W9-AHH-500

SpringerWienNewYork

CISM COURSES AND LECTURES

The series presents lecture notes, monographs, edited works and proceedings in the field of Mechanics, Engineering, Computer Science and Applied Mathematics.
Purpose of the series is to make known in the international scientific and technical community results obtained in some of the activities organized by CISM, the International Centre for Mechanical Sciences.

INTERNATIONAL CENTRE FOR MECHANICAL SCIENCES

COURSES AND LECTURES - No. 499

DYNAMIC METHODS FOR DAMAGE DETECTION IN STRUCTURES

EDITED BY

ANTONINO MORASSI
UNIVERSITY OF UDINE, ITALY

FABRIZIO VESTRONI
UNIVERSITY OF ROMA LA SAPIENZA, ITALY

SpringerWienNewYork

This volume contains 108 illustrations

Printed in Italy
SPIN 12244355

All contributions have been typeset by the authors.

ISBN 978-3-211-78776-2 SpringerWienNewYork

PREFACE

Non-destructive testing aimed at monitoring, structural identification and diagnostics is of strategic importance in many branches of civil and mechanical engineering. This type of tests is widely practiced and directly affects topical issues regarding the design of new buildings and the repair and monitoring of existing ones. The load-bearing capacity of a structure can now be evaluated using well-established mechanical modelling methods aided by computing facilities of great capability. However, to ensure reliable results, models must be calibrated with accurate information on the characteristics of materials and structural components. To this end, non-destructive techniques are a useful tool from several points of view. Particularly, by measuring structural response, they provide guidance on the validation of structural descriptions or of the mathematical models of material behaviour.

Diagnostic engineering is a crucial area for the application of non-destructive testing methods. Repeated tests over time can indicate the emergence of possible damage occurring during the structure's lifetime and provide quantitative estimates of the level of residual safety.

Of the many non-destructive testing techniques now available, dynamic methods enjoy growing focus among the engineering community. Conventional diagnostic methods, such as those based on visual inspection, thermal or ultrasonic analysis, are local by nature. To be effective these require direct accessibility of the region to be inspected and a good preliminary knowledge of the position of the defective area. Techniques based on the study of the dynamic response of the structure or wave propagation, on the contrary, are a potentially effective diagnostic tool. These can operate on a global scale and do not require a priori information on the damaged area.

Recent technological progress has generated extremely accurate and reliable experimental methods, enabling a good estimate of changes in the dynamic behaviour of a structural system caused by possible damage. Although experimental techniques are now well-established, the interpretation of measurements still lags somewhat behind. This particularly concerns identification and structural diagnostics due to their nature of inverse problems. Indeed, in these applications one wishes to determine some mechanical properties of a system on the basis of measurements of its response, in part exchanging the role of the unknowns and data compared to the direct problems of structural analysis.

Hence, concerns typical of inverse problems arise, such as high nonlinearity, non-uniqueness or non-continuous dependence of the solution on the data. When identification techniques are applied to the study of real-world structures, additional obstacles arise given the complexity of structural modelling, the inaccuracy of the analytical models used to interpret experiments, measurement errors and incomplete field data. Furthermore, the results of the theoretical mathematical formulation of problems of identification and diagnostics, given the present state-of-knowledge in the field, focus on quality, while practical needs often require more specific and quantitative estimates of quantities to be identified. To overcome these obstacles, standard procedures often do not suffice and an individual approach must be applied to tackle the intrinsic nature of the problem, using specific experimental, theoretical and numerical methods. It is for these reasons that use of damage identification techniques still involves delicate issues that are only now being clarified in international scientific literature.

The CISM Course "Dynamic Methods for Damage Detection in Structures" was an opportunity to present an updated state-of-the-art overview. The aim was to tackle both theoretical and experimental aspects of dynamic non-destructive methods, with special emphasis on advanced research in the field today.

The opening chapter by Vestroni and Pau describes basic concepts for the dynamic characterization of discrete vibrating systems. Chapter 2, by Friswell, gives an overview of the use of inverse methods in damage detection and location, using measured vibration data. Regularisation techniques to reduce ill-conditioning effects are presented and problems discussed relating to the inverse approach to structural health monitoring, such as modelling errors, environmental effects, damage models and sensor validation. Chapter 3, by Betti, presents a methodology to identify mass, stiffness and damping coefficients of a discrete vibrating system based on the measurement of input/output time histories. Using this approach, structural damage can be assessed by comparing the undamaged and damaged estimates of the physical parameters. Cases of partial/limited instrumentation and the effect of model reduction are also discussed. Chapter 4, by Vestroni, deals with the analysis of structural identification techniques based on parametric models. A numerical code, that implements a variational procedure for the identification of linear finite element models based on modal quantities, is presented and applied for modal updating and damage detection purposes. Pseudo-experimental and experimental cases are solved. Ill-conditioning and other peculiarities of the method are also investigated. Chapter 5, by Vestroni, deals with damage detection in beam structures via natural frequency measurements. Cases of single, multiple and interacting cracks are considered in detail. Attention is particularly focussed on the consequences that certain peculiarities, such as the limited number of unknowns (e.g., locations and stiffness reduction of damaged sections), have on the inverse problem solution. The analysis of damage identification in vibrating beams is continued in Chapter 6 by Morassi. Damage analysis is formulated as a reconstruction problem and it is shown that frequency shifts caused by damage contain information on certain Fourier coefficients of the unknown stiffness variation. The rest of the chapter is devoted to the identification of localized damage in beams from a minimal set of natural frequency measurements. Closed form solutions for certain crack identification problems in vibrating rods and beams are presented. Applications based on changes in the nodes of the mode shapes and on antiresonant data are also discussed. Chapter 7, by Testa, is on the localization of concentrated damage in beam structures based on frequency changes caused by the damage. A second application deals with a crack closure that may develop in fatigue and the potential impact on damage detection. Chapter 8 proposes a paper by Cawley on the use of guided waves for long-range inspection and the integrity assessment of pipes. The aim is to determine the reflection coefficients from cracks and notches of varying depth, circumferential and axial extent when the fundamental torsional mode is travelling in the pipe. Chapter 9, by Vestroni and Vidoli, discusses a technique to enhance sensitivity of the dynamic response to local variations of the mechanical characteristics of a vibrating system based on coupling with an auxiliary system. An application to a beam-like structure coupled to a network of piezoelectric patches is discussed in detail to illustrate the approach.

Antonino Morassi
Fabrizio Vestroni

CONTENTS

[illegible] the RMS [illegible] for locating the damaged element, the [illegible] model [illegible] clear [illegible]. Table 1 gives the RMS values of the deviations of the six frequency change ratios [illegible] on the beam. For both the closed and open crack conditions, RMS [illegible] indicates that the [illegible] element of the beam is the damaged one. Although the frequency effects of the crack were considerably obscured by crack closure, the simple crack [illegible] model [illegible] sensitivity in this example.

[illegible] frequency changes

Crack Location Beam Element	RMS Deviation Closed Crack	RMS Deviation Open Crack
[illegible]	0.34	0.35
[illegible]	[illegible]	[illegible]
[illegible]	0.30	0.31
4	0.09	0.13

[illegible]

[illegible] a simple model that can give information on damage location even [illegible] a structure that one seeks in the general problem of system identification. The model outlined here has the requisite simplicity and it shows sensitivity in an application where the damage effects are obscured by crack closure, something that may well be encountered in a structure with fatigue cracking. The simplification introduced in the description of damage rendered the model independent of the severity of the damage at a location, a feature that [illegible] observed in the application to the cracked beam.

[illegible] modeled the effects of a crack in a beam, usually treating the crack as a slot. While the results presented here show a great potential difference between the slot and a crack if there is closure [illegible] effects on frequency and damping are concerned, there is not such [illegible] distinction between the two in the results of the application of the present crack [illegible].

[illegible] of crack closure on stiffness, which appears in Figure 2 to give a bilinear load [illegible] which is actually nonlinear near the knee of the curve, there are also [illegible] response. These have been addressed further in ([illegible] and Zhang [illegible]).

Further work on the application of this simplified model requires consideration of the proper selection of [illegible] to be monitored for a specific diagnostic, as well as questions of uniqueness.

Elements of Experimental Modal Analysis

Fabrizio Vestroni* and Annamaria Pau*

* Dipartimento di Ingegneria Strutturale e Geotecnica, University of Roma *La Sapienza*, Italy

Abstract Fundamental concepts for the characterization of the dynamical response of SDOF and NDOF systems are provided. A description is given of the main techniques to represent the response in the frequency domain and its experimental characterization. Two classical procedures of modal parameter identification are outlined and selected numerical and experimental examples are reported.

1 Dynamic Characterization of a SDOF

The experimental study of a structure provides an insight into the real behavior of the system. In particular, the study of its dynamic response, exploiting vibration phenomena, aims to determine the dynamic properties closely connected to the geometrical and mechanical characteristics of the system. Hence, some concepts of structural dynamics will be briefly summarized. It is assumed that the reader has had some exposure to the matter (Craig, 1981; Meirovitch, 1997; Ewins, 2000; Braun et al., 2001).

The classical model of a single degree-of-freedom (SDOF) system is the spring-mass-dashpot model of Figure 1, where the equation of motion and the steady-state solution is reported. Assuming a harmonic excitation, the frequency response function (FRF) $H(\omega)$ can be defined as the ratio between the amplitude of the steady-state response and the load intensity. The FRF shows that in a small range of the ratio ω/ω_0, when the frequency of the excitation approaches the natural frequency of the system, the response amplitude is much larger than the static response. This is called resonance. Furthermore, the amplitude of the steady-state response is linearly dependent on both p_0 and $H(\omega)$. By knowing $H(\omega)$ the response of a SDOF system to a harmonic excitation can be estimated.

In the real world, forces are not simply harmonic, being frequently periodic or approximated closely by periodic forces. A periodic function $p(t)$ having period T_1 can be represented as a series of harmonic components by means of its Fourier series expansion. As an example, in Figure 2, the Fourier series expansion is applied to a square wave. The Fourier series is convergent, i.e. the more terms used, the better the approximation obtained.

Since the response of a SDOF system to a harmonic force is known and a periodic forcing function $p(t)$ can be represented as a sum of harmonic forces, the response of the system $u(t)$ to a periodic excitation can be obtained by exploiting the principle of effect superposition:

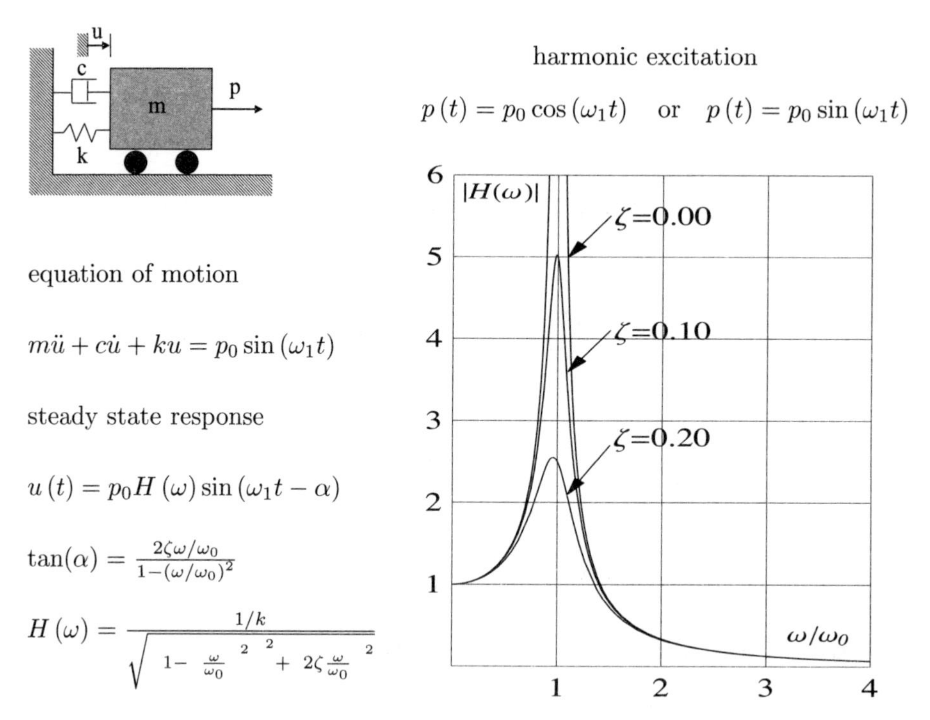

Figure 1. Response of a SDOF system to a harmonic excitation.

$$p(t) = \sum_{n=1}^{\infty} p_n \cos(\omega_n t + \varphi_n), \quad \omega_n = n\omega_1 \tag{1.1}$$

$$u(t) = \sum_{n=1}^{\infty} U_n \cos(\omega_n t + \varphi_n - \alpha_n) \tag{1.2}$$

$$U_n = \frac{p_n/k}{\sqrt{(1-r_n^2)^2 + (2\zeta r_n)^2}} = p_n H(\omega_n, \zeta), \quad tg\alpha_n = \frac{2\zeta r_n}{1-r_n^2}, \quad r_n = \frac{\omega_n}{\omega_0}. \tag{1.3}$$

In this case, too, a knowledge of $H(\omega)$ is sufficient to predict the response of the system.

The steady-state response of a SDOF system to a harmonic force can also be written in complex form, where the bar denotes complex quantities:

$$\overline{u}(t) = \overline{U}(\omega) e^{i\omega t} = \overline{H}(\omega) p_0 e^{i\omega t} \quad \text{and} \quad \overline{H}(\omega) = \frac{1/k}{\left[1-(r)^2\right] + i(2\zeta r)}. \tag{1.4}$$

It is clear that the amplitude and phase of the steady-state response are determined from the amplitude and phase of the complex FRF.

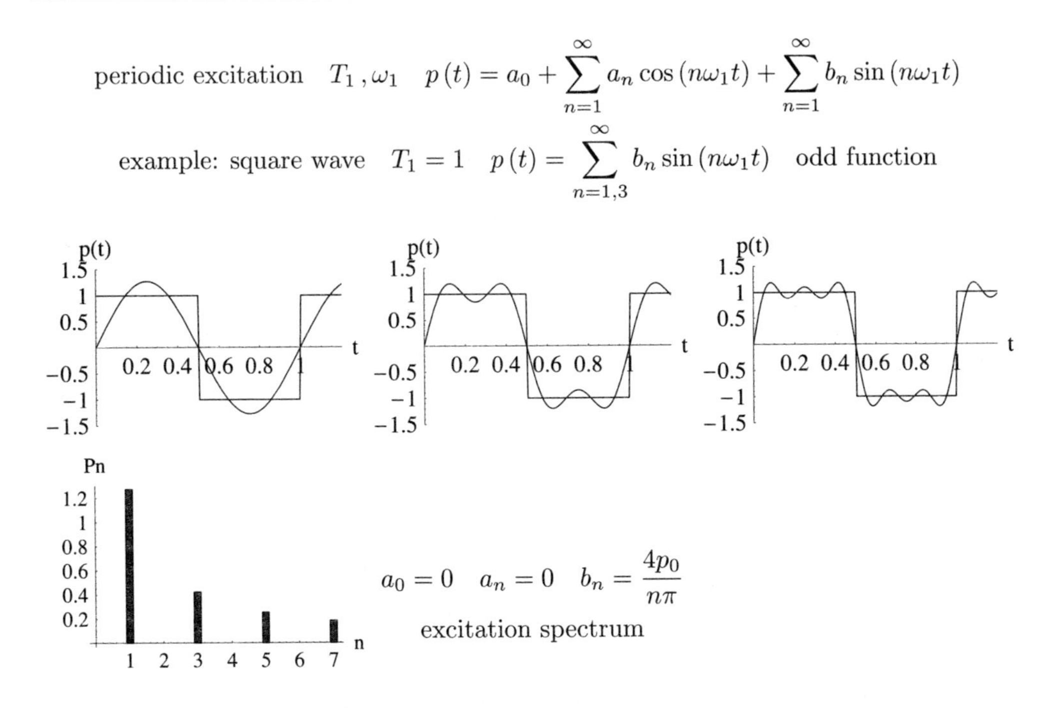

periodic excitation T_1, ω_1 $p(t) = a_0 + \sum_{n=1}^{\infty} a_n \cos(n\omega_1 t) + \sum_{n=1}^{\infty} b_n \sin(n\omega_1 t)$
example: square wave $T_1 = 1$ $p(t) = \sum_{n=1,3}^{\infty} b_n \sin(n\omega_1 t)$ odd function
p(t)
t
Pn
n
$a_0 = 0$ $a_n = 0$ $b_n = \frac{4p_0}{n\pi}$
excitation spectrum

Figure 2. Fourier series expansion of a square wave.

If the forcing function is periodic, the response can be written as:

$$\overline{u}(t) = \sum_{n=-\infty}^{\infty} \overline{U}_n \, e^{in\omega t} = \sum_{n=-\infty}^{\infty} \overline{H}(\omega_n) \, \overline{p}_n \, e^{in\omega t}. \tag{1.5}$$

When the excitation is non periodic, it can be represented by a Fourier integral, which is obtained from the Fourier series by letting the period T_1 approach infinity. Let us define:

$$T_1 = \frac{2\pi}{\omega_1}, \quad \omega_1 = \Delta\omega, \quad n\omega_1 = \omega_n. \tag{1.6}$$

In the Fourier series

$$p(t) = \sum_{n=-\infty}^{\infty} \overline{p}_n(\omega_n) \, e^{i\omega_n t}, \quad \overline{p}_n(\omega_n) = \frac{1}{T_1} \int_{T_1} p(t) \, e^{-i\omega_n t} dt \tag{1.7}$$

since T_1 tends to infinity, p_n is newly defined as:

$$\overline{p}_n(\omega_n) = T_1 \overline{p}_n(\omega_n) \quad \text{and} \quad p(t) = \sum_{n=-\infty}^{\infty} \frac{\Delta\omega}{2\pi} \overline{p}_n(\omega_n) \, e^{in\Delta\omega t}. \tag{1.8}$$

When $T_1 \longrightarrow \infty$, $n\Delta\omega = \omega_n = \omega$ becomes a continuous variable and $\Delta\omega$ becomes the differential $d\omega$, then a Fourier transform pair is obtained:

$$\bar{P}(\omega) = \int_{-\infty}^{\infty} p(t) e^{-i\omega t} dt \quad \text{direct Fourier transform} \tag{1.9}$$

$$p(t) = \int_{-\infty}^{\infty} \frac{1}{2\pi} \overline{P}(\omega) e^{i\omega t} d\omega \quad \text{inverse Fourier transform.} \tag{1.10}$$

When the forcing function is non periodic, a relationship in the frequency domain between the response and the force can be obtained by applying the Fourier transform to each term of the motion equation:

$$\left(-m\omega^2 + ic\omega + k\right) \bar{U}(\omega) = \bar{P}(\omega) \tag{1.11}$$

where use is made of the following properties:

$$\dot{\bar{U}}(\omega) = \frac{i\omega}{2\pi} \int_{-\infty}^{\infty} u(t)\, e^{-i2\pi f t} dt = \frac{i\omega}{2\pi} \bar{U}(\omega) \tag{1.12}$$

$$\ddot{\bar{U}}(\omega) = \frac{-\omega^2}{2\pi} \int_{-\infty}^{\infty} u(t)\, e^{-i2\pi f t} dt = \frac{-\omega^2}{2\pi} \bar{U}(\omega). \tag{1.13}$$

The Fourier transform of the response is obtained as the product of the complex FRF and the Fourier transform of the excitation

$$\bar{U}(\omega) = \bar{H}(\omega)\, \bar{P}(\omega). \tag{1.14}$$

Once $U(\omega)$ is known, the response in the time domain is given by the inverse Fourier transform:

$$\overline{u}(t) = \int_{-\infty}^{\infty} \frac{1}{2\pi} \overline{U}(\omega)\, e^{i2\pi f t} df. \tag{1.15}$$

In this case also, by knowing $H(\omega)$, the response to a generic excitation can be estimated.

A significant relationship exists between the $H(\omega)$ and the unit impulse response $h(t)$. The latter defines the SDOF response in the time domain to a forcing function equal to a Dirac delta. Since the Fourier transform of an impulse $p(t) = \delta(0)$ is

$$p(\omega) = \frac{1}{2\pi}, \tag{1.16}$$

the impulse response can be written as

$$u(t) = h(t) = \int_{-\infty}^{\infty} \overline{H}(\omega)\, \overline{p}(\omega) e^{i\omega t} d\omega = \frac{1}{2\pi} \int_{-\infty}^{\infty} \overline{H}(\omega) e^{i\omega t} d\omega. \tag{1.17}$$

In other words, the time domain response to a Dirac delta is the inverse Fourier transform of the FRF. The FRF and the impulse response function form a couple of Fourier transforms. It is possible to refer both to $H(\omega)$ or to $h(t)$ to characterize the system and to provide a predictive model.

2 Display of a FRF

As a complex quantity, the FRF contains information regarding both the amplitude and the phase of the oscillation.

The real and imaginary components of $H(\omega)$ are:

$$H_R(\omega) = \frac{(1-r^2)/k}{(1-r^2)^2 + (2\zeta r)^2} \quad H_I(\omega) = \frac{-2\zeta r/k}{(1-r^2)^2 + (2\zeta r)^2}. \tag{2.1}$$

The three most common forms of representation of $H(\omega)$ are reported in Figure 3-4 for a SDOF with $\zeta = 0.125$.

(1) Real and Imaginary parts of $H(\omega)$ vs r (Figure 3). The real part of $H(\omega)$ crosses the frequency axis at resonance, while, at the same frequency, the imaginary part reaches a minimum.

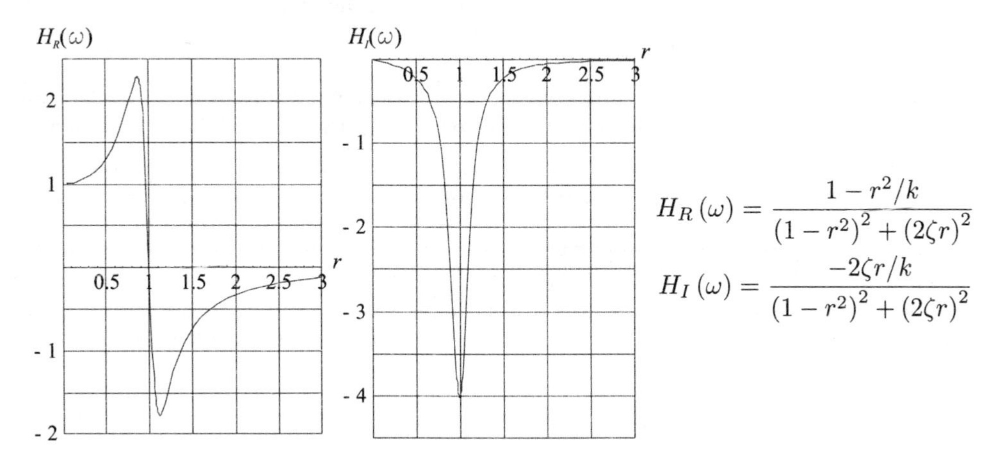

Figure 3. Real and Imaginary part of the FRF vs r.

(2) Modulus of FRF and phase vs r (Figure 4). Resonance is pointed out by a maximum in the modulus of the FRF and by a phase change from 0 to $-\pi$.

(3) Real part vs Imaginary part in the Argand plane (Figure 4). This is a circular loop that contains all the information and enhances the region close to resonance, which is practically coincident with the intersection of the circle with the y axis.

The dynamic properties of a system can be expressed in terms of any convenient response characteristics: FRF can be presented in terms of displacement (receptance), velocity (mobility) or acceleration (intertance). Mobility and inertance are obtained from receptance by multiplying by $i\omega$ and $(i\omega)^2$, respectively.

From the analytical relationships, previously reported, it is clear that the FRF can be experimentally evaluated mainly by two different methods (Ewins, 2000; Maia and Silva, 1997). The former involves the steady-state response to a harmonic force P at different assigned frequencies, which implies the use of an exciter connected to the structure that

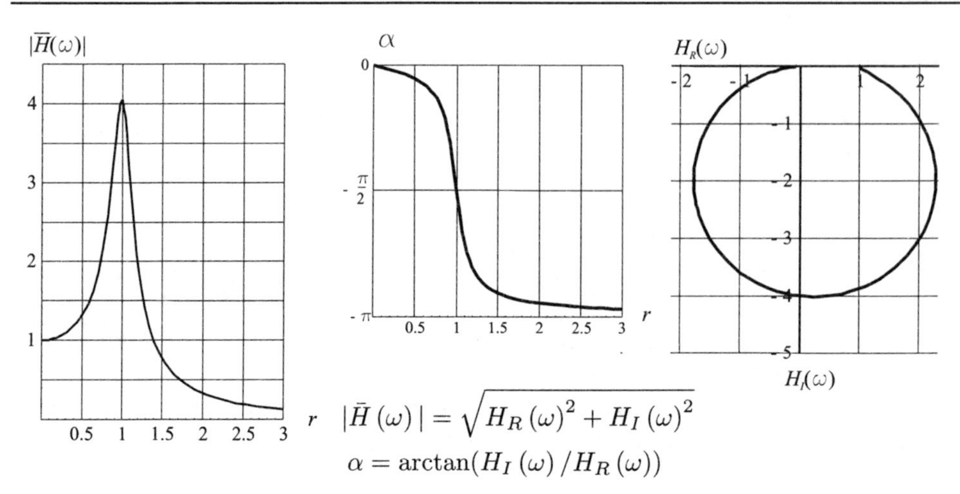

Figure 4. Modulus of FRF and phase vs r and Argand plane.

generates a harmonic force. The amplitude U of the stationary response is recorded and the values of the FRF at the discrete frequencies ω_j of the applied force are directly obtained from the ratio:

$$H(\omega_j) = \frac{U}{P}. \tag{2.2}$$

The other method involves free oscillations of the structure and includes devices which are able to exert impulsive force, such as an instrumented hammer or a sharp release of a static force. The FRF is obtained as the ratio between the response Fourier transform and the input Fourier transform:

$$H(\omega) = \frac{U(\omega)}{P(\omega)}. \tag{2.3}$$

Through this relationship, $H(\omega)$ can only be obtained in the frequency band contained in the forcing function. In particular, the input generated by a real impulsive force has a wide spectrum which is flat within a frequency band $0 - f_{max}$, then rapidly approaches zero. This implies that f_{max} is not infinite, as in the theoretical Dirac delta, but that in the real world depends on the impulse duration T.

In the example of Figure 5, the duration of an experimental impulse is about 0.0006s and the related Fourier transform is constant only up to about 1000 Hz, then rapidly decreases and reaches zero at 3000 Hz, which can be approximately estimated as $2/T$.

3 Dynamic Characterization of Multidegree-of-Freedom Systems

The multidegree-of-freedom systems are representative of discrete systems or discretized continuous systems. The governing equations of motion of a linear system with N-degrees-of-freedom (NDOF) can be written in matrix form as (Craig, 1981; Meirovitch, 1997)

$$\mathbf{M}\,\ddot{u}(t) + \mathbf{C}\,\dot{u}(t) + \mathbf{K}\,u(t) = p(t) \tag{3.1}$$

where **M**, **C** and **K** are respectively the mass, damping and stiffness matrices, with dimensions $N \times N$ and $p(t)$ is the force vector. If the damping matrix **C** is proportional

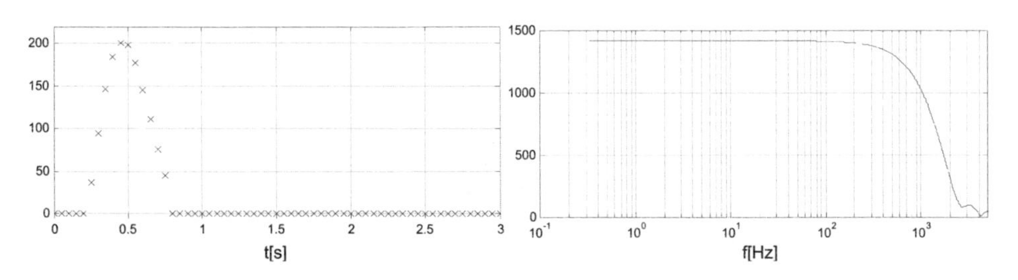

Figure 5. Experimental impulse (left) and its Fourier transform (right).

to a linear combination of **M** and **K**, the damped system will have real eigenvectors coincident with that of the undamped system (classical damping), so that the equations of motion can be decoupled by describing the displacement motion in terms of eigenvectors weighted by the modal coordinates $q(t)$

$$u\left(t\right)=\Phi q\left(t\right)=\sum_{r}\Phi_{r}q_{r}\left(t\right). \tag{3.2}$$

By substituting this expression in the equations of motion, then pre-multiplying by the matrix Φ^{T} and using the orthogonality eigenvector conditions, N uncoupled modal equations are obtained:

$$\ddot{q}_{r}\left(t\right)+2\zeta_{r}\omega_{r}\dot{q}_{r}\left(t\right)+k_{r}q_{r}\left(t\right)=p_{r}\left(t\right) \qquad r=1,N. \tag{3.3}$$

Hence, the system behaviour is represented as a linear superposition of the response of N SDOF systems. In this case, the FRF is a matrix $N\times N$; the displacement in the node i caused by a force applied in node j is:

$$H_{ij}\left(\omega\right)=\sum_{r=1}^{N}\frac{\Phi_{ir}\Phi_{jr}}{k_{r}\left[\left(1-r_{r}^{2}\right)+i2\zeta_{r}r_{r}\right]}. \tag{3.4}$$

An NDOF systems has N natural frequencies ω_r and N related mode shapes Φ_r along with the modal damping coefficients ζ_r. These are the dynamic characteristics of the structure. When the structure is excited by a harmonic force with a frequency coincident to one of its natural frequencies, the response is amplified. As an example, Figure 6 reports the modulus of $H_{33}(\omega)$ and the mode shapes of a 3DOF shear-type frame. The FRF exhibits three sharp peaks at natural frequencies ω_1, ω_2, ω_3. H_{13}, H_{23}, H_{33} may be obtained by the response measured at nodes 1, 2 and 3 to a forcing function applied in the node 3.

With regard to a 3DOF system, Figure 7 reports H_{11} and H_{21}, represented in two of the three possible forms described for a SDOF system. From these curves, the resonances can be obtained as for the SDOF system, as well as other data needed to define the mode shapes. When dealing with a NDOF system, it is mandatory to define if two points oscillate in phase. The difference of phase between two points in a mode can be directly read from the phase plot. As an example, nodes 1 and 2 oscillate in-phase in mode 1 and

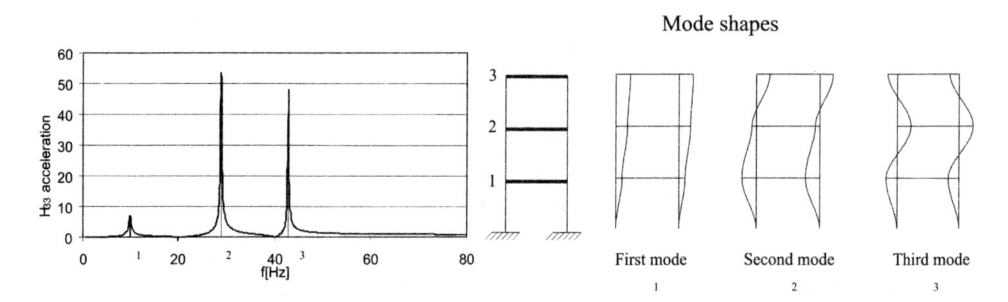

Figure 6. FRF and mode shapes of a 3DOF shear-type frame.

2. In fact, the phase difference between the two responses is zero. By contrast, these two nodes oscillate out-of-phase in mode 3, as shown by a π phase difference. The information can also be deduced from the Nyquist plot: a change in the quadrant occupied by the circles pinpoints a π phase difference. This is indicated in Figure 7 for the third mode. The results are in agreement with the mode shapes of Figure 7.

Figure 8 shows the Fourier transform of acceleration response of a simply supported continuous beam measured in CH2 obtained by means of a static force in CH3 sharply removed (left) or of a hammer impulse (right) at the same point. As for the NDOF, each peak corresponds to a natural frequency. The impulsive force is more effective in exciting a large number of frequencies with respect to the imposed initial displacement.

4 Modal Parameter Identification

Once the registration of the response is available, there are various procedures in the time domain and in the frequency domain to determine the modal properties of the structure. Two classical procedures are presented here. The first algorithm presented is based on a multimodal estimate in the frequency domain, following the pioneering work by Goyder (1980). In the neighborhood of the *r-th* resonance, the inertance $H_{ij}(\omega)$ can be regarded as the sum of two terms:

$$H_{ij}(\omega) = \frac{-\omega^2 \Phi_{ir} \Phi_{rj}}{\omega_r^2 - \omega^2 + 2i\zeta_r \omega_r \omega} + \sum_{s \neq r = 1}^{N} \frac{-\omega^2 \Phi_{is} \Phi_{js}}{\omega_s^2 - \omega^2 + 2i\zeta_s \omega_s \omega}. \tag{4.1}$$

The former is the prevailing resonant term and the latter is the contribution of all the other modes. By comparing the experimental (e) and the analytical inertance functions the error at the generic frequency ω_k around ω_r can be defined:

$$E_k = H_{ij}\left(\omega_k\right) - H_{ij}^{(e)}\left(\omega_k\right) = \frac{-\omega_k^2 \Phi_{ir} \Phi_{jr}}{\omega_r^2 - \omega_k^2 + 2i\zeta_r \omega_r \omega_k} - C_k \tag{4.2}$$

where the constant C_k is given by the difference between the experimental value of the FRF and the modal contributions of all the other modes $s \neq r$. By introducing the following modal quantities as variables of the problem:

$$a_r = \omega_r^2, \quad b_r = 2\zeta_r \omega_r, \quad c_{ijr} = \phi_{ir} \phi_{jr} \tag{4.3}$$

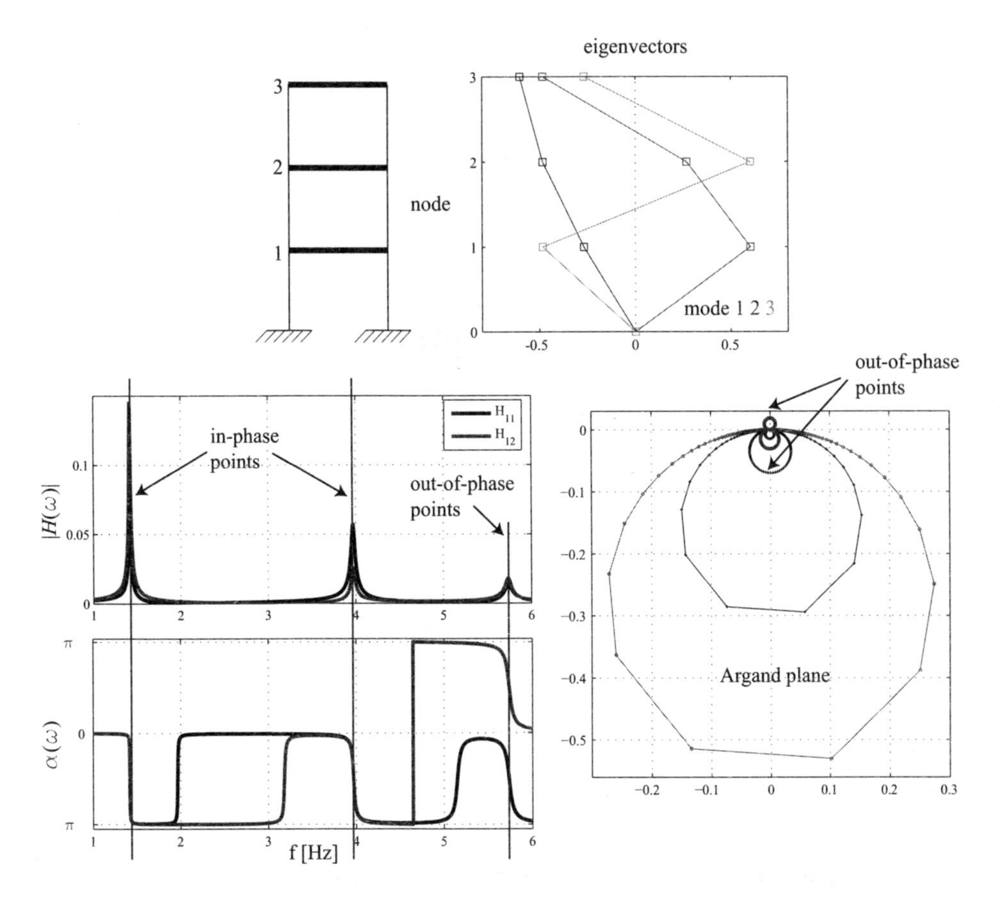

Figure 7. Display of the FRF H_{33} of a 3DOF system.

and a suitable weight function

$$w_k = \frac{-1}{a_r - \omega_k^2 + i b_r \omega_k}, \tag{4.4}$$

the error function can be linearized with respect to the assumed variables:

$$E_k = \frac{-\omega_k^2 \Phi_{ir} \Phi_{jr}}{\omega_r^2 - \omega_k^2 + 2i\zeta_r \omega_r \omega_k} - C_k = \omega_k^2 w_k c_{ijr} + w_k C_k \left(a_r - \omega_k^2 + i b_r \omega_k \right). \tag{4.5}$$

The procedure follows an iterative scheme: at the *h-th* iteration, the error is expressed as:

$$E_{k,h} = \omega_k^2 w_{k,h-1} c_{ijr,h} + w_{k,h-1} C_{k,h-1} \left(a_{r,h} - \omega_k^2 + i b_{r,h} \omega_k \right) \tag{4.6}$$

which is linear with respect to the three *r-th* modal unknowns since w_k and C_k are determined by the values at the step $h - 1$. The objective function is obtained by summing the square error in a frequency range in the neighborhood of ω_r. On the basis

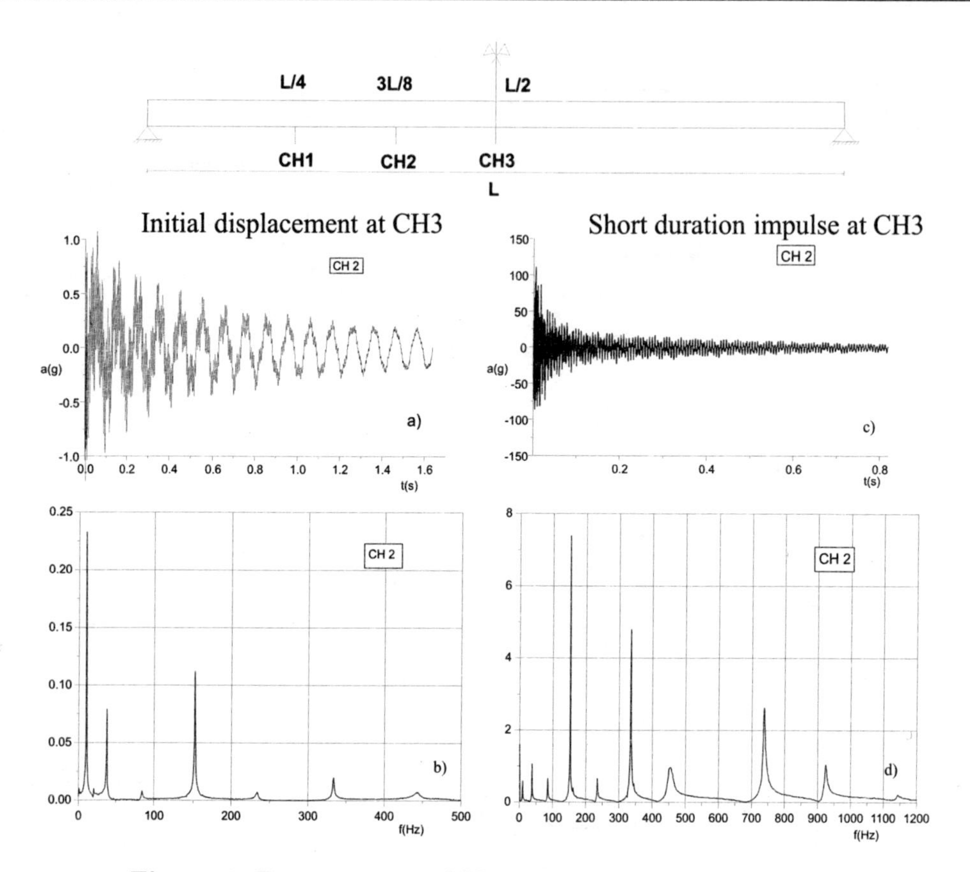

Figure 8. Free response and FRF of a simply supported beam.

of the least-squares method, the modal unknowns are determined. At each step, this procedure is extended to all meaningful resonances which appear in the response and, at the end of the process, all the modal parameters in the range of frequencies investigated are determined (Beolchini and Vestroni, 1994; Genovese and Vestroni, 1998; De Sortis et al., 2005).

The second method presented is a frequency domain decomposition method and relies on the response to ambient excitation sources when the output only is available. The method is based on the singular value decomposition of the spectral matrix (Brincker et al., 2001). It exploits the relationship:

$$\mathbf{G}yy(\omega) = \bar{\mathbf{H}}(\omega)\,\mathbf{G}xx(\omega)\,\mathbf{H}^T(\omega) \tag{4.7}$$

where $\mathbf{G}xx(\omega)$ (RxR, R number of inputs) and $\mathbf{G}yy(\omega)$ (MxM, M number of measured responses) are respectively the input and output power spectral density matrices, and $\mathbf{H}(\omega)$ is the frequency response function matrix (MxR).

Supposing the inputs at the different points to be completely uncorrelated and white noise, $\mathbf{G}xx$ is a constant diagonal matrix $\mathbf{G}$, independent of ω. Thus:

$$\mathbf{G}yy(\omega) = G\,\mathbf{H}(\omega)\,\mathbf{H}^T(\omega) \tag{4.8}$$

whose term jk can be written as, by omitting the constant G:

$$Gyy_{jk}(\omega) = \sum_{r=1}^{R}\left(\sum_{p=1}^{N}\frac{\phi_{jp}\phi_{rp}}{\bar{\lambda}_p^2-\omega^2}\right)\left(\sum_{q=1}^{N}\frac{\phi_{kq}\phi_{rq}}{\lambda_q^2-\omega^2}\right). \tag{4.9}$$

In the neighborhood of the *i-th* resonance, the previous equation can be written as:

$$Gyy_{jk}(\omega) \cong \sum_{r=1}^{R}\frac{\phi_{ji}\phi_{ri}}{\bar{\lambda}_i^2-\omega^2}\frac{\phi_{ki}\phi_{ri}}{\lambda_i^2-\omega^2} = \frac{\phi_{ji}\phi_{ki}}{\left(\bar{\lambda}_i^2-\omega^2\right)\left(\lambda_i^2-\omega^2\right)}\sum_{r=1}^{R}\phi_{ri}^2 \tag{4.10}$$

where $\sum_{r=1}^{R}\phi_{ri}^2$ is a constant. By ignoring this term, the matrix $\mathbf{G}yy$ can thus be expressed as the product of three matrices:

$$\mathbf{G}yy(\omega) = \mathbf{\Phi}\mathbf{\Lambda}_i\mathbf{\Phi}^T \tag{4.11}$$

which represents a singular value decomposition of the matrix $\mathbf{G}yy$, where:

$$\mathbf{\Lambda}_i = \begin{bmatrix} \frac{1}{\left(\lambda_i^2-\omega^2\right)\left(\bar{\lambda}_i^2-\omega^2\right)} & 0... & 0 \\ 0 & 0... & 0 \\ 0 & 0... & 0 \end{bmatrix}. \tag{4.12}$$

The peaks of the first singular values indicate the natural frequencies of the system. In the neighborhood of the *i-th* peak of the first singular value, the first singular vector is coincident with the *i-th* eigenvector. This occurs at each *j-th* resonance, when the prevailing contribution is given by the *j-th* mode. This procedure, which had recently a great diffusion, was implemented in a commercial code (ARTeMIS).

5 Conclusions

For SDOF and NDOF systems the knowledge of $H(\omega)$ provides a predictive model of the mechanical system in evaluating the response to any excitation. Moreover, it is possible to obtain experimental values of some components of $H(\omega)$ and then extract experimental values of modal parameters which are characteristic dynamic properties of the structure. Several methods in frequency and time domain are available to evaluate modal parameters from measured response to known and unknown excitation.

Bibliography

ARTeMIS. *modal software, www.svibs.com.*

G. C. Beolchini and F. Vestroni. Identification of dynamic characteristics of base-isolated and conventional buildings. In *Proceedings of the 10th European Conference on Earthquake Engineering*, 1994.

S. Braun, D. Ewins, and S. S. Rao, editors. *Encyclopedia of Vibration*, 2001. Academic Press, San Diego.

R. Brincker, L. Zhang, and P. Andersen. Modal identification of output-only systems using frequency domain decomposition. *Smart Materials and Structures*, 10:441–445, 2001.

R. R. Craig. *Structural Dynamics.* John Wiley & Sons, 1981.

A. De Sortis, E. Antonacci, and F. Vestroni. Dynamic identification of a masonry building using forced vibration tests. *Engineering Structures*, 27:155–165, 2005.

D. J. Ewins. *Modal Testing: theory, practice and application.* Reaserch Studies Press, 2000.

F. Genovese and F. Vestroni. Identification of dynamic characteristics of a masonry building. In *Proceedings of the 1th European Conference on Earthquake Engineering*, 1998.

H. G. D. Goyder. Methods and applications of structural modelling from measured frequency response data. *Journal of Sound and Vibration*, 68:209–230, 1980.

N.M.M. Maia and J.M.N. Silva. *Theoretical and Experimental Modal Analysis.* Reaserch Studies Press, 1997.

L. Meirovitch. *Principles and Techniques of Vibration.* Prentice-Hall, 1997.

Damage Identification using Inverse Methods

Michael I. Friswell
Department of Aerospace Engineering, University of Bristol, Bristol BS8 1TR, UK.
m.i.friswell@bristol.ac.uk

Abstract This chapter gives an overview of the use of inverse methods in damage detection and location, using measured vibration data. Inverse problems require the use of a model and the identification of uncertain parameters of this model. Damage is often local in nature and although the effect of the loss of stiffness may require only a small number of parameters, the lack of knowledge of the location means that a large number of candidate parameters must be included. This leads to potential ill-conditioning problems, and this topic is reviewed in this chapter. This chapter then goes on to discuss a number of problems that exist with the inverse approach to structural health monitoring, including modelling errors, environmental effects, damage localisation, regularisation, models of damage and sensor validation.

1 Introduction to Inverse Methods

Inverse methods combine an initial model of the structure and measured data to improve the model or test an hypothesis. In practice the model is based on finite element analysis and the measurements are acceleration and force data, often in the form of a modal database, although frequency response function (FRF) data may also be used. The estimation techniques are often based on the methods of model updating, which have had some success in improving models and understanding the underlying dynamics, especially for joints (Friswell and Mottershead, 1995; Mottershead and Friswell, 1993). Model updating methods may be classified as sensitivity or direct methods. Sensitivity type methods rely on a parametric model of the structure and the minimisation of some penalty function based on the error between the measured data and the predictions from the model. These methods offer a wide range of parameters to update that have physical meaning and allow a degree of control over the optimisation process. The alternative is direct updating methods that change complete mass and/or stiffness matrices, although the updated models obtained are often difficult to interpret for health monitoring applications. These methods will be considered in more detail later. However it should be emphasised that a huge number of papers have been written on the application of inverse methods to damage identification, and this chapter aims to give an overview of the approaches rather than a complete literature review. This chapter will also consider some of the difficulties that occur when inverse methods are used for damage identification (Friswell, 2007; Doebling et al., 1998).

The four stages of damage estimation, first given by Rytter (1993), are now well established as detection, location, quantification and prognosis. Detection is readily performed by pattern recognition methods or novelty detection (Worden, 1997; Worden et al., 2000). The key issue for inverse methods is location, which is equivalent to error localisation in model updating. Once the damage is located, it may be parameterised with a limited set of parameters and quantification, in terms of the local change in stiffness, is readily estimated. Prognosis requires that the underlying damage mechanism is determined, which may be possible using inverse methods using hypothesis testing among several candidate mechanisms. This questions is considered in more detail later in the chapter. However, once the damage mechanism is determined, the associated model is available for prognosis, and this is a great advantage of model based inverse methods.

1.1 Objective Functions

Friswell and Mottershead (1995) discussed sensitivity based methods in detail. The approach minimises the difference between modal quantities (usually natural frequencies and less often mode shapes) of the measured data and model predictions. This problem may be expressed as the minimization of J, where

$$J(\theta) = \|\mathbf{z}_m - \mathbf{z}(\theta)\|^2 = \epsilon^T \epsilon \tag{1.1}$$

and

$$\epsilon = \mathbf{z}_m - \mathbf{z}(\theta). \tag{1.2}$$

Here $\mathbf{z}_m$ and $\mathbf{z}(\theta)$ are the measured and computed modal vectors, θ is a vector of all unknown parameters, and ϵ is the modal residual vector. The modal vectors may consist of both natural frequencies and mode shapes, although often mode shapes are only used to pair individual modes. If mode shapes are included then they must be carefully normalised, the sensor locations must be carefuly matched to the finite element degrees of freedom and weighting should be applied to Equation (1.1).

Frequency response functions may also be used, although a model of damping is required, and the penalty function is often a very complicated function of the parameters with many local minima, making the optimisation very difficult. dos Santos et al. (2005) presents an example of such a method for damage in a composite structure.

1.2 Sensitivity Methods

Sensitivity based methods allow a wide choice of physically meaningful parameters and these advantages has led to their widespread use in model updating. The approach is very general and relies on minimising a penalty function, which usually consists of the error between the measured quantities and the corresponding predictions from the model. Parameters are then chosen that are assumed uncertain, and these are usually estimated by approximating the penalty function using a truncated Taylor series and iterating to obtain a converged solution. If there are sufficient measurements and a restricted set of parameters then the identification may be well-conditioned. Often some form of regularisation must be applied, and this is considered in detail later. Other optimisation methods may be used, such as quadratic programming, simulated annealing or genetic algorithms, but these are not considered further in this chapter. Problems will also arise if an incorrect or incomplete set of parameters is chosen, or even worse, if the structure of the model is wrong.

The modal residual in Equation (1.1) is a non-linear function of the parameters and the minimization is solved using a truncated linear Taylor series and iteration. Thus the Taylor series is

$$\mathbf{z}_m = \mathbf{z}_j + \mathbf{S}_j \delta\theta_j + \text{higher order terms} \tag{1.3}$$

where

$$\mathbf{z}_j = \mathbf{z}(\theta_j), \quad \mathbf{S}_j = \mathbf{S}(\theta_j), \quad \delta\theta_j = \theta_m - \theta_j. \tag{1.4}$$

The matrix $\mathbf{S}_j$ consists of the first derivatives of the modal quantities with respect to the model parameters, index j denotes the jth iteration and θ_m is the parameter vector that gives the measured outputs. Standard methods exist to calculate the modal derivatives required (Friswell and Mottershead, 1995; Adhikari and Friswell, 2001). By neglecting higher order terms in Equation (1.3), an iterative scheme may be derived, using the linear approximation,

$$\delta\mathbf{z}_j = \mathbf{S}_j \delta\theta_j \tag{1.5}$$

where $\delta\mathbf{z}_j = \mathbf{z}_m - \mathbf{z}_j$ and $\delta\theta_j = \theta_{j+1} - \theta_j$. Often, for damage location studies, only the residual and sensitivity matrix for the initial model are used. Avoiding iteration reduces the computation required, particularly where multiple parameter sets have to be estimated. However, particularly if the damage is severe, there is a risk that the wrong location is identified.

As indicated above, one of the problems with sensitivity methods is the need for a parameteric model of the damage. Mottershead et al. (1999) proposed an approach where the system was constrained so that unknown stiffnesses are replaced with rigid connections. The constraint is not imposed physically but the behaviour inferred from the unconstrained measurements. The best fit between the measured and predicted data is obtained when the damage is located in the substructure that is made rigid.

1.3 Model Parameters

One of the key aspects of a model based identification method is the parameterisation of the candidate damage. Since inverse approaches rely on a model of the damage, the success of the estimation is dependent on the quality of the model used. The type of model used will depend on the type of structure and the damage mechanism, which leads to an increase either in local or distributed flexibility. The damage model may be simple or complex. For example, a cracked beam may be modelled as a reduction in stiffness in a large finite element or substructure, or alternatively using a very detailed model from fracture mechanics. Whether such a detailed model is justified will often depend on the requirements of the estimation procedure and the quality of the measured data. Using a measured modal model consisting of the lower natural frequencies and associated mode shapes will mean that only a coarse model of the damage may be identified. The simple example used for illustration will use element stiffnesses as the parameters and is the simplest form of equivalent model for the damage. More detailed models of damage will be considered in Section 3.

1.4 Optimisation Procedures and Ill-Conditioning

When the parameters of a model are unknown, they must be estimated using measured data. Usually the measured response will be a non-linear function of the parameters. In these cases, minimizing the error between the measured and predicted response will produce a non-linear optimisation problem, with the usual questions about convergence and local minima. The most common approach is to linearise the residuals, obtain a least squares solution and iterate. If the identification problem is well posed then this simple approach will be adequate. The usual response to problems encountered in the optimisation is to try more advanced algorithms, but often the issue is that the estimation problem has not been posed correctly, and including some physical insight into the problem provides a much better solution.

Probably the most important difficulty in parameter estimation is ill-conditioning. In the worst case this can mean that there is no unique solution to the estimation problem, and many sets of parameters are able to fit the data. Many optimization procedures result in the solution of linear equations for the unknown parameters. The use of the singular value decomposition (SVD) (Golub and van Loan, 1996) for these linear equations enables ill-conditioning to be identified and quantified. The options are then to increase the available data, which is often difficult and costly, or to provide extra conditions on the parameters. These can take the form of smoothness conditions (for example, the truncated SVD), minimum norm parameter values (Tikhonov regularization) or minimum changes from the initial estimates of the parameters (Hansen, 1992, 1994).

Black box methods are often not considered as model based approaches. However any simulation of an input-output relationship must make some assumptions about the underlying process, and hence essentially has an underlying model. For example a neural network is essentially a very sophisticated curve fitting algorithm, and ill-conditioning is a major problem, evidenced by over-fitting and a lack of generalisation. The advantage of neural networks is that the class of input-output relationships that may be fitted is huge. However, better results will always be obtained if physical insight is used to guide the modelling and estimation process. Indeed there is often a need to reduce the number of input nodes to present to a neural network, and understanding is vital to obtain the correct feature extraction and data reduction. Another use of physical models is the generation of training or test data for these identification schemes. Typically experimental data for a sufficient range of events is difficult or expensive to obtain. Since running a model many times is relatively easy and cheap, these simulations may be used to increase the quantity of the test data. However it is vital that this simulated data correctly reproduces the important features of the real structure, and hence requires a validated and, if necessary, updated model.

Neural networks and genetic algorithms have been viewed as potential saviours for the solution of the difficult problems in damage location. Although these methods may be useful in some circumstances they do not deal with the root cause of the problem. Genetic algorithms have some advantage in finding a global minimum in very difficult optimisation problems, particularly where there are many local minima as is often the case in damage location. That said, the method still requires that the dynamics of the structure changes sufficiently and predictably enough for the optimisation to be meaningful. The crucial decision and difficulty is what to optimise, not the optimisation method used.

Neural networks are able to treat damage mechanisms implicitly, so that it is not necessary to model the structure in so much detail. The method can also deal with non-linear damage mechanisms easily. Models are still required to provide the training cases for the networks, and this is their major problem. There will always be systematic errors between the model used for training and the actual structure. For success, neural networks require that the essential features in the damaged structure were represented in the training data. The robustness of networks to these errors has not been tested sufficiently. Of course, the other major problem with both genetic algorithms and neural networks is that they require a huge amount of computation for structures of practical complexity, although these methods are well suited to parallel computation.

1.5 Problems and Errors in Damage Identification

The discussion thus far has indicated some of the problems with damage identification. There are always errors in the measured data and the numerical model that affect all of the algorithms. These errors, and the adequacy of the data, are now discussed. Damage identification algorithms should always be tested on realistic experimental examples, as many methods that work well on simulated data often fail due to the problems highlighted in this section. As a first step, methods may be tested using simulated data, but even then realistic systematic errors should be incorporated.

Modelling errors One of the major problems in damage location is the reliance on the finite element model. This model is also an important strength because the very incomplete set of measured data requires extra information from the model to be able to identify damage location. There will undoubtedly be errors even in the model of the undamaged structure. Thus if the measurements on the damaged structure are used to identify damage locations, the methods will have great difficulty in distinguishing between the actual damage sites and the location of errors in the original model. If suitable parameters are not included to allow for the undamaged model errors then the result will be a systematic error between the model and the data. Identification schemes generally have considerable difficulty with systematic errors. It is very likely that the original errors in the model will produce frequency changes that are far greater than those produced by the damage. There are two basic approaches to reducing this problem, although both rely on having measured data from an undamaged structure. The first is to update the finite element model of the undamaged structure to produce a reliable model (Friswell and Mottershead, 1995). Obviously the quality of the damage location assessment is critically dependent upon the updated model being physically meaningful (Friswell et al., 2001; Link and Friswell, 2003). Generally, this requires model validation using a control set of data not used for the updating. The second alternative uses differences between the damaged and undamaged response data in the damage location algorithm (Parloo et al., 2003; Titurus et al., 2003b). To first order, any error in the undamaged model of the structure that is also present in the damaged structure will be removed. This does rely on the structure remaining unchanged, except for the damage, between the two sets of measurements.

Another potential source of error is the mismach between the measurement locations and the model degrees of freedom. Such a mismatch makes the direct comparison of frequency response functions and mode shapes impossible, and the generation of residuals, inaccurate. The magnitude of the errors involved will depend on the mesh density in the sensor region and the complexity of the mode shapes. The best solution is to ensure nodes in the model exist at the sensor locations. Alternatively interpolation techniques may be used.

Environmental and other non-stationary effects One very difficult aspect of damage assessment is the change in the measured data due to environmental effects. This is one undesirable non-stationary effect and makes damage location very difficult. Of course progressive damage is also a non-stationary phenomenon, and damage can be difficult to identify if other non-stationary effects are also present. Typical environmental effects are demonstrated by highway bridges, especially those constructed using concrete, which have been the subject of many studies in damage location. For example, temperature changes can cause the stiffness properties of a bridge to change significantly, and the difficulty is to predict the effects of temperature from readily available measurements. Peeters and de Roeck (2001) reported on measurements of the Z24 bridge over a whole year and suggested a *black box* model to predict the temperature variation. Sohn et al. (1999) considered the effect of temperature on the Alamosa Canyon Bridge. Sohn et al. (2002) used a combination of time series analysis, neural networks and statistical inference to determine damage state for structures affected by the environmental conditions. Mickens et al. (2003) corrected frequency response function measurements by assuming the temperature affected the global stiffness of the structure. On a highway bridge, the changing traffic conditions cause different mass loading effects that can change the natural frequencies by as much as 1% (Zhang et al., 2002). There are further difficulties with highway bridges because they are highly damped with low natural frequencies. They are in a noisy environment and are difficult to excite. The frequency resolution in the measurements is invariably quite low, leading to considerable difficulties in detecting small frequency changes due to damage.

Typical of environmental effects are those in highway bridges. These bridges have been the subject of many studies in damage location, but in the UK, where most bridges are constructed using concrete, such identification has considerable problems with changes due to environmental factors (Wood, 1992). For example, concrete absorbs considerable moisture during damp weather, which considerably increases the mass of the bridge. Temperature changes the stiffness properties of the road surface, known as the *black-top*, significantly. On a hot summer's day in the UK, the road surface will provide little stiffness, but on a cold winter's day the stiffness contribution is considerable. The difficulty is trying to predict the effects of temperature and moisture absorption from readily available measurements. Figure 1 shows the variation of the first 4 natural frequencies of a concrete highway bridge in Birmingham, UK with soffit temperature (Wood, 1992). Soffit temperature is the variable that correlates best with the frequencies, but even then relatively large, unexplained variations in frequency occur.

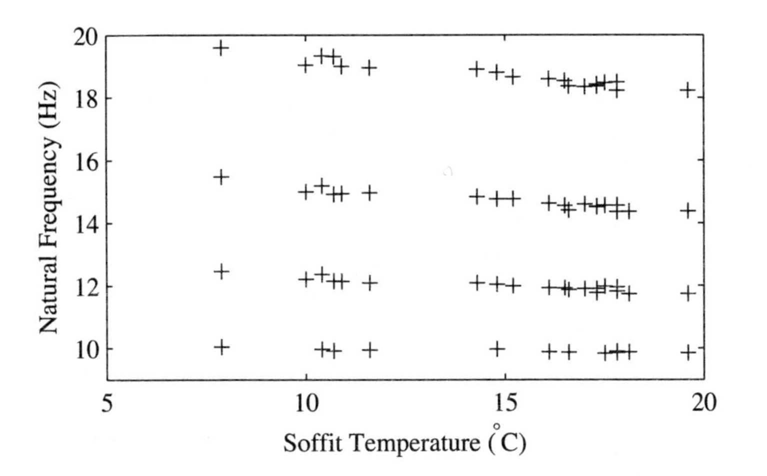

Figure 1. Variation of the 4 lowest natural frequencies of the New Haymills Bridge, Birmingham, UK.

The effect of frequency range The range of frequencies employed in damage location has a great influence on the resolution of the results and also the physical range of application. The great advantage in using low frequency vibration measurements is that the low frequency modes are generally global and so the vibration sensors may be mounted remotely from the damage site. Equally fewer sensors may be used. The problem with low frequency modes is that the spatial wavelengths of the modes are large, and typically are far larger than the extent of the damage. The spatial resolution of the damage identification scheme requires that there is a significant change in response between two adjacent potential damage sites. If low frequency modes are used then this resolution is closely related to the spatial wavelengths of the modes. Using high frequency excitation uses very local modes which are able to accurately locate damage, but only very close to the sensor and actuator position. Estimating accurate models at these high frequency ranges is also very difficult, and often changes in the response are used for damage identification. For example, Park et al. (2000, 2001) used changes in measured impedance to identify damage in civil structures and pipeline systems. Schulz et al. (1999) used high frequency transmissability to detect delaminations in composite structures.

Moving to even higher frequencies can also yield good results. Acoustic emission (Rogers, 2001) is a transient elastic wave, typically in the region of 50 to 500kHz, and is able, for example, to detect the energy released when cracks propagate. One approach to damage assessment is to use physical models to deduce quantitative relationships between measured acoustic emission signals and the damage mechanism that cause them. Significant research has been undertaken to obtain a physical understanding of various source mechanisms (Scruby and Buttle, 1991) and the radiation pattern of bulk shear and longitudinal acoustic waves that they produce (Ono, 1991). The difficulty in using these models in inverse estimation procedures is the accuracy of these high frequency models, and the huge computational requirements. In recent years a pattern recognition philosophy has dominated, that relies on using large databases of empirical data from which correlations between measured acoustic emission signals and damage mechanisms are inferred. Many advanced signal processing algorithms have been employed to interpret experimental data. Damage location is often determined using time of flight methods, that require the events to be well separated in time, the wave speed to be approximately equal in all parts of the structure, and the effects of reflection and refraction to be insignificant. Examples of the use of acoustic emission for health monitoring are given by Rogers (2001), Atherton et al. (2004) and (Holford et al., 2001).

Damage magnitude A frequent problem that arises in model-based vibration-based damage detection, whether parametric or non-parametric, is the need for a very accurate mathematical model, so that it correctly captures the actual structural dynamic behaviour in some predetermined frequency range. Often in structural health monitoring the changes in the measured quantities caused by structural damage are smaller than those observed between the healthy (i.e. undamaged) structure and the mathematical model. Consequently, it becomes almost impossible to discern between inadequate modelling and actual changes due to damage. There are two alternative approaches to this problem. The first is to update the healthy model so that the correlation between the model and the measured data is improved. This approach requires that the errors that remain after updating are smaller in magnitude than the changes due to the damage. Furthermore the changes to the model should be physically meaningful, so that the updating process corrects actual model errors, and doesn't merely reproduce the measured data. The second approach is based on the use of (relative) differences between data measured on healthy and potentially damaged structure. In this case, assuming that the only changes in the structure are due to damage, the problem may be reduced to finding those parameters that reproduce the measured changes.

Non-linearity Many forms of damage cause a change in the stiffness non-linearity that qualitatively and quantitatively affects the dynamic response of a structure. For example, Nichols et al. (2003a,b) used the features of the chaotic response of a structure to detect changes in a joint. Adams and Nataraju (2002) gave a variety of features based on the non-linear dynamic response. Kerschen et al. (2003) considered model based estimation methods and identified the form of nonlinearity that is most likely present in the measured data. Meyer and Link (2003) identified a parametric non-linear model using harmonic balance and a model updating approach. A breathing crack, which opens and closes, can produce interesting and complicated non-linear dynamics. Brandon (1998) and Kisa and Brandon (2000) gave an overview of some of the techniques that may be applied. Many techniques to analyse the resulting non-linear dynamics are based on approximating the bilinear stiffness when the crack opens and closes. Linear approaches to damage estimation approximates a local reduction in the stiffness matrix of the beam. Since the non-linearity introduced by a crack is often weak, many of the common testing techniques will tend to linearise the response (Friswell and Penny, 2002). Sinusoidal forcing will tend to emphasise the non-linearity, and damage detection methods based on detecting harmonics of the forcing frequency have been proposed (Shen, 1998). In rotor dynamic applications these approaches are useful because the forcing is inherently sinusoidal (Dimarogonas, 1996). However in structural health monitoring applications this approach requires considerable hardware and software to implement, and also requires a lengthy experiment. Johnson et al. (2004) used a transmissibility approach that was insensitive to boundary condition non-linearities. Neild et al. (2003) investigated the potential of a time frequency analysis procedure to identify damage in concrete beams.

Although using the non-linear response has a huge potential in health monitoring, model based inverse approaches have a number of difficulties because of the high number of degrees of freedom required, and therefore the computational burden imposed. In practice, any realistic multi degree of freedom non-linear analysis would have to be based on a reduced order model of the structure. Furthermore, many of the difficulties outlined in this section for linear systems, are also a problem for non-linear systems.

Strength vs. stiffness The philosophy of damage detection using measured vibration data is based on the premise that the damage will change the stiffness of the structure. In some instances there is a significant difference between strength and stiffness. Indeed, estimating the remaining useful life of a component based on conclusions from a dynamic analysis is very difficult. For example, a concrete highway bridge will have steel reinforcement cables running in channels in the concrete. The cables are tensioned, either before or after the concrete has set, to ensure that the concrete remains in compression. One major failure mechanism is by the corrosion of these cables. Once the cables have failed the concrete has no strength in tension and so the bridge is liable to collapse. Unfortunately the stiffness of the bridge is mainly due to the concrete, and so the progressive corrosion of the cables is very difficult to identify from stiffness changes. Essentially the dynamics of the bridge do not change until it collapses.

1.6 The Role of Simulation and Physical Testing

Many of the algorithms suggested for damage location are tested on simulated data. It is necessary to fully test any method on both simulated and real data. The simulated tests are able to fully exercise the location methods, with the benefit that the answer is known. In simulation, far more damage cases may be used and the effect of errors may be fully investigated. The need for real testing arises because experimental work always produces errors and problems that are unexpected. For simulation to be useful, the errors that might be expected in real structures must be simulated. Thus, adding random noise to a model of the structure and then using the same model to identify the damage in not enough! Most identification schemes are able to cope very well with random noise, and although such simulations are important parts of the overall performance assessment of an algorithm, they are not sufficient. It is vital that systematic type errors are included in the simulation. Thus, discretisation errors may be included by generating the simulated measurements using a fine finite element model; the damage mechanism introduced to generate the measurements may be different to those modelled for the identification; or boundary conditions on the structure could be changed between the measured data set and the identification.

1.7 A Simple Cantilever Beam Example

A simulated cantilever beam example will be used to demonstrate some of the problems. Although the example is somewhat artificial it will highlight how easily methods fail even on very simple structures. Any practical method would have to be robust and should therefore succeed on simple structures, even though some systematic errors are included. This example also demonstrates the use of simulation in damage identification. Not all of the methods are tried and this example is not supposed to represent an extensive scientific evaluation of the methods. Its purpose is for illustration.

The beam has a cross section of 25mm × 50mm, a length of 1m and is assumed to be rigidly clamped at one end. Only motion in the plane of the thinner beam dimension is considered. The beam has a Young's modulus of 210GN/m^2 and a mass density of 7800kg/m^3.

The first test of any method is its application to a simulated example with no noise or systematic errors. Any parameter changes in the model should be identified exactly. The simulated *measurements* are assumed to be the relative changes in the lower natural frequencies of the beam and are taken from a model with 20 elements. The undamaged natural frequencies are taken from the uniform beam, whilst the damaged frequencies are derived from a model where the stiffness of element 4 has been reduced by 30%. Table 1 gives the damaged and undamaged natural frequencies, showing that the 30% damage only results in a 2.4% change in natural frequency at most. These small frequency changes are typical in damage location and are one of the major difficulties in the identification of the location of damage. Measurement noise, environmental factors and structure non-stationarity can easily lead to incorrect conclusions on damage location.

Table 1. Natural frequencies of the simulated undamaged and damaged beam.

Mode No.	Undamaged (Hz)	Damaged (Hz)	Difference(%)
1	20.96	20.45	2.39
2	131.3	131.1	0.15
3	367.7	366.6	0.31
4	720.6	711.3	1.29
5	1191	1172	1.61
6	1780	1762	1.02
7	2487	2479	0.32
8	3313	3303	0.30

The standard sensitivity approach based on modal data will now be used to identify the damage. The set of candidate parameters is chosen to be relatively large and consists of the stiffness of each of the 20 elements. If the relative changes to the first 8 natural frequencies are used as the measurements then the identification of the parameters is underdetermined. In this case some form of regularisation must be employed. Figure 2 shows the change in element stiffness required to reproduce the damaged natural frequencies, using a minimum norm constraint on the parameter changes. Although the largest stiffness change occurs at element 4 the identified damage is spread over the whole beam, and there are some significant increases in stiffness. Note that this is the ideal case with no measurement noise or modelling errors. Suppose that, by some means, the damage is known to be somewhere in the eight elements closest to the fixed end. The number of parameters is now reduced to eight, the same as the number of natural frequencies. There is now a unique solution to the estimation problem and this solution is given in Figure 3. Note that the stiffness of elements 9 to 20 cannot change, but are included in Figure 3 for easy comparison. The damage has clearly been correctly located to element 4. However the magnitude of the damage is incorrect because the estimation is based on the sensitivity matrix which is a linear approximation to the residual. The other seven parameters are non-zero for the same reason.

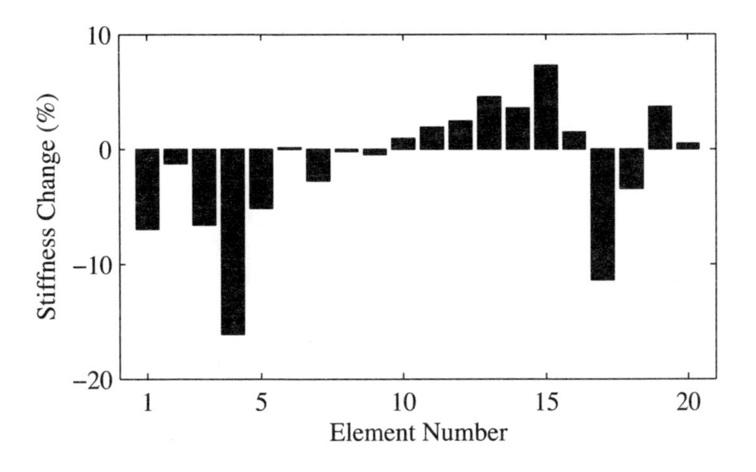

Figure 2. The change in element stiffness estimated for the cantilever beam example with no noise and a minimum norm constraint.

2 Regularisation

The advantages of sensitivity type model updating methods have been highlighted in this chapter. However there are significant differences in the application of these methods in model updating and damage location, which necessitates different methods of regularisation. In both cases the number of potential parameters is very large and the estimation process is likely to be ill-conditioned unless the physical understanding can be used to introduce extra information.

In model updating, the number of parameters may be reduced by only including those parameters that are likely to be in error. Thus if a frame structure is updated, the beams are likely to be modelled accurately but the joints are more difficult to model. It would therefore be sensible to concentrate the uncertain parameters to those associated with the joints. Even so, a large number of potential parameters may be generated, the measurements may still be reproduced and the parameters are unlikely to be identified

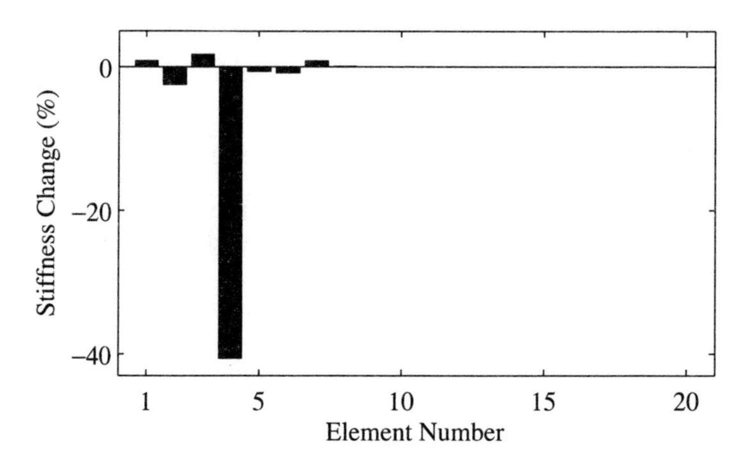

Figure 3. The change in element stiffness estimated for the cantilever beam example with no noise, but only 8 non-zero parameters.

uniquely. In this situation all the parameters are changed, and regularisation must be applied to generate a unique solution (Friswell et al., 2001). Regularisation generally applies extra constraints to the parameter estimation problem to ensure a unique solution. Applying the standard Moore-Penrose pseudo inverse is a type of regularisation where the parameter vector with the minimum norm is chosen. The parameter changes may be weighted separately to give a weighted least squares problem, where the penalty function is a weighted sum of squares of the measurement errors and the parameter changes. Such weighting may also be extended to include minimising the difference between equivalent parameters that are nominally equal in different substructures such as joints. Although using parametric models can reduce the number of parameters considerably, for damage location there will still be a large number of parameters. Most regularisation techniques rely on minimum norm type solutions that will tend to spread the identified damage over a large number of parameters. Using subset selection, where only the optimum subset of the parameters are used for the estimation (Friswell et al., 1997), has been used for model updating and also for damage location.

2.1 Tikhonov Regularisation

The treatment of ill-conditioned, noisy systems of equations is a problem central to finite element model updating (Ahmadian et al., 1998). Such equations often arise in the correction of finite element models by using vibration measurements. The regularisation problem centres around the linear equation,

$$\mathbf{A}\mathbf{x} = \mathbf{b} \tag{2.1}$$

where $\mathbf{x}$ is a vector of the m parameter changes we wish to determine, and $\mathbf{b}$ is a vector of n residual quantities derived from the measured data and the current estimate of the model. Note that the iteration index j has been dropped from the expressions in Section 1.2, and $\delta\theta$ has been replaced by $\mathbf{x}$. In model updating the relationship between the measured output (for example, natural frequencies, mode shapes, or the frequency response function) is generally non-linear. In this case the problem is linearised using a Taylor series expansion and iteration performed until convergence. When $\mathbf{b}$ is contaminated with additive, independent random noise with zero mean, it is well known that the least-squares solution, $\mathbf{x}_{LS}$, is unique and unbiased provided that $\operatorname{rank}(\mathbf{A}) = m$. When $\mathbf{A}$ is close to being rank deficient then small levels of noise may lead to a large deviation in the estimated parameters from its exact value. The solution is said to be unstable and Equation (2.1) is ill-conditioned

A different problem occurs when $m > n$ so that Equation (2.1) is under-determined and there are an infinite number of solutions. The Moore-Penrose pseudo-inverse in the form,

$$\mathbf{x}_{LS} = \mathbf{A}^\top \left[\mathbf{A}\mathbf{A}^\top\right]^{-1} \mathbf{b} \tag{2.2}$$

provides the solution of minimum norm, as does singular value decomposition (SVD). For the case when $\operatorname{rank}(\mathbf{A}) = r < \min(m, n)$, the SVD will again result in the minimum norm solution. This is a form of regularisation which has been widely applied in the model updating community. Unfortunately minimum norm solutions rarely lead to physically meaningful updated parameters.

Side Constraints Model updating often leads to an ill-conditioned parameter estimation problem, and an effective form of regularisation is to place constraints on the parameters. This could be that the deviation between the parameters of the updated and the initial model are minimised, or differences between parameters could be minimised. For example, in a frame structure a number of 'T' joints may exist that are nominally identical. Due to manufacturing tolerances the parameters of these joints will be slightly different, although these differences should be small. Therefore a side constraint is placed on the parameters, so that both the residual and the differences between nominally identical parameters are minimised. Thus if Equation (2.1) generates the residual, the parameter is sought which minimises the quadratic cost function,

$$J(\mathbf{x}) = \|\mathbf{A}\mathbf{x} - \mathbf{b}\|^2 + \lambda^2 \|\mathbf{C}\mathbf{x} - \mathbf{d}\|^2 \tag{2.3}$$

for some matrix $\mathbf{C}$, vector $\mathbf{d}$ and regularisation parameter λ. The regularisation parameter is chosen to give a suitable balance between the residual and the side constraint. For example, if there were only two parameters, which were nominally equal, then

$$\mathbf{C} = [\,1 \ \ -1\,] \tag{2.4}$$

Minimising Equation (2.3) is equivalent to minimising the residual of

$$\begin{bmatrix} \mathbf{A} \\ \lambda\mathbf{C} \end{bmatrix} \mathbf{x} = \begin{Bmatrix} \mathbf{b} \\ \lambda\mathbf{d} \end{Bmatrix}. \tag{2.5}$$

Equation (2.5) then replaces Equation (2.1), although with the significant difference that Equation (2.5) is generally over-determined, whereas Equation (2.1) if often under-determined. The constraints should be chosen to satisfy Morozov's complementation condition

$$\text{rank}\begin{bmatrix} \mathbf{A} \\ \mathbf{C} \end{bmatrix} = m \tag{2.6}$$

which ensures that the coefficient matrix in Equation (2.5) is full rank.

The Singular Value Decomposition The singular value decomposition (SVD) of $\mathbf{A}$ may be written in the form,

$$\mathbf{A} = \mathbf{U}\mathbf{\Sigma}\mathbf{V}^\top = \sum_{i=1}^{m} \sigma_i \mathbf{u}_i \mathbf{v}_i^\top \tag{2.7}$$

where $\mathbf{U} = [\mathbf{u}_1\mathbf{u}_2 \ldots \mathbf{u}_n]$ and $\mathbf{V} = [\mathbf{v}_1\mathbf{v}_2 \ldots \mathbf{v}_m]$ are $n \times n$ and $m \times m$ orthogonal matrices and

$$\mathbf{\Sigma} = \text{diag}\,(\sigma_1, \sigma_2, \ldots, \sigma_m) \tag{2.8}$$

where the singular values, σ_i, are arranged in descending order ($\sigma_1 \geq \sigma_2 \geq \ldots \geq \sigma_m$). In ill-posed problems two commonly occurring characteristics of the singular values have been observed; the singular values decay steadily to zero with no particular gap in the spectrum, and the left and right singular vectors $\mathbf{u}_i$ and $\mathbf{v}_i$ tend to have more sign changes in their elements as the index i increases.

The solution for the parameters using the SVD is

$$\mathbf{x} = \sum_{i=1}^{m} \frac{\mathbf{u}_i^\top \mathbf{b}}{\sigma_i} \mathbf{v}_i. \tag{2.9}$$

Thus the components of $\mathbf{A}$ corresponding to the low singular values have only a small contribution to $\mathbf{A}$ but a large contribution to the estimated parameters. The elements of these singular vectors (corresponding to the low singular values) are also generally highly oscillatory. Equation (2.9) shows the noise will be amplified when $\sigma_i < \mathbf{u}_i^\top \mathbf{b}$, and this may be used to decide where to truncate the singular values. If $\mathbf{A}$ does not contain noise then the singular values will decay to zero whereas the $\mathbf{u}_i^\top \mathbf{b}$ terms will decay to the noise level. Ahmadian et al. (1998); Hemez and Farhat (1995) consider this approach in more detail.

The standard SVD is incapable of taking account of the side constraint, as this requires the generalised SVD. Space does not permit a full explanation of the generalised SVD, and the reader is referred to Hansen (1994) for more complete detail of the decomposition. In Equation (2.5), $\mathbf{A}$ and $\mathbf{C}$ are decomposed as

$$\mathbf{A} = \mathbf{U}\begin{bmatrix} \mathbf{I} & \mathbf{0} \\ \mathbf{0} & \mathbf{\Sigma} \end{bmatrix}\mathbf{X}^{-1} \qquad \mathbf{C} = \mathbf{V}\,[\,\mathbf{0}\;\;\mathbf{M}\,]\,\mathbf{X}^{-1} \tag{2.10}$$

where $\mathbf{X}$ is a non-singular $m \times m$ matrix. $\mathbf{U}$ and $\mathbf{V}$ are $n \times m$ and $p \times p$ respectively, and their columns are orthogonal (but they are not related to the matrices $\mathbf{U}$ and $\mathbf{V}$ of the standard SVD) and $n \geq m \geq p$. The matrices $\mathbf{\Sigma}$ and $\mathbf{M}$ are

$$\mathbf{\Sigma} = \text{diag}\,(\sigma_1, \sigma_2, \ldots, \sigma_p) \qquad \mathbf{M} = \text{diag}\,(\mu_1, \mu_2, \ldots, \mu_p) \tag{2.11}$$

where $1 \geq \sigma_1 \geq \sigma_2 \geq \ldots \geq \sigma_p \geq 0$ and $0 \leq \mu_1 \leq \mu_2 \leq \ldots \leq \mu_p \leq 1$, and σ_i and μ_i are normalised so that,

$$\sigma_i^2 + \mu_i^2 = 1. \tag{2.12}$$

The p generalised singular values of $\begin{bmatrix} \mathbf{A} \\ \mathbf{C} \end{bmatrix}$, in decreasing order, are then

$$\gamma_i = \frac{\sigma_i}{\mu_i}. \tag{2.13}$$

The solution to Equation (2.5) is then

$$\mathbf{x} = \sum_{i=1}^{p} \frac{\gamma_i^2}{\gamma_i^2 + \lambda^2} \frac{\mathbf{u}_i^\top \mathbf{b}}{\sigma_i} \mathbf{v}_i + \sum_{i=p+1}^{m} \left(\mathbf{u}_i^\top \mathbf{b}\right) \mathbf{v}_i. \tag{2.14}$$

The regularisation parameter, λ, has the effect of damping the effect of the lower singular values (lower than about λ) and thus smoothing the solution. The expansion in terms of the SVD, (2.14), may also be used to specify a solution as a truncated SVD. If, instead of specifying λ, the series is truncated by only keeping the largest k singular values, then the solution is

$$\mathbf{x} = \sum_{i=1}^{k} \frac{\mathbf{u}_i^\top \mathbf{b}}{\sigma_i} \mathbf{v}_i + \sum_{i=p+1}^{m} \left(\mathbf{u}_i^\top \mathbf{b}\right) \mathbf{v}_i. \tag{2.15}$$

Picard's condition may be used to choose k, and the expansion is truncated when $\frac{\mathbf{u}_i^\top \mathbf{b}}{\sigma_i}$ becomes large.

'L' Curves One way of obtaining the optimum value of the regularisation parameter in the presence of correlated noise is to define an upper bound for the side constraint and minimise the residue,

$$\min_{\mathbf{x}} \|\mathbf{A}\mathbf{x} - \mathbf{b}\| \quad \text{subject to} \quad \|\mathbf{C}\mathbf{x} - \mathbf{d}\| \leq \gamma, \tag{2.16}$$

or alternatively to set a limit for the residue and minimise the deviation from the side constraint,

$$\min_{\mathbf{x}} \|\mathbf{C}\mathbf{x} - \mathbf{d}\| \quad \text{subject to} \quad \|\mathbf{A}\mathbf{x} - \mathbf{b}\| \leq \varepsilon. \tag{2.17}$$

Of course the success of this approach is highly dependent on the physical insight of the analyst in determining the allowable constraint violation or measurement error (residue magnitude).

A different approach is to plot the norm of the side constraint, $\|\mathbf{C}\mathbf{x} - \mathbf{d}\|$, against the norm of the residue, $\|\mathbf{A}\mathbf{x} - \mathbf{b}\|$, obtained by minimising the penalty function Equation (2.3) for different values of λ. Hansen (1992) showed that the norm of the side constraint is a monotonically decreasing function of the norm of the residue, and any point (ε, γ) on the curve is a solution to the two constrained least-squares problems

Equations (2.16) and (2.17). He pointed out that for a reasonable signal-to-noise ratio and the satisfaction of the Picard condition, the curve is approximately vertical for $\lambda < \lambda_{\text{opt}}$, and soon becomes a horizontal line when $\lambda > \lambda_{\text{opt}}$, with a corner near the optimal regularisation parameter λ_{opt}. The curve is called the 'L'-curve because of this behaviour. The optimum value of the regularisation parameter, λ_{opt}, corresponds to the point with maximum curvature at the corner of the log-log plot of the 'L'-curve. This point represents a balance between confidence in the measurements and the analyst's intuition.

Cross-Validation The idea of cross-validation is to maximise the predictability of the model by choice of the regularisation parameter λ. A predictability test can be arranged by omitting one data point, b_k, at a time and determining the best parameter estimate using the other data points, by minimising Equation (2.3). Then for each of the estimates, predict the missing data and find the value of λ that on average predicts the b_k best, in the sense of minimising the cross-validation function

$$V_0(\lambda) = \frac{1}{n}\sum_{k=1}^{n}\left(b_k - \tilde{b}_k(\lambda)\right)^2 \tag{2.18}$$

where $\tilde{b}_k(\lambda)$ is the estimate of b_k obtained from the remaining data. This is the method of cross-validation. Equation (2.18) is equivalent to (Ahmadian et al., 1998),

$$V_0(\lambda) = \frac{1}{n}\left\|[\text{diag}(\mathbf{I} - \mathbf{R}(\lambda))]^{-1}[\mathbf{A}\mathbf{x}(\lambda) - \mathbf{b}]\right\|^2 \tag{2.19}$$

where

$$\mathbf{R}(\lambda) = \mathbf{A}\left[\mathbf{A}^\top\mathbf{A} + \lambda^2\mathbf{C}^\top\mathbf{C}\right]^{-1}\mathbf{A}^\top \tag{2.20}$$

and diag denotes the matrix with zeros assigned to the off-diagonal terms.

2.2 Subset Selection

One solution to the problem of ill-conditioning is to select only a subset of the parameters for updating (Friswell et al., 1997). The parameters that are chosen are those to which the response data is sensitive, but the parameters must also be able to correct the errors in the model. Parameter subset selection is a technique that selects the best subset of parameters from a candidate set, utilising some application dependent cost function that provides a measure of goodness of each subset. Often, these techniques only obtain a sub-optimal estimate of the best subsets in some sense due to the excessive computational burden posed by the original problem. These techniques are firmly rooted in statistics and related fields (Millar, 1990), although recently applications in structural mechanics have appeared. Friswell et al. (1997) gave an overview of subset selection and also proposed the use of this technique for damage detection. They suggested an approach based on forward parameter subset selection, which is especially suited to local damage, and applied the method to a simulated cantilever beam example with physical parameters corresponding to either element or node properties. Different selection and iteration strategies were evaluated, and the case where multiple measurement sets are available was handled by computing the principal angles between two vector subspaces. Fritzen et al. (1998) used a orthogonalisation scheme for subset selection.

In damage location statistical methods and performance measures have been used that work on a similar principle (Cawley and Adams, 1979; Cawley et al., 1978; Friswell et al., 1994). Only a limited number of sites are assumed to be damaged, and the model updated based on the reduced number of parameters. This process is repeated for all possible combinations of damage site, and possibly even damage mechanism. The results from all the updated models are compared and the one that best matches the measured data is chosen.

The major problem with both subset selection and the statistical type approach, is that many smaller model updating exercises have to be performed. To optimally derive the best set of parameters, or the best damage location, requires the evaluation of many subsets of parameters. With a large number of parameters evaluating all subsets of even 2 or 3 parameters can become daunting. Thus sub-optimal methods must be used to derive good, but not necessarily the best, subsets of the parameters. In the forward approach parameters are chosen one at a time, and the parameters selected previously are retained. However there is no guarantee that the optimal subset will be found. The number of candidate damage locations may be controlled based on the expected reduction in the residual (Millar, 1990). The addition of a parameter to a previously selected subset inevitably reduces the residual terms, and thus there is a trade off between the number of parameters selected and the magnitude of the residual. Often only a single damage location will be required, in which case the optimal parameter may be determined. Often a reasonable number of parameter subsets (say between 3 and 20) are selected for more detailed study (Millar, 1990). Friswell et al. (1997) reviewed the relationship between subset selection and matrix decomposition, and also expanded the methods to parameter groups using subspace angles. Titurus et al. (2003b) considered the weighting requirements within the inner product defining the subspace angles, following the work of Knyazev and Argentati (2002).

The process of subset selection will now be described. It should be highlighted that the standard approach to subset selection is not iterative, but only uses Equation (2.1), evaluated at the initial parameter values. It would be possible to update each candidate parameter set until convergence, and then compare the performance of the different subsets, although it practice the computational cost is prohibitive. As the model parameters are usually local in nature and may also allow for different damage mechanisms, parameter subset selection selects parameters from $\mathbf{x}$ that identify both the damage location and mechanism. This formulation requires the selection of the optimum parameter subset from $\mathbf{x}$. The most straightforward approach is to use an exhaustive search where all $(2^m - 1)$ possible cases have to be searched. The number of cases renders this approach computational intensive and thus impractical in many real situations. Consequently sub-optimal schemes have to be used. An additional problem is that the addition of a parameter to a previously selected subset inevitably reduces the residual generated by Equation (2.1). Thus there is a trade off between the number of parameters selected and the magnitude of the residual.

Equation (2.1) may be written as

$$\mathbf{A}\mathbf{x} = [\mathbf{a}_1, \mathbf{a}_2, \ldots, \mathbf{a}_m]\,\mathbf{x} = \mathbf{b}. \tag{2.21}$$

The case of a single damage location leads to a simplified version of above philosophy. When only one parameter is selected, the optimum parameter is that which best fits the changes due to damage characterised by the vector $\mathbf{b}$ in Equation (2.21). Thus, the goal is to find the column $\mathbf{a}_j$ of matrix $\mathbf{A}$ that minimises

$$J = \|\mathbf{b} - \mathbf{a}_j \hat{x}_j\|^2 \tag{2.22}$$

where $\hat{x}_j$ is the least squares estimate of the jth parameter in $\mathbf{x}$. Friswell et al. (1997) showed that minimising Equation (2.22) is equivalent to finding the column of $\mathbf{A}$ that minimises the angle with $\mathbf{b}$. Hence the best parameter is the jth and found by

$$\min\left(\{\psi_1, \psi_2, \ldots, \psi_m\}\right) \quad \Rightarrow \quad \hat{x}_j, \mathbf{a}_j \tag{2.23}$$

where

$$\cos^2 \psi_i = \frac{\left(\mathbf{a}_i^\top \mathbf{b}\right)^2}{\left(\mathbf{a}_i^\top \mathbf{a}_i\right)\left(\mathbf{b}^\top \mathbf{b}\right)}, \quad i = 1, 2, \ldots, m \tag{2.24}$$

and ψ_i is the angle between vectors $\mathbf{a}_i$ and $\mathbf{b}$. This step is part of a general technique used in damage detection (Friswell et al., 1997) and is called forward parameter subset selection. This is a sub-optimal technique of subset selection, starting with the above step and continuing by additional parameter searches where the already selected parameters are retained. For subsequent steps a new modified problem is created, respecting the previous parameter selections. Suppose the single parameter with index j_1 has been chosen, then the parameter estimate $\hat{x}_{j_1}$ and residual ε are

$$\hat{x}_{j_1} = \frac{\mathbf{a}_{j_1}^\top \mathbf{b}}{\mathbf{a}_{j_1}^\top \mathbf{a}_{j_1}} \quad \Rightarrow \quad \varepsilon = \mathbf{b} - \hat{x}_{j_1} \mathbf{a}_{j_1}. \tag{2.25}$$

Note that ε is orthogonal to $\mathbf{a}_{j_1}$. A new parameter is then sought by considering the subspace defined by columns of $\mathbf{A}$, but orthogonal to $\mathbf{a}_{j_1}$. The modified problem is defined as (Friswell et al., 1997),

$$\mathbf{a}_j \to \mathbf{a}_j - \alpha_j \mathbf{a}_{j_1}, \qquad \mathbf{b} \to \mathbf{b} - \hat{x}_{j_1} \mathbf{a}_{j_1} \tag{2.26}$$

where

$$\alpha_j = \left(\mathbf{a}_{j_1}^\top \mathbf{a}_j\right) / \left(\mathbf{a}_{j_1}^\top \mathbf{a}_{j_1}\right). \tag{2.27}$$

A second parameter may now be selected by means of the modified problem defined by Equation (2.26), where $j \neq j_1$. Further parameters may be selected in the same way. An algorithm is thus created to search for the best parameter subset, denoted by $\left[\hat{x}_{j_1}, \hat{x}_{j_2}, \ldots, \hat{x}_{j_p}\right]$, that minimises the cost function,

$$J_p = \left\| \mathbf{b} - \sum_{i=1}^{p} \hat{x}_{j_i} \mathbf{a}_{j_i} \right\|^2. \tag{2.28}$$

This cost function is also employed in Efroymson's algorithm for forward subset selection, which focuses on adding or removing parameter selections from chosen subsets. Thus the number of candidate damage locations may be controlled based on the expected reduction in the residual (Millar, 1990; Friswell et al., 1997). Since only single damage location cases will be examined in detail here this subject will not be considered further.

Weighting The final theoretical aspect is the need for weighting when Equation (1.5) is used for damage location, and two types of weighting will be considered. First, weighting is needed to handle the different numerical values corresponding to the different modal quantities. Thus, only relative, or percentage changes in the modal quantities, due to damage, will be employed. The second type of weighting arises as a result of combining two different entities in the sensitivity matrix, namely natural frequencies and mode shapes for complete damage location. The experimental origin of the measurements means that the errors in mode shape estimation are usually greater than the errors in natural frequency estimation. This weighting is employed in the calculation of the subspace angles between the vector $\delta\mathbf{z}$ and the columns of the matrix $\mathbf{A}$, Equation (2.23), and is based on the weighted scalar product (Knyazev and Argentati, 2002). The procedure is also called the scalar A-based product and has its origins in statistics (note that the A in the name of this product has nothing to do with the matrix $\mathbf{A}$ in Equation (2.21)). Knyazev and Argentati (2002) studied this scalar product in the context of the numerically stable computation of principal angles between two linear subspaces. The scalar A-based inner product is defined as

$$(\mathbf{x}, \mathbf{y})_A = (\mathbf{x}, \mathbf{A}_W \mathbf{y}) = \mathbf{y}^\top \mathbf{A}_W \mathbf{x} \tag{2.29}$$

where $\mathbf{x}, \mathbf{y} \in \Re^n$ are vectors, and $\mathbf{A}_W \in \Re^{n \times n}$ is a symmetric, positive definite matrix. A-based vector and matrix norms, $\|\ldots\|_A$, may be defined as

$$\|\mathbf{x}\|_A = \sqrt{(\mathbf{x}, \mathbf{x})_A} = \left\| \mathbf{A}_W^{1/2} \mathbf{x} \right\|, \qquad \|\mathbf{B}\|_A = \left\| \mathbf{A}_W^{1/2} \mathbf{B} \mathbf{A}_W^{-1/2} \right\| \tag{2.30}$$

where $\mathbf{B} \in \Re^{n \times n}$ is an arbitrary matrix.

The applicability of this type of product for damage location based on the additional use of mode shape sensitivities in $\mathbf{S}$ and mode shape differences due to damage will be studied for an experimental, geometrically symmetric structure in a later section. Since $\delta\mathbf{z}$ is derived from experimental data, and assuming that the mass distribution does not change with damage, no additional scaling of individual mode shapes, with respect to other modes, will be employed. Since the natural frequencies are measured much more accurately than the mode shapes, the natural frequencies should be used to determine the candidate damage locations. The A weighting on the mode shapes is then used for geometrically symmetric structures to ensure that the most likely damage location from among the candidate locations identified from the natural frequencies is chosen. The weighting of the mode shapes is increased until a perceptible difference occurs between these candidate locations, but is kept as low as possible to reduce the effect of the noise on the mode shapes.

2.3 The Simple Cantilever Beam Example Revisited

The simple cantilever beam example of Section 1.7 will be used to demonstrate some of the properties of the methods given in this section. Candidate parameters now include element mass, and discrete mass and springs, as well as the element stiffness. Figure 4 shows the angles between the columns of the sensitivity matrix of the initial finite element model and the vector of the relative changes in the first 8 natural frequencies due to the damage. Clearly the column relating to the stiffness of element 4 has a small angle, although it is not zero because the method is based on a first order approximation and the extent of the damage (30%) is large. Changing the mass of element 17 is also able to model the *measured* changes accurately. This is a problem that relates to the symmetry of the beam, and the fact than no spatial information is incorporated into the measurements. Mode shapes could also be incorporated into the measurement vector, although the accuracy with which they could be measured may be insufficient to show a change in mode shape due to damage. This is an example of the more general problem, where damage or changes of parameters at more than one location causes the same changes in the lower natural frequencies.

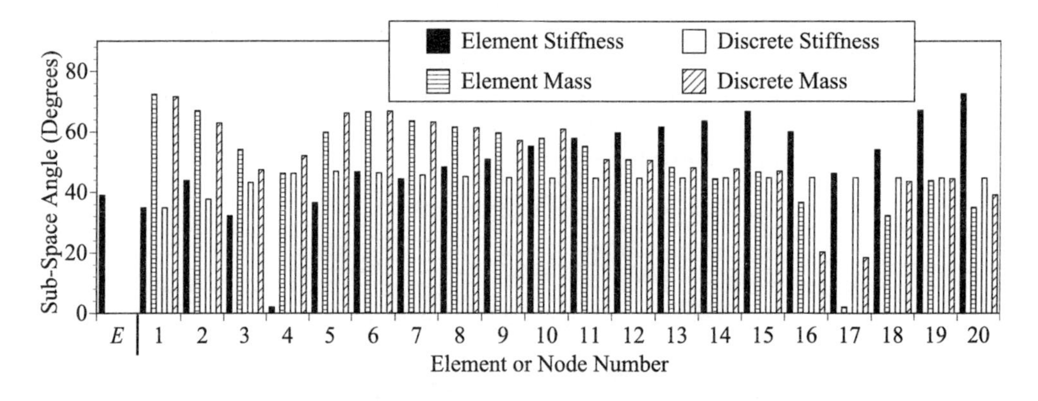

Figure 4. Subspace angles for a 30% change to the stiffness of element 4.

Subset selection is next demonstrated on an example where damage is introduced at 2 locations. A 0.1kg mass is added to node 12, in addition to the 30% stiffness change to element 4. This example does not include any measurement noise or modelling errors. Table 2 shows the results when the best subsets of 1, 2 and 3 parameters are chosen. The parameters are specified by type and element or node number. Thus $(\rho A)_{17}$ is the mass / unit length of element 17, $(EI)_4$ is the stiffness of element 4, k_1 is a discrete spring at node 1 and m_{12} is a discrete mass at node 12. At each stage the 2 best parameters are chosen. The residuals under the first 2 parameters relate to the values when a subset of size 1 or 2 is selected. Also shown are the residuals after convergence based on optimising the values of the chosen parameters. From the values of the residuals, it is clear that the two correct parameters should be selected. A number of other parameter subsets have small residuals and the addition of random noise would make the selection of the best subset more difficult.

Table 2. The selection of three parameters for the beam example.

Parameter 1			Parameter 2			Parameter 3		
	Res-idual	Conv-erged		Res-idual	Conv-erged		Res-idual	Conv-erged
$(\rho A)_{17}$	154.6	160.4	m_{12}	1.49	0.782	m_8	1.21	0.701
						$(\rho A)_{12}$	1.48	0.286
			m_8	8.70	4.76	m_{12}	1.21	0.701
						$(\rho A)_{12}$	8.68	4.75
$(EI)_4$	154.7	160.5	m_{12}	1.49	0.000	m_8	1.20	0.000
						$(\rho A)_{12}$	1.48	0.000
			m_8	8.70	5.40	m_{12}	1.20	0.000
						$(\rho A)_{12}$	8.69	5.35

3 Parameterisation of Models of Damage

Damage usually causes a reduction in the local stiffness of the structures. One option is to model this a reduction in stiffness at the element or substructure level. This equivalent modelling approach is often sufficient for the identification of local damage using low frequency vibration measurements. This section considers more detailed models of damage that have parameters that may be identified using inverse methods.

3.1 Crack Models

The modelling of cracks in beam structures and rotating shafts has been a significant research topic. The models fall into three main categories; local stiffness reduction, discrete spring models, and complex models in two or three dimensions. Dimarogonas (1996); Ostachowicz and Krawczuk (2001) gave comprehensive surveys of crack modelling approaches. The simplest methods for finite element models reduce the stiffness locally, for example by reducing a complete element stiffness to simulate a small crack in that element (Mayes and Davies, 1984). This approach suffers from problems in matching damage severity to crack depth, and is affected by the mesh density. An improved method introduces local flexibility based on physically based stiffness reductions, where the crack position may be used as a parameter for identification purposes. The second class of methods divides a beam type structure into two parts that are pinned at the crack location and the crack is simulated by the addition of a rotational spring. These approaches are a gross simplification of the crack dynamics and do not involve the crack size and location directly. The alternative, using beam theory, is to model the dynamics close to the crack more accurately, for example producing a closed form solution giving the natural frequencies and mode shapes of cracked beam directly or using differential equations with compatible boundary conditions satisfying the crack conditions (Christides and Barr, 1984; Sinha et al., 2002; Lee and Chung, 2001). Friswell and Penny (2002) compared several of the simple cracks models that may be used for health monitoring, for both the linear and non-linear response. Alternatively two or three dimensional finite element meshes for beam type structures with a crack may be used. Meshless approaches may also be used, but are more suited to crack propagation studies. No element connectivity is required and so the task of remeshing as the crack grows is avoided, and a growing crack is modelled by extending the free surfaces corresponding to the crack (Belytschko et al., 1995). However the compuational cost of these meshless methods generally exceeds that of conventional finite element analysis (FEA). Rao and Rahman (2001) avoided this difficulty by coupling a meshless region near the crack with an FEA model in the remainder of the structure. The two and three dimensional approaches produce detailed and accurate models but are a complicated and computational intensive approach to model simple structures like beams, and are unlikely to lead to practical algorithms for damage identification.

Models of open cracks Although the geometry of a crack can be very complicated, the contention in this paper is that for low frequency vibration only an effective reduction in stiffness is required. Thus, for comparison, a simple model of an open crack, which is essentially a saw cut, will be used. This will allow the comparison of models using beam elements, with those using plate elements. Only a selection of beam models will be used, that illustrate the fact that many beam models are able to model the effect of the crack at low frequencies.

Two standard approaches using beam elements are shown in Figure 5. In the first approach, the stiffness of a single element is reduced, which requires a fine mesh, and also the derivation of the effect of a crack on the element stiffness. In the second approach, the beam is separated into two halves at the crack location. The beam sections are then pinned together and a rotational spring used to model the increased flexibility due to the crack. Translational springs may also be used in place of the pinned constraint. The major difficulties with this approach is that a finite element node must be place at the crack location, requiring remeshing for health monitoring applications, and the relationship between the spring stiffness and crack depth needs to be derived.

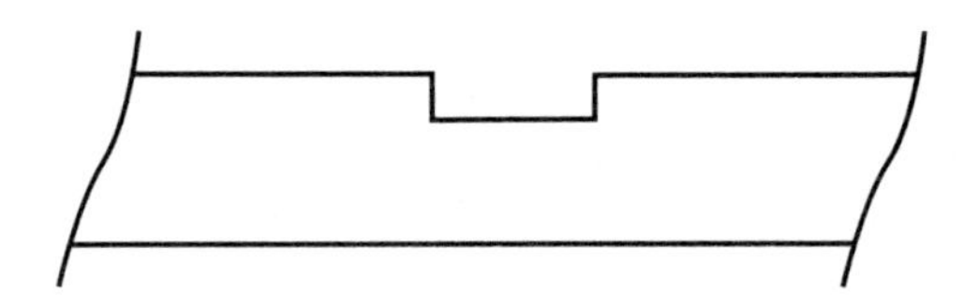

Reduction in Element Stiffness

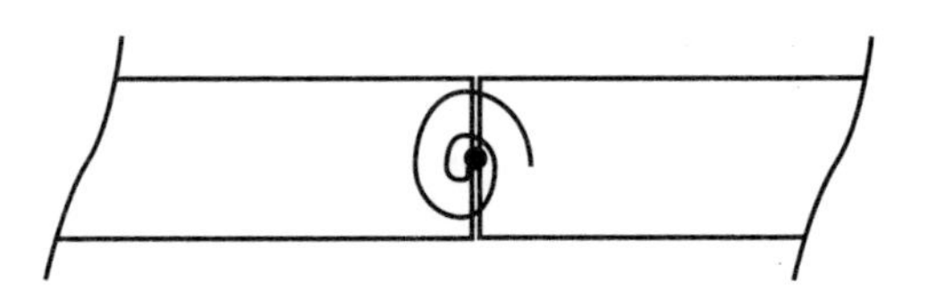

Pinned Joint at Crack Location

Figure 5. Simple crack models for beam elements.

For illustration, the open crack will be modeled using plate elements. The geometry is modeled by removing elements where the crack is located. Figure 6 shows this in the case of plate elements, and shows the side view of the mesh used. Clearly more complex methods may be used, and the review papers quoted earlier give further details.

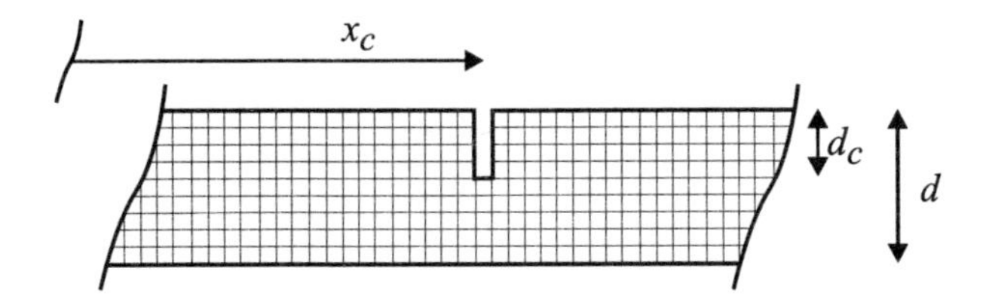

Figure 6. A simple crack model using plate elements.

The approach of Christides and Barr Clearly some of the material adjacent to the crack will not be stressed and thus will offer only a limited contribution to the stiffness. The actual form of this increased flexibility is quite complicated, but in this paper we approximate this phenomenon as a variation in the local flexibility. In reality, for a crack on one side of a beam, the neutral axis will change in the vicinity of the crack, but this will not be considered here. Shen and Pierre (1994); Carneiro and Inman (2001, 2002) have extended this approach to consider single edge cracks. Christides and Barr (1984) considered the effect of a crack in a continuous beam and calculated the stiffness, EI, for a rectangular beam to involve an exponential function given by

$$EI(x) = \frac{EI_0}{1 + C\exp\left(-2\alpha\left|x - x_c\right|/d\right)} \tag{3.1}$$

where $C = (I_0 - I_c)/I_c$. $I_0 = \dfrac{wd^3}{12}$ and $I_c = \dfrac{w(d-d_c)^3}{12}$ are the second moment of areas of the undamaged beam and at the crack. w and d are the width and depth of the undamaged beam, and d_c is the crack depth. x is the position along the beam, and x_c the position of the crack. α is a constant that Christides and Barr estimated from experiments to be 0.667. The inclusion of the stiffness reduction of Christides and Barr (1984) into a finite element model of a structure, using beam elements, is complicated because the flexibility is not local to one or two elements, and thus the integration required to produce the stiffness matrix for the beam would have to be performed numerically every time the crack position changed. Furthermore, for complex structures, without uniform long beams, Equation (3.1) would only be approximate. Sinha et al. (2002) used a simplified approach, where the stiffness reduction of Christides and Barr was approximated by a triangular reduction in stiffness. An example of this approximation is shown in Figure 7, for a crack of depth 5%, located at $x = 0$. The advantage of this simplified model is that the stiffness reduction is now local, and the stiffness matrix may be written as an explicit function of the crack location and depth. For cracks of small depth a good approximation to the length of the beam influenced by the crack is $2d/\alpha$.

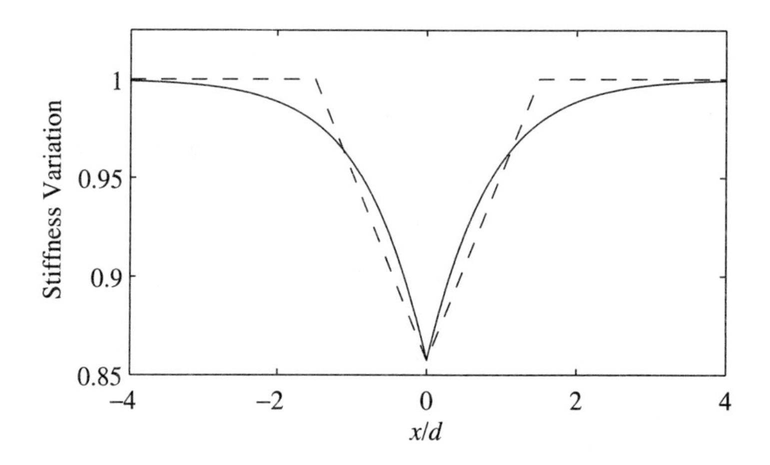

Figure 7. The variation in beam stiffness for the approaches of Christides and Barr (1984) (solid line) and Sinha et al. (2002) (dashed line).

Fracture mechanics approach An alternative approach is to estimate the increased flexibility caused by the crack, using empirical expressions of stress intensity factors from fracture mechanics. Lee and Chung (2001) gave such an approach based on the relationships given by Tada et al. (1973). Only a summary of the relevant equations will be given here. The element stiffness matrix is given by

$$\mathbf{K}_c = \mathbf{T}^\top \mathbf{C}^{-1} \mathbf{T} \tag{3.2}$$

where the transformation, $\mathbf{T}$, is

$$\mathbf{T} = \begin{bmatrix} -1 & 0 & 1 & 0 \\ 0 & -1 & 0 & 1 \end{bmatrix}. \tag{3.3}$$

The flexibility matrix, $\mathbf{C}$, for an element containing the crack in the middle, is given by

$$\mathbf{C} = \frac{1}{6EI} \begin{bmatrix} 2\ell_e^3 & 3\ell_e^2 \\ 3\ell_e^2 & 6\ell_e \end{bmatrix} + \frac{18\pi\left(1-\nu^2\right)}{Ewd^2} \begin{bmatrix} \ell_e^2 & 2\ell_e \\ 2\ell_e & 4 \end{bmatrix} \int_0^{d_c/d} \beta F_I^2(\beta)\, \mathrm{d}\beta. \tag{3.4}$$

where ℓ_e is the element length and ν is Poisson's ratio. $F_I(\beta)$ is the correction factor for the stress intensity factor, and may be approximated as

$$F_I(\beta) = \sqrt{\frac{\tan(\pi\beta/2)}{\pi\beta/2}} \frac{0.923 + 0.199\left[1 - \sin(\pi\beta/2)\right]^4}{\cos(\pi\beta/2)}. \tag{3.5}$$

This formulation does give the stiffness matrix of the element containing the crack explicitly in terms of the crack depth. There are two difficulties with using this approach for structural health monitoring. The main problem is that the crack is located at the centre of the element, requiring that the finite element mesh be redefined as the crack moves. Furthermore the stiffness matrix of the crack is a complicated function of the crack depth, and does not depend on the crack location explicitly.

A numerical comparison of the models The approaches to crack modelling will be compared using a simple example of a steel cantilever beam 1m long, with cross section 25 × 50mm. Bending in the more flexible plane is considered. The crack is assumed to be located at a distance 200mm from the fixed end, and has a constant depth of 10mm across the beam width.

The beam is modelled using 20 Euler-Bernoulli beam elements, and gives the natural frequencies shown in Table 3. For the plate elements the length is split into 401 elements and the depth into 10 elements. Thus the elements are approximately 2.5mm square. A large number of elements is required because an element with linear shape functions is used. Table 3 shows the estimated natural frequencies using the Quad4 element in the Structural Dynamics Toolbox (Balmes, 2000).

Table 3. Natural frequencies (in Hz) for the undamaged beam.

		Beam	Plate
Number DoF		40	13233
Modes	1	20.709	20.707
	2	129.78	129.39
	3	363.40	360.62
	4	712.16	701.96
	5	1177.4	1150.6

The damaged beam was also modelled using the approaches discussed earlier, and the results are shown in Table 4. The beam models all contain 20 elements, and the nodes are arranged such that the crack occurs in the middle of an element. Of course in the case of the discrete rotational spring a node is placed at the crack location. The reduction in the element stiffness is adjusted so that the percentage change in the first natural frequency is the same as that for the plate model. The other beam models are adjusted in a similar way. In the plate model, the crack is simulated by removing 4 elements and thus represents a saw cut 10 mm deep. The row of elements below the crack is also made thinner, so that the crack has negligible width. The differences in the lower natural frequencies are very similar for all models, and these differences are smaller than the changes that would occur due to small modelling errors, or changes due to environmental effects. Of course the accuracy at higher frequencies becomes less since the modes are influenced more by local stiffness variations.

Comparison with experimental results The previous section has shown that the natural frequencies predicted from different models are very close. Of course the question is whether the differences in these predictions are smaller than the measurement errors. As a demonstration the example of Rizos et al. (1990) will be used. Kam and Lee (1992) and Lee and Chung (2001) also used these results. The example is a steel cantilever beam of cross-section 20 × 20mm and length 300mm. Table 5 shows the measured and predicted frequencies of the uncracked beam. Rizos et al. (1990) propagated cracks at a number of different positions and depths, but here only a crack 80mm from the cantilever root, and depths of 2 and 6mm will be considered. Table 5 also shows the measured

Table 4. The percentage changes in the natural frequencies for the damaged beam.

		Beam				Plate
		Element Stiffness Reduction	Discrete Spring	Sinha et al. (2002)	Lee and Chung (2001)	
Modes	1	4.18	4.18	4.18	4.18	4.18
	2	0.07	0.04	0.08	0.04	0.04
	3	1.24	1.23	1.24	1.20	1.22
	4	2.99	3.08	2.98	2.99	3.07
	5	2.37	2.45	2.37	2.34	2.69

natural frequencies for these crack depths. The damaged cantilever beam is modelled using the beam methods described earlier. The depth of the crack is optimised so that the percentage change in the first mode matches the experimental result, to allow for possible errors in measuring the crack depth. Tables 6 and 7 show the measured and predicted frequency changes for the 2 crack depths. The results clearly show that the differences in the natural frequencies predicted by the models are smaller than the measurement errors. Thus the simple models for cracked beams may be used with confidence in health monitoring applications.

Table 5. The natural frequencies (in Hz) for the experimental cantilever beam example.

		FE Model	Experimental		
		Undamaged	Undamaged	2 mm Crack	6 mm Crack
Modes	1	185.1	185.2	184.0	174.7
	2	1159.9	1160.6	1160.0	1155.3
	3	3247.6	3259.1	3245.0	3134.8

Table 6. The percentage changes in the natural frequencies for the damaged beam with a 2 mm crack.

		Element Stiffness Reduction	Discrete Spring	Sinha et al. (2002)	Lee and Chung (2001)	Experi- mental
Modes	1	0.648	0.648	0.648	0.648	0.648
	2	0.065	0.063	0.130	0.063	0.052
	3	0.606	0.610	0.604	0.606	0.433

3.2 Composite Structures

Composite structures have an excellent performance, although this deteriorates significantly with damage. Unfortunately damage, due to impact events for example, are difficult to detect visually, and hence some method of non-destructive testing of these structures is required. Zou et al. (2000) reviewed the vibration based methods that are available to monitor composite structures. Since this paper considers inverse methods for

Table 7. The percentage changes in the natural frequencies for the damaged beam with a 6 mm crack.

		Element Stiffness Reduction	Discrete Spring	Sinha et al. (2002)	Lee and Chung (2001)	Experi-mental
	1	5.67	5.67	5.67	5.67	5.67
Modes	2	0.56	0.54	0.88	0.54	0.46
	3	4.92	4.95	4.49	4.92	3.81

damage estimation, this section will only consider the parameterisation of the damage in composite structures, and in particular the modelling of delaminations. Although composite structures have other modes of failure, such as matrix cracking, fibre breakage or fibre-matrix debonding (Ostachowicz and Krawczuk, 2001), these damage mechanisms produce similar changes in the vibration response to that obtained for damage in metallic structures. However delamination is a serious problem in composite structures, and has no parallel to damage mechanisms in other materials. Once the damage is parameterised then inverse methods, such as sensitivity analysis, may be applied.

Zou et al. (2000) reviewed methods to model delaminations, and here we will concentrate on simple models. For example, if a structure is modelled with beam or plate elements, then only beam or plates elements should be used to model the structure with delaminations. Delamination occurs when adjacent plies in a laminated composite debond. For beam structures the simpliest case of a through width delamination, parallel to the beam surface, was modelled using four beam segments (Majumdar and Suryanarayan, 1988; Tracy and Pardoen, 1989). Separate beam elements were used above and below the delamination, and the constraints to join these elements to those of the undamaged parts of the beam needed to be applied carefully. Zou et al. (2000) detailed further development of these models. One difficulty with using these models for parameter based identification is that changing the length and position of a delamination requires the model to be remeshed, and care must be exercised in calculating the associated sensitivity matrices. The techniques detailed by Sinha et al. (2002) for the position of cracks might be extended to this case. Paolozzi and Peroni (1990) highlighted that the most sensitive modes are those whose wavelength is approximately the same size as the delamination. Luo and Hanagud (1995) used a sensitivity based method to detect delaminations, and they also discovered that some modes split to give two closely spaced natural frequencies.

3.3 Joint Models and Generic Elements

One major difficulty in parametric approaches is that a model is required that accurately reflects the effect of damage on the mass and stiffness matrices. To some extent the situation is helped when low frequency vibration measurements are used because any local stiffness reduction will have a very similar effect on the dynamic response. Thus it is possible to use equivalent parameters, such as element stiffnesses, to model the damage. Generic elements (Gladwell and Ahmadian, 1995; Friswell et al., 2001) take this approach further by allowing changes to the eigenvalues and eigenvectors of the stiffness matrices of structural elements or substructures. These changes are usually constrained so that properties such as the rigid body modes and the geometric symmetry are retained.

Generic elements introduce flexibility into the joint in a controlled way. Other equivalent models, such as discrete rotational springs, offset parameters or changing element properties may also be used, although generic parameters do have advantages (Friswell et al., 2001). In particular, all models prejudge how the damage will affect the full model of the structure, whereas the generic element approach automatically finds the likely low frequency motion of the joint. Consider a two dimensional T joint constructed from three beam elements. Each node has three degrees of freedom and, since the substructure has four nodes, the substructure stiffness matrix has three rigid body eigenvectors and nine flexible eigenvectors (Titurus et al., 2003a). The lower eigenvectors have much simpler deformation shapes that are more likely to represent the motion the substructure would undergo in many of the global modes of the structure. Thus reducing the eigenvalues corresponding to these eigenvectors makes the joint substructure more flexible in the frequency range of the global dynamics, and may be used to model damage. Higher frequency eigenvectors of the substructure may also be included if the motion of the joint is more complex, however the lower eigenvectors of the joint are likely to adequately characterise the low frequency dynamics of the structure.

Generic elements have been developed for use in model updating and may be considered as equivalent models of elements or substructures (Gladwell and Ahmadian, 1995). Law et al. (2001) applied generic elements to the finite element model updating of the Tsing Ma bridge in Hong Kong. Wang et al. (1999) used generic elements in damage detection, dealing with the simulated problem of damage detection in a frame structure with flexible L-shaped and T-shaped structural joints.

The form of generic element parameterisation assumes that the damage only influences the stiffness properties and that the mass properties are modelled correctly. Thus only changes in the stiffness matrices are allowed. The eigenvalue problem for any selected sub-structure or element stiffness matrix can be written as

$$\left(\mathbf{K}^{\mathrm{SUB}} - \lambda_i \mathbf{I}\right)\phi_i = \mathbf{0}, \qquad \left(\mathbf{\Phi}^{\mathrm{SUB}}\right)^{\mathrm{T}} \mathbf{K}^{\mathrm{SUB}} \mathbf{\Phi}^{\mathrm{SUB}} = \begin{bmatrix} \mathbf{0} & \mathbf{0} \\ \mathbf{0} & \mathbf{\Lambda}_S \end{bmatrix} \tag{3.6}$$

where

$$\mathbf{\Phi}^{\mathrm{SUB}} = [\phi_1, \ldots, \phi_{n_R}, \phi_{n_R+1}, \ldots, \phi_{n_{SUB}}] = [\mathbf{\Phi}_R, \mathbf{\Phi}_S] \in \Re^{n_{SUB} \times n_{SUB}}, \tag{3.7}$$

and $n_R \leq 6$. $\mathbf{K}^{\mathrm{SUB}}$ is a sub-structure stiffness matrix, $\mathbf{\Phi}^{\mathrm{SUB}}$ is the eigenvector matrix of $\mathbf{K}^{\mathrm{SUB}}$, λ_i and ϕ_i are the ith eigenvalue and eigenvector of matrix $\mathbf{K}^{\mathrm{SUB}}$, respectively. Sub-matrix $\mathbf{\Lambda}_S$ is a diagonal matrix of non-zero eigenvalues of matrix $\mathbf{K}^{\mathrm{SUB}}$. The dimensions of these matrices depend on the size of the chosen sub-structure, where n_{SUB} is a number of degrees of freedom of substructure and $n_R \leq 6$ is the number of rigid body modes, $\mathbf{\Phi}_R, \mathbf{\Phi}_S$ are sub-matrices of $\mathbf{\Phi}^{\mathrm{SUB}}$ corresponding to the rigid and structural modes, respectively.

A modified set of sub-structure eigenvectors may be obtained by a linear transformation, as

$$[\mathbf{\Phi}_{0R}, \mathbf{\Phi}_{0S}] = [\mathbf{\Phi}_R, \mathbf{\Phi}_S] \begin{bmatrix} \mathbf{S}_R & \mathbf{S}_{RS} \\ \mathbf{0} & \mathbf{S}_S \end{bmatrix} \tag{3.8}$$

where the index 0 denotes the original quantities and matrices without index 0 represent modified quantities. Notice that in Equation (3.8) the modified rigid body modes do not contain any of the structural modes. By rearranging Equation (3.6) and using Equation (3.8), the modified sub-structure stiffness matrix may be written as

$$\mathbf{K}^{\mathrm{SUB}} = \mathbf{\Phi}_{0S}\mathbf{S}_S^\top \mathbf{\Lambda}_S \mathbf{S}_S \mathbf{\Phi}_{0S}^\top = \mathbf{\Phi}_{0S} \begin{bmatrix} \kappa_{1,1} & \cdots & \kappa_{1,(n_{SUB}-n_R)} \\ & \ddots & \vdots \\ \mathrm{SYM} & & \kappa_{(n_{SUB}-n_R),(n_{SUB}-n_R)} \end{bmatrix} \mathbf{\Phi}_{0S}^\top. \quad (3.9)$$

Equation (3.9) is the basis for generic element parameterisation for damage detection. $\kappa_{1,1}, \ldots, \kappa_{(n_{SUB}-n_R),(n_{SUB}-n_R)}$ are the most general parameters for this parameterisation. Employing additional assumptions related to the geometric symmetry or anti-symmetry of the corresponding eigenvectors will significantly reduce the total number of parameters. The sensitivity of natural frequencies with respect to these parameters is

$$\frac{\partial \lambda_i}{\partial x_j} = \phi_i^\top \frac{\partial}{\partial x_j}\left(\mathbf{K}_0 + \sum_{l=1}^{N_P} \mathbf{K}_l(\mathbf{x}_l)\right)\phi_i = \phi_i^\top \frac{\partial \mathbf{K}_r(\mathbf{x}_r)}{\partial x_j}\phi_i \quad (3.10)$$

where N_P is a number of parameterised substructures or elements, $\mathbf{x}_l$ is a group of parameters corresponding to lth substructure or element with corresponding stiffness matrix $\mathbf{K}_l$, $\mathbf{x}$ is a vector of all parameters, $\mathbf{K}_0$ is non-parameterised part of the global stiffness matrix, λ_i is the ith eigenvalue and x_j is jth parameter of a chosen parameterisation that is associated with the rth substructure or element.

Consider a two dimensional T joint constructed from three beam elements. Each node has three degrees of freedom and, since the substructure has four nodes, the substructure stiffness matrix has three rigid body eigenvectors and nine flexible eigenvectors. Figure 8 shows the nine flexible eigenvectors for this substructure, where the circles and dots represent the nodes and the dotted line is the undeformed joint. The finite element shape functions have been used to produce smooth deformation shapes. The lower eigenvectors have much simpler deformation shapes that are more likely to represent the motion the substructure would undergo in many of the global modes of the structure. Thus reducing the eigenvalues corresponding to these eigenvectors makes the joint substructure more flexible in the frequency range of the global dynamics. Higher frequency eigenvectors of the substructure may also be included if the motion of the joint is more complex, however the first two eigenvectors of the T joint were found to characterise the dynamics of the frame structure considered later. Gladwell and Ahmadian (1995) gave further explanation of the physical meaning of generic elements.

3.4 Distributed Damage

Teughels et al. (2002) presented a sensitivity-based finite element updating method for damage assessment that minimised differences between the experimental and predicted modal data. The parameterisation of the damage (both localisation and quantification) was represented by a reduction factor of the element bending stiffness. The number of unknown variables was reduced to obtain a physically meaningful result, by using a set of

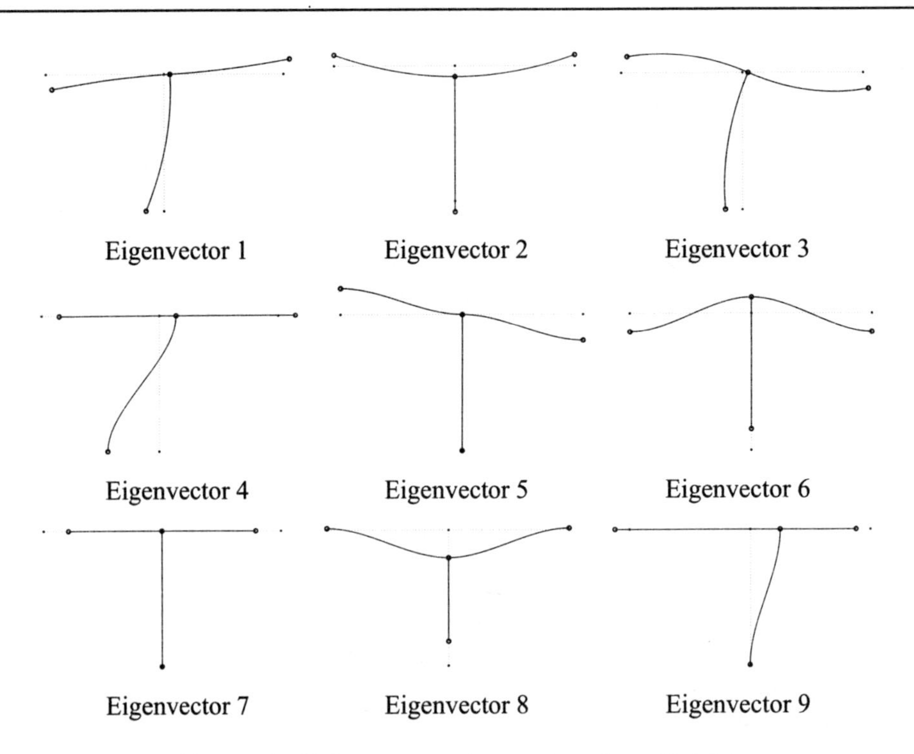

Figure 8. Substructure eigenvectors for a T joint.

damage functions to determine the spatial bending stiffness distribution. The updating parameters were then the multiplication factors of the damage functions. The procedure was illustrated on a reinforced concrete beam and on a highway bridge (Teughels and Roeck, 2004).

4 An Example of Subset Selection using Generic Elements

The proposed strategy is evaluated on a structure consisting of four thin-walled tubes connected to each other by four fillet welds. These joints were intentionally manipulated to produce one healthy and six damage cases. Titurus et al. (2003a) gave a detailed discussion of the identification results for the healthy/undamaged structure and Titurus et al. (2003b) described the estimation of the damage cases. Figure 9 shows the experimental structure, and Figure 10 shows the discretisation and experimental (EMA) measurement locations (the response was measured at the FE nodes). The finite element (FEM) nodes were placed at the measurement locations. Thus 32 degrees of freedom were measured, whereas the FE model contained 96 degrees of freedom (three degrees of freedom per node). The in-plane dynamics of the structure were measured, and the structure was supported in the free-free condition by elastic bands.

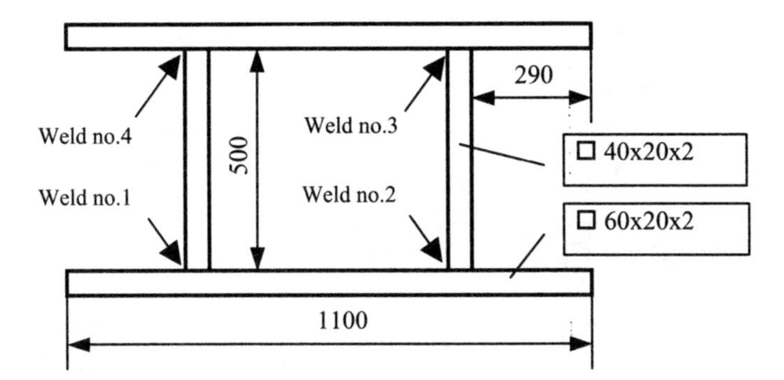

Figure 9. The outline of the H-frame structure (dimensions in mm).

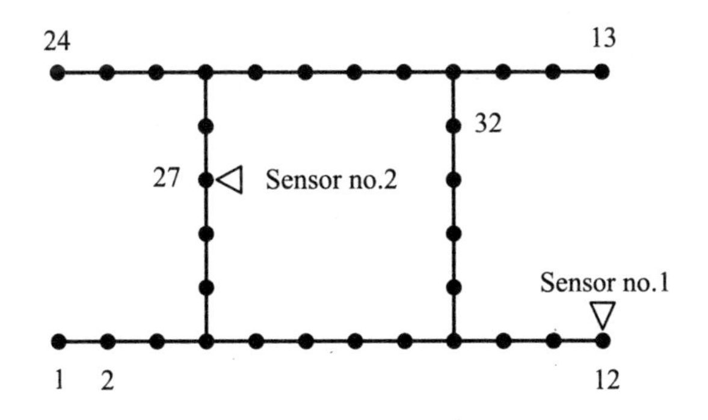

Figure 10. The discretisation of the H-frame structure.

4.1 Damage cases

Figure 11 gives a detailed description of all of the damage cases. These cases were produced by the intentional incompleteness of one or more of the fillet welds used to interconnect the four tubular parts of the structure. The shaded lines show the welds completed for each of the four joints, for each damage state. Note that State VII has all welds in place and hence is the undamaged structure. A distinctive feature of this structure is its geometrical symmetry, which is likely to cause problems for damage location based on measured natural frequencies alone. Two different approaches will be evaluated; the first assumes that the influence of the transducer mass will be sufficient to break the symmetry, whilst the second uses the measured mode shapes and their associated sensitivities. However, partial damage location will be tried, based on the use of natural frequencies alone.

These damage cases were selected to give a reasonable coverage of all possible combinations of damage cases, within practical constraints. This section concentrates on the single location damage case, and so only State VII, State VI and State V from Figure 11 will be studied in detail.

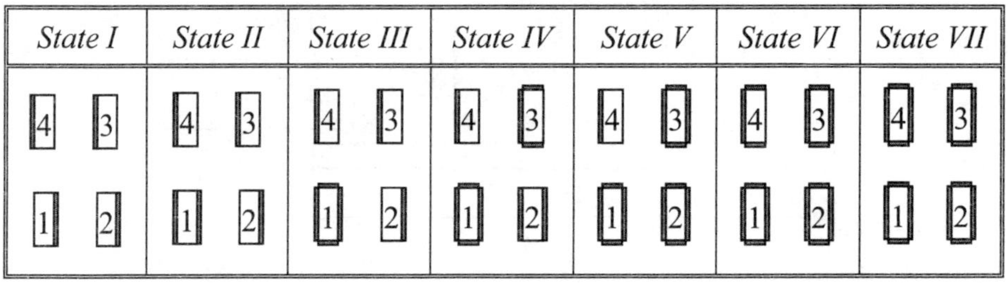

State I	*State II*	*State III*	*State IV*	*State V*	*State VI*	*State VII*
4 3	4 3	4 3	4 3	4 3	4 3	4 3
1 2	1 2	1 2	1 2	1 2	1 2	1 2

Figure 11. An overview of the damage cases considered.

4.2 Identification results for the damage cases

A full modal test was performed for each of the damage cases shown in Figure 11, however as the number of results is large only a selection will be considered here. The measurements were performed in the frequency range from 0 to 625 Hz. Table 8 gives the first nine measured natural frequencies for all of the damage cases. The fifth and sixth modes swap order between damage states III and IV. The last column corresponds to the undamaged/healthy structure, that is the structure with fully welded joints. Generally, the natural frequencies decrease with increasing level of damage, as a result of the decreasing stiffness of the structure. However, some small increases were observed in some natural frequencies from one case to another. One possible reason might be a small decrease in mass due to the absence of some weld material. Alternatively, taking the structure from the free-free suspension to undertake the welding may give small frequency changes due to slightly different suspension conditions.

Table 8. The natural frequencies (Hz) of the healthy (State VII) and damaged (State I to VI) structure. Note that modes 5 and 6 swap between States III and IV due to the damage.

Mode	State I	State II	State III	State IV	State V	State VI	State VII
1	27.63	33.57	34.18	48.60	50.26	60.06	60.57
2	118.64	120.94	120.74	125.04	124.82	126.60	126.53
3	126.38	129.92	130.64	138.97	139.63	147.86	147.05
4	169.77	172.28	172.32	174.77	175.42	175.86	175.89
5	264.77	275.73	275.42	280.09	280.33	280.81	280.76
6	279.64	280.38	280.24	300.44	301.72	319.60	320.56
7	298.91	312.58	317.15	342.10	347.97	359.55	360.70
8	393.61	396.78	399.06	418.68	420.05	436.24	437.72
9	550.55	551.78	552.71	560.06	560.63	565.93	566.52

Figure 12 shows the modal assurance criteria (MAC) matrices between the reference mode shapes of the healthy structure and mode shapes corresponding to the damaged cases. It is clear that the fifth and sixth modes interact and swap over between damage states I and VII. Another interesting feature shown by the MAC matrices is the relative insensitivity of the mode shapes to increasing damage, despite large changes in the natural frequencies. Titurus et al. (2003a) gave other experimental results, in particular the mode shapes corresponding to the healthy structure and further discussion of modelling issues.

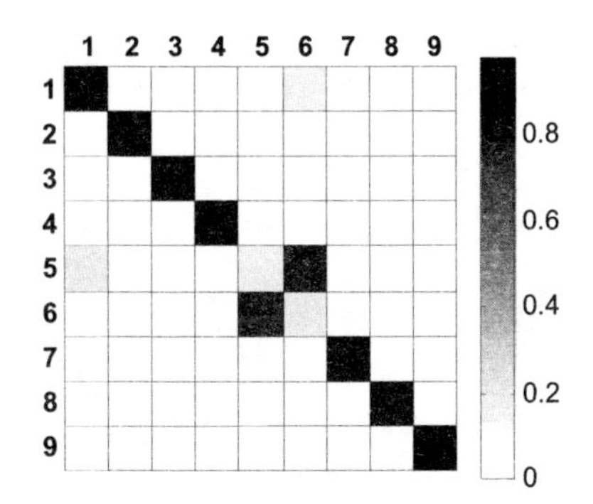

Figure 12. MAC criterion between State VII and damage State I. The rows of each MAC matrix correspond to mode shapes of State VII while columns correspond to mode shapes of State I.

4.3 Parameterisation overview

Section 3.3 provided a detailed explanation of parameterisations to be used for damage location, however, for the sake of completeness, a summary is provided here. Parameterisation A is expressed in terms of two groups of generic elements. The first group consists of one generic substructure that models the parts of the structure containing the fillet welds, and two parameters are required for each substructure, as shown in Figure 13. The other group consists of three different generic elements, each requiring one parameter, as shown in Figure 13. Thus, parameterisation A requires the parameter vector $\mathbf{x}$ given by

$$\mathbf{x} = [x_1, x_2, x_3, x_4, x_5]^\top = \left[\kappa_{11}^1, \kappa_{22}^1, \kappa_{11}^2, \kappa_{11}^3, \kappa_{11}^4\right]^\top \tag{4.1}$$

where κ_{jk}^i denotes the (j,k) element of the matrix of the ith element/substructure, based on generic elements as detailed in Equation (3.9). The values of these parameters may be determined by model updating. This parameterisation allows partial localisation to the type of region where damage has occurred.

Parameterisation B allows similar elements or substructures to have independent values of the corresponding generic parameters, to enable complete damage localisation. Parameterisation B requires 28 parameters for the H-shaped structure, as shown in Fig-

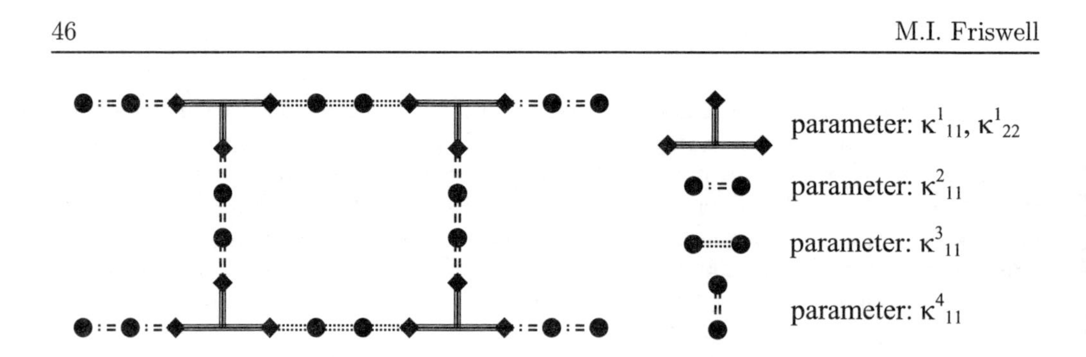

Figure 13. Parameterisation A of the baseline model of the thin-walled H structure.

ure 14, and these are defined as,

$$\begin{aligned} x_{2i-1} &= \kappa_{11}^{i}, \quad x_{2i} = \kappa_{22}^{i}, \qquad i = 1, 2, 3, 4, \\ x_{j+4} &= \kappa_{11}^{j}, \qquad\qquad\qquad\; j = 5, 6, \ldots, 24. \end{aligned} \tag{4.2}$$

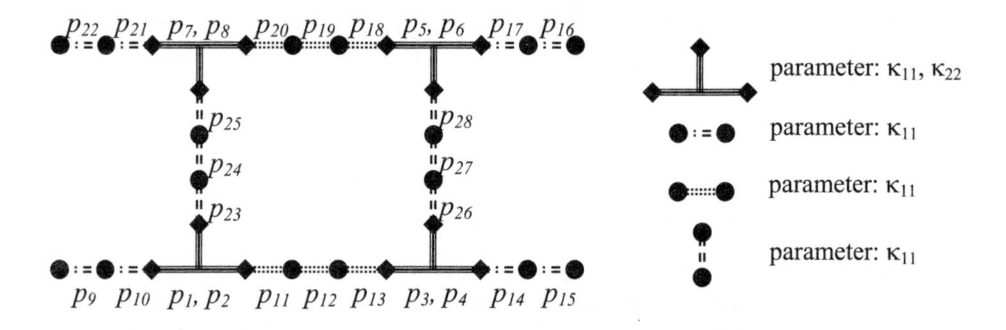

Figure 14. Parameterisation B of the baseline model of the thin-walled H structure.

4.4 Partial damage location - natural frequencies alone

In this section, parameterisation A will be used for partial damage localisation, using only the measured natural frequencies. This simplified form of damage localisation is chosen as a first step in the damage detection of a geometrically symmetric structure. The first seven natural frequencies corresponding to the healthy and damaged structures, as well as the sensitivity matrix **S**, determined at the updated parameters values (Titurus et al., 2003a), were used to test the subset selection approach proposed in Section 2.2.

Both the sensitivity matrix and the measured frequency differences for the considered damage cases, were normalised by the corresponding measured natual frequencies. Since single-location damage states are of primary interest, State VI and State V will be used, as they represent two levels of damage in one fillet weld. State IV and State II will also be considered, as these are multi-location damage states with different levels of damage (see Figure 11).

The results of damage location, in the form of subspace angles, are shown in Figure 15. Individual groups of columns correspond to particular model parameters. Within each group, corresponding to each parameter, the columns represent different comparisons of the damage cases (State II, IV, V, VI) with the healthy structure (State VII). Figure 15 suggests that the damage corresponds to parameter $x_1 = \kappa_{11}^1$, which corresponds to the generic substructures containing the welded joints. Thus the damage is correctly localised to the welded joint. An important feature of this study is that even relatively small damage corresponding to State VI is readily observable and clearly identifiable. An increasing level of damage, represented here by State V, leads to improved and clearer identification of damage location or damage type.

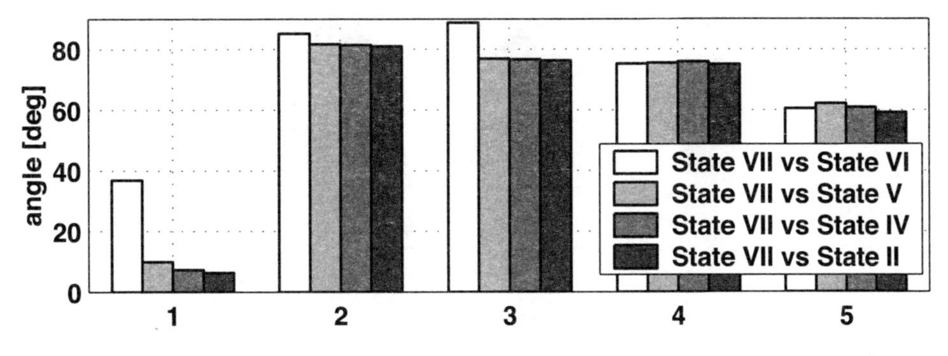

Figure 15. Subspace angles for partial localisation, parameterisation A.

The results disagree with the expectation that the increasing level of damage should lead to increasing subspace angles and consequently to a deteriorating quality of damage localisation, due to the increasing error in the linearisation of the original non-linear problem. However the level of damage for State VI is relatively small (the maximum difference between State VII and State VI is -0.83% for the first natural frequency, see Table 8) and therefore susceptible to measurement error. State V is characterised by a larger extent of damage, and therefore large differences in the natural frequencies (the maximum difference for this combination is -17.02% for the first natural frequency), produces smaller subspace angles. Figure 15 also gives the subspace angles for State II and State IV, and the angles corresponding to parameter 1 are smaller still. Although these are multi-damage cases, the damage still lies in the joints. Another noticeable and beneficial feature of the results is the insensitivity of the subspace angles from other parameters due to damage in the welded joints, reducing the possibility of false alarms.

Figure 16 provides additional information in terms of a selection tree. A selection tree is a representation of the forward parameter subset selection where each node of the tree corresponds to a selected subset and its colour represents the numerical value of the residual. The root of the tree represents the initial system, Equation (4.2). The branching factor and the depth of the tree are decided in advance and in our case results in binary selection trees with three levels corresponding to the selected parameter subsets. All three figures are determined using the first seven natural frequencies. The second best single parameter would be x_5, a parameter that also effectively monitors the stiffness in the regions connected with welded joints. The important relative indicators of damage level for a given situation are the absolute values of the residuals provided by the amplitude bar on the left of the Figure 16, as the ability to reproduce the measurement vector decreases with increasing damage level and consequently the magnitude of residuals also increases.

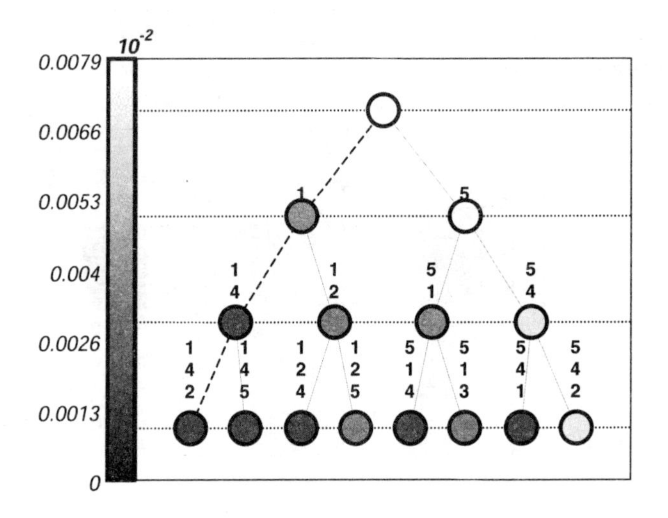

Figure 16. Binary tree representing forward selection, State VII vs. State VI.

4.5 Complete damage detection - natural frequencies and mode shapes

The only way to deal with damage localisation for geometrically symmetric structures is to use spatial information in the form of mode shapes. Once again the analysis was limited to the single location damage case, i.e. State VII (healthy), State VI (level 1 damage at welded joint 4, see Figure 9) and State V (level 2 damage at welded joint 4). The subspace angles corresponding to the individual parameters (parameterisation B) were computed by the techniques presented in Section 2.2. However, since the vector $\delta\mathbf{z}$ and the sensitivity matrix $\mathbf{S}$ contain elements corresponding to both the natural frequencies and the mode shapes, additional weighting must be included to represent the relative importance of the natural frequency and mode shape information, as presented in Section 2.2.

There are problems in using mode shape information, particularly since the accuracy of their estimation from measured data is worse than for natural frequencies. This is compounded since the proposed approach uses the differences between the measured damaged and measured undamaged mode shapes. Thus only mode shapes that are sensitive to the candidate damage sites should be chosen. Table 8 shows the changes in the natural frequencies for the different damage states, and gives a good indication of this sensitivity. However the table shows the sensitivity of the natural frequencies, which is not necessarily the same as the sensitivity of the mode shapes. Certainly, if the natural frequencies change very little with damage, then the corresponding mode shapes will not be sensitive. Thus, of the first seven modes, modes 2, 3, 4 and 5 are unlikely to give useful spatial information (note that mode 3 has been excluded because of the slight increase in the natural frequency in State VI). The sensitivity of the mode shapes to damage also increases with mode number, as the mode shapes corresponding to higher frequencies, contain more local deformation. Mode 1 is a global mode and therefore its shape is insensitive to damage.

The proposed approach using mode shapes will be demonstrated using spatial information from mode 6. Relative errors in the first seven natural frequencies and the difference in the mode shape elements were used. No further weighting was included, as similar results were obtained with other weighting values. Figure 17 shows the subspace angles corresponding to damage State V, and provides the correct indication of damage location, corresponding to parameter x_7. This parameter belongs to the fourth generic T substructure, representing fillet weld number 4 (see Figure 14). Other significant parameters indicated by the subspace angles are parameters x_{23} to x_{28}, which are located on the crossbar neighbouring the damaged region. However the results from the frequency only estimation clearly indicated that the damage is located in the joints. Thus damage in welded joint 4 may be confidently predicted.

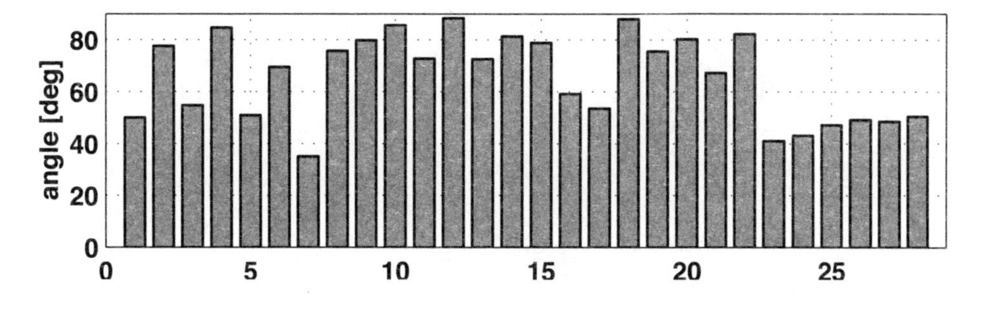

Figure 17. Subspace angles corresponding to State V, using seven natural frequencies and mode shape 6.

5 Methods using Mode Shapes

This section considers a different approach where changes in the measured modes shapes are used directly to detect and locate damage. Farrar and Jauregui (1998a,b) compared several of these methods, such as the damage index method (Stubbs et al., 1992), mode shape curvatures (Pandey et al., 1991), the change in flexibilities (Pandey and Biswas, 1994) or the change in stiffness (Zimmerman and Kaouk, 1994). The example used was a road bridge with a concrete deck and steel supports. Different levels of damage were introduced, but the damage was only clearly located with most methods at the most severe level where the first natural frequency changed by over 7%, and the mode shapes changed significantly. The damage index method (basically a measure of changes in modal strain energy) was found to be the most promising. Other methods based on pattern recognition, often using neural networks, are also popular (Sohn et al., 2001, 2002; Trendafilova and Heylen, 2003). These methods essentially provide curve fits using interpolation functions and are not based on physical models. The lack of a physical model also limits the scope for damage prognosis.

One of the major problems with methods using mode shapes is the incompleteness of the mode shapes. The number of measured degrees of freedom is always far smaller that the number of analytical degrees of freedom. Thus the mode shapes must be expanded or the analytical model must be reduced. In either case errors will be introduced because the model without damage is generally used for this reduction or expansion. The other source of incompleteness is the measured frequency range, which means that only a small number of modes will be measured. If damage causes a change in the stiffness matrix, this is further complicated because the high frequency modes have most effect on the elements of the stiffness matrix, but the lower frequency modes are generally measured. A huge number of different methods have been proposed and only a selection of the available methods will now be described.

5.1 COMAC

Perhaps the simplest method is the Coordinate Modal Assurance Criterion (COMAC). The usual Modal Assurance Criterion (MAC) correlates modes shapes by summing over the measured degrees of freedom. The COMAC sums over the modes and thus gives information about the correlation of the degrees of freedom (Lieven, 1988). If $\phi_{ui,j}$ is the jth element of the ith mode shape vector for the undamaged structure, and $\phi_{di,j}$ is the corresponding quantity for the damaged structure, then the COMAC for degree of freedom j is

$$\text{COMAC}_j = \frac{\left\|\sum_i \phi_{ui,j}\phi_{di,j}\right\|^2}{\left\|\sum_i \phi_{ui,j}\phi_{ui,j}\right\| \left\|\sum_i \phi_{di,j}\phi_{di,j}\right\|}. \tag{5.1}$$

Damage location is determined by those degrees of freedom with a low correlation between the healthy and damaged states. Note that local damage will affect all of the degrees of freedom to some extent, and so the changes in the mode shapes will not necessarily be local.

5.2 Balancing the Eigenvalue Equation

Suppose that the first r natural frequencies and mode shapes are measured. If λ_{di} are the eigenvalues and ϕ_{di} the mode shapes of the damaged structure, then

$$\left[\lambda_{di}^2\mathbf{M}_d + \lambda_{di}\mathbf{D}_d + \mathbf{K}_d\right]\phi_{di} = \mathbf{0}, \qquad i = 1,\ldots,r, \tag{5.2}$$

where $\mathbf{M}_d$, $\mathbf{D}_d$ and $\mathbf{K}_d$ are the mass, damping and stiffness matrices of the damaged structure, which are of course unknown. Suppose that damage only affects the stiffness matrix, then $\mathbf{M}_d = \mathbf{M}_u$, $\mathbf{D}_d = \mathbf{D}_u$ and $\mathbf{K}_d = \mathbf{K}_u - \delta\mathbf{K}$ where $\mathbf{M}_u$, $\mathbf{D}_u$ and $\mathbf{K}_u$ are the mass, damping and stiffness matrices of the undamaged structure, which are assumed to be known. In practice model updating may be used to ensure that the model of the undamaged structure accurately represents the measured dynamics. $\delta\mathbf{K}$ is the unknown change in the stiffness matrix due to damage. Equation (5.2) then becomes,

$$\mathbf{L}_i = \left[\lambda_{di}^2 \mathbf{M}_u + \lambda_{di}\mathbf{D}_u + \mathbf{K}_u\right] \phi_{di} = \delta\mathbf{K}\phi_{di}, \qquad i = 1, \dots, r. \tag{5.3}$$

In Equation (5.3) the vector $\mathbf{L}_i$ on the left side of the equation may be calculated. If the change in the stiffness matrix due to damage is local then only those parts of the stiffness matrix corresponding to the affected degrees of freedom will be non-zero, and hence the damage may be localised by the non-zero elements of the vector $\mathbf{L}_i$. If more than one mode is measured then the contribution from each mode may be averaged using the root mean square for each degree of freedom.

5.3 Modal Strain Energy

Zhang et al. (1998) presented a method which compared the modal strain energy within the elements. For element j, with corresponding element stiffness $\mathbf{K}_j$, the element modal strain energy ratio for the ith mode (SER_{ij}) is

$$SER_{ij} = \frac{\phi_i^\top \mathbf{K}_j \phi_i}{\phi_i^\top \mathbf{K} \phi_i} = \frac{\phi_i^\top \mathbf{K}_j \phi_i}{\omega_i^2} \tag{5.4}$$

where ϕ_i is the ith mass normalised mode shape, ω_i is the ith natural frequency and $\mathbf{K}$ is the stiffness matrix of the structure. A damage indicator β_{ij} is then defined as the difference of the element modal strain energy ratio before and after damage as,

$$\beta_{ij} = \frac{\phi_{di}^\top \mathbf{K}_j \phi_{di}}{\omega_{di}^2} - \frac{\phi_{ui}^\top \mathbf{K}_j \phi_{ui}}{\omega_{ui}^2} \tag{5.5}$$

where the subscripts u and d denote the healthy and damaged structures.

5.4 Direct and Minimum Rank Update Methods

Consider first the direct updating methods. The goal is often to reproduce the measured data (usually the modal model), by changing the stiffness matrix as little as possible (in some minimum norm sense). Historically these method were among the earliest in model updating (Friswell and Mottershead, 1995) and a number of generalisations have been proposed, depending on what is considered to be the *reference* quantities (Kenigsbuch and Halevi, 1998). A number of problems exist with the direct methods. There is no guarantee that the resulting matrices are positive definite (or semi-definite for structures with free-free modes), and extra modes may be introduced into the frequency range of interest. The standard methods do not enforce the connectivity of the structure, represented by the bandedness of the matrices and the pattern of zero terms, although Kabé

(1985) gave a method that enforced the expected connectivity. More fundamental is that forcing the model to reproduce the data does not allow for the errors that will be present in the measured data. Mode shapes, in particular, can only be measured with a limited accuracy. The major problem for damage location, and indeed for error location in model updating, is that all elements in the matrices may be changed. If only a small number of sites are modelled incorrectly (or are damaged) then only a small number of the matrix elements will be changed. Generally, because of the minimum norm optimisation in the updating method, all the matrix elements would be changed a little, rather than a small number of elements changed substantially. Thus the effect of any damage present would be spread over all the degrees of freedom making location difficult.

Zimmerman and Kaouk (1994) and Kaouk and Zimmerman (1994) proposed that the change in the stiffness matrix should be low rank. This does not ensure that the change in stiffness will be local, as the stiffness change could be global but low rank. The method requires the rank of the stiffness change to be less than or equal to the number of measured modes used in the update. Zimmerman et al. (1995) gave an overview of this approach, and discussed issues such as the number of measured modes to use. Doebling (1996) extended the method by updating the elemental parameter vector rather than the global stiffness matrix. Abdalla et al. (1998, 2000) developed methods by minimising the change in the stiffness matrix, while enforcing constraints such as symmetry, sparsity and positive definiteness.

The development begins by combining Equation (5.3) for all r measured modes to give,

$$\delta \mathbf{K} \mathbf{V}_d = \mathbf{M}_u \mathbf{V}_d \Lambda_d^2 + \mathbf{D}_u \mathbf{V}_d \Lambda_d + \mathbf{K}_u \mathbf{V}_d = \mathbf{B}, \tag{5.6}$$

where $\mathbf{V}_d = [\phi_{d1}\ \phi_{d2} \ldots \phi_{dr}]$ and $\Lambda_d = \operatorname{diag}[\lambda_{d1}\ \lambda_{d2} \ldots \lambda_{dr}]$

It may be proved (Zimmerman and Kaouk, 1994; Kaouk and Zimmerman, 1994) that the minimum rank of $\delta\mathbf{K}$ is r, and that this minimum rank solution to Equation (5.6) is

$$\delta \mathbf{K} = \mathbf{B} \left[\mathbf{B}^\top \mathbf{V}_d\right]^{-1} \mathbf{B}^\top. \tag{5.7}$$

5.5 Change in Flexibility

The flexibility matrix is the inverse of the stiffness matrix. In terms of the mode shapes the flexibility matrix $\mathbf{C}$ is

$$\mathbf{C} = \sum_{i=1}^{n} \frac{1}{\omega_i^2} \phi_i \phi_i^\top \tag{5.8}$$

where λ_i and ϕ_i are the ith natural frequency and mass normalised mode shape and n is the number of degrees of freedom in the model. Note that the lower (measured modes) have the largest influence on the flexibility matrix. The flexibility method (Pandey and Biswas, 1994) compares the flexibility matrices for the healthy and damaged structure, based on the r measured modes, as

$$\delta \mathbf{C} = \mathbf{C}_u - \mathbf{C}_d \tag{5.9}$$

where

$$\mathbf{C}_u = \sum_{i=1}^{r} \frac{1}{\omega_{ui}^2} \phi_{ui} \phi_{ui}^\top, \qquad \mathbf{C}_d = \sum_{i=1}^{r} \frac{1}{\omega_{di}^2} \phi_{di} \phi_{di}^\top. \tag{5.10}$$

A measure in terms of degrees of freedom is obtained by taking the maximum along each column of $\delta\mathbf{C}$.

6 Sensor Validation

The correct functioning of structural health monitoring systems requires that the sensors be functioning. Errors introduced by faulty sensors can cause undamaged areas to be identified as damaged. In many civil structures applications for health monitoring (such as bridges), ambient loads must be used for excitation. These loads are not known and may be measured or estimated as part of the health monitoring algorithm, which requires a large number of sensors. Sensor validation, where the sensors are confirmed to be functioning during operation, seems to have received little attention. The critical aspect in structural health monitoring is that there are usually more sensors than excited modes. This redundancy may be used, together with a modal model of the structure, to validate the sensor functionality.

The control and chemical engineering community have considered the sensor validation problem, and have used models and sensor redundancy to good effect. However, these approaches usually use the faulty sensor to predict the response and look for errors between predictions and measurement. Clearly using the faulty sensor in the prediction process will propagate errors to the predicted responses. Often neural networks, or artificial intelligence approaches are used for the analysis.

Friswell and Inman (1999) assumed that only the lower modes of the structure are usually excited, producing a large redundancy in the data. This has similarities to the principal component analysis used in chemical plant (Dunia et al., 1996; Dunia and Qin, 1998). Moreover, the approach seems to work only under the assumption of additive faults while giving erroneous results for the multiplicative faults case. Physically additive faults might arise from DC offsets in the electronic equipment and multiplicative faults might arise from calibration errors. The alternative used here, is to generate new residuals using the modal filtering approach which has similarities to the approach of Friswell and Inman. It is shown that these new residuals have interesting fault isolation properties. The approach is demonstrated on a subframe structure, although the method is completely general and may be applied to any structure for which a modal model is available. If necessary, such a model could be obtained from an identification experiment. For fault isolation a correlation index is proposed which is shown to correctly identify the faulty sensor.

Faults may cause a variety of changes in the dynamic response of a sensor, and many of these are difficult to model. However the two most common faults, namely additive and multiplicative faults, are relatively straightforward to model. Physically additive faults might arise from DC offsets in the electronic equipment and multiplicative faults might arise from calibration errors. In this section the sensor faults are assumed to be additive and modelled as a constant signal added to the sensor response. The problem of detecting sensor faults is then transformed into the problem of the detection of the change in the mean of a Gaussian variable with known covariance matrix, which switches from zero under the no-fault condition to a mean value with unknown magnitude under the fault condition. This problem may be solved using a likelihood ratio test resulting in a χ^2 distributed variable which is then compared to a threshold. In order to decide which sensor or subset of sensors is most likely to be responsible for the fault, the so-called sensitivity tests are computed, which are also χ^2 distributed.

6.1 Sensor Validation Concepts

Although there is redundancy in the data, based on the number of sensors and the number of modes excited, it is still not straightforward to identify those sensors that are damaged. When all sensors are working it is possible to estimate the modal contributions to the response and therefore produce a predicted response that will give some idea of the accuracy of the model of structure and the extent of the measurement noise. However if a sensor is damaged, then using data from this sensor to estimate the modal participation factors will propagate the errors from the faulty channel through the estimate of the modal response to the estimate of the response in all channels. Thus to predict faulty sensors the sensors are split into two groups. If S represents the set of all sensors then these two groups are,

$$\begin{aligned} S_f &= \{\text{sensors assumed to be faulty}\} \\ S_w &= \{\text{sensors assumed to be working}\} \end{aligned} \tag{6.1}$$

Note that these two sets are disjoint so that

$$S_f \cap S_w = \{\ \}, \qquad S_f \cup S_w = S. \tag{6.2}$$

Note that the distribution of faulty and working sensors seems to have been determined at the outset. In practice which sensors will be faulty is unknown and so every potential subset of faulty sensors must be tried. This approach has parallels with the subset selection technique in parameter estimation (Friswell et al., 1997; Millar, 1990). The difficulty in sensor validation, as in parameter estimation, is to determine which sensor or parameter subset is optimal. Note that for sensor validation, the number of assumed working sensors should be at least as great as the number of modes of interest.

6.2 Validation via Modal Filtering

Central to the proposed strategy for sensor validation is a modal model of the structure and also the estimation of the modal participation factors during operation. At any time instant, t_k, the measured output is

$$\mathbf{y}(t_k) = \mathbf{y}_k = \mathbf{H\Phi q}(t_k) = \mathbf{H\Phi}_r\mathbf{q}_r(t_k) + \mathbf{H\Phi}_d\mathbf{q}_d(t_k) \tag{6.3}$$

where the modes have been split into those that are retained, $\mathbf{\Phi}_r$, and those that will be discarded, $\mathbf{\Phi}_d$. If $\mathbf{H}_w$ picks out those outputs that are assumed to be working (i.e. are elements of S_w), then we need to estimate $\mathbf{q}_{r,k} = \mathbf{q}_r(t_k)$ from

$$\mathbf{y}_{w,k} = \mathbf{H}_w\mathbf{\Phi}_r\mathbf{q}_{r,k} + \mathbf{H}_w\mathbf{\Phi}_d\mathbf{q}_{d,k} \tag{6.4}$$

where $\mathbf{y}_{w,k}$ denotes the response at the fully functioning sensors at time t_k and $\mathbf{q}_{d,k} = \mathbf{q}_d(t_k)$. Clearly the discarded modes in Equation (6.4) could be neglected and the pseudo inverse used to estimate $\mathbf{q}_{r,k}$ from the resulting over-determined set of equations, as

$$\hat{\mathbf{q}}_{r,k} = (\mathbf{H}_w \boldsymbol{\Phi}_r)^{\dagger} \mathbf{y}_{w,k} \tag{6.5}$$

where $()^{\dagger}$ denotes the usual Moore-Penrose pseudo inverse. This gives an estimate of the response at the functioning sensors as

$$\hat{\mathbf{y}}_{w,k} = \mathbf{H}_w \boldsymbol{\Phi}_r (\mathbf{H}_w \boldsymbol{\Phi}_r)^{\dagger} \mathbf{y}_{w,k} = \mathbf{P}\mathbf{y}_{w,k}. \tag{6.6}$$

There will be an error introduced because

$$(\mathbf{H}_w \boldsymbol{\Phi}_r)^{\dagger} \mathbf{H}_w \Phi_d \neq \mathbf{0} \tag{6.7}$$

and a better estimate may be obtained by using the orthogonality of the modes as

$$\hat{\mathbf{q}}_{r,k} = \left(\boldsymbol{\Phi}_r^{\top} \mathbf{H}_w^{\top} \mathbf{M}_{w,r} \mathbf{H}_w \boldsymbol{\Phi}_r\right)^{-1} \boldsymbol{\Phi}_r^{T} \mathbf{H}_w^{\top} \mathbf{M}_{w,r} \mathbf{y}_{w,k} \tag{6.8}$$

where $\mathbf{M}_{w,r}$ is the mass matrix reduced to the degrees of freedom corresponding to the functioning sensors in the set S_w. Given that the mode shapes are assumed known, SEREP would be the most appropriate reduction method (O'Callahan et al., 1989). However, if the discarded modes lie outside the frequency range of interest then the estimator based on the pseudo inverse, Equation (6.5), will be adequate. The corresponding estimate of the response is

$$\hat{\mathbf{y}}_{w,k} = \mathbf{H}_w \boldsymbol{\Phi}_r \left(\boldsymbol{\Phi}_r^{\top} \mathbf{H}_w^{\top} \mathbf{M}_{w,r} \mathbf{H}_w \boldsymbol{\Phi}_r\right)^{-1} \boldsymbol{\Phi}_r^{\top} \mathbf{H}_w^{\top} \mathbf{M}_{w,r} \mathbf{y}_{w,k} = \mathbf{P}\mathbf{y}_{w,k}. \tag{6.9}$$

Both approaches give a projector matrix $\mathbf{P}$ from the response space to the space of the lower modes. The quality of the model may be determined by reconstructing the response at the functioning sensors and producing the error as

$$\varepsilon_{w,k} = (\mathbf{I} - \mathbf{P})\, \mathbf{y}_{w,k}. \tag{6.10}$$

Reconstructing the responses of the faulty sensors gives the error as

$$\varepsilon_{f,k} = \mathbf{y}_{f,k} - \mathbf{H}_f \boldsymbol{\Phi}_r \hat{\mathbf{q}}_{r,k} \tag{6.11}$$

where $\mathbf{H}_f$ picks out those outputs that are assumed to be faulty (i.e. are elements of S_f).

In practice we do not know which sensors are working and which are faulty. Therefore the errors in Equations (6.10) and (6.11) are generated for all possible sets S_w and S_f. Of course the estimation of the modal participation factors has been performed at every time step, and so the errors will be produced at every time step. The average error over the time range of interest may be easily computed. The projector matrix, $\mathbf{P}$, is constant for a particular choice of sets S_w and S_f and only needs to be computed once. Those sets where the error in the faulty sensor(s) is much greater than the error in the functioning sensors are then used to locate the faulty sensors.

6.3 The Parity Space Approach

Abdelghani and Friswell (2001) introduced a different approach to treating the residuals, that performs better on systems with multiplicative sensor errors. Three residuals are required. The first is related to the modal residuals given above, and is essentially the negative of the residual in Equation (6.10), and is

$$\gamma_k = \left[\mathbf{H}_w \mathbf{\Phi}_r \left(\mathbf{H}_w \mathbf{\Phi}_r \right)^\dagger - \mathbf{I} \right] \mathbf{y}_{w,k}. \tag{6.12}$$

The second residual is similar, but the complete set of sensors (including any faulty sensor) is used to calculate the modal quantities. Thus

$$\gamma_k^0 = \mathbf{H}_w \mathbf{\Phi}_r \left(\mathbf{\Phi}_r \right)^\dagger \mathbf{y}_k - \mathbf{y}_{w,k}. \tag{6.13}$$

The final residual is the difference between the two, namely

$$\xi_k = \gamma_k^0 - \gamma_k. \tag{6.14}$$

The damage correlation index is then given by

$$\rho = \frac{\mathrm{E}\left[\xi^\top \gamma^0 \right]}{\sqrt{\mathrm{E}\left[\xi^\top \xi \right]}} \tag{6.15}$$

where the expected value is over the time index k. This correlation index may be computed for each potentially faulty sensor, and hence any faulty sensor determined.

6.4 Example

The structure considered in this study consists of a suspended steel subframe used extensively in modal identification studies (Abdelghani et al., 1997). The structure was excited at two different locations using random noise inputs, and 28 accelerometers were used to measure the time response. The analysis was performed in the 0 – 500 Hz frequency range and 32000 data points per channel were collected at 1024 Hz sampling frequency.

All 28 sensors were used to identify the experimental natural frequencies, damping ratios and mode shapes from the first 3000 data samples. The Balanced Realization algorithm using data correlations was used for the identification (Abdelghani et al., 1999). Only the 5 first modes were retained (up to 300 Hz) and the natural frequencies and damping ratios are given in Table 9. The corresponding real mode shapes were then used to generate the residuals. Faults are added to the measured signals to simulate realistic behaviour. For all cases the data samples 3000-4000 were used to generate the residuals.

Suppose that an additive fault is simulated, where of 50% of the maximum response is added to sensor 6. Figure 18 shows the results using modal residuals and demonstrates that additive faults may be detected.

Next a multiplicative fault to sensor 6 is introduced: the time responses are multiplied by a factor of 1.5. The modal residuals perform poorly on multiplicative faults. Figure 19

Table 9. Frequencies and damping ratios identified using the Balanced Realisation algorithm.

Frequency (Hz)	Damping (%)
60.72	0.13
156.32	0.17
190.66	0.16
229.19	0.20
287.11	0.10

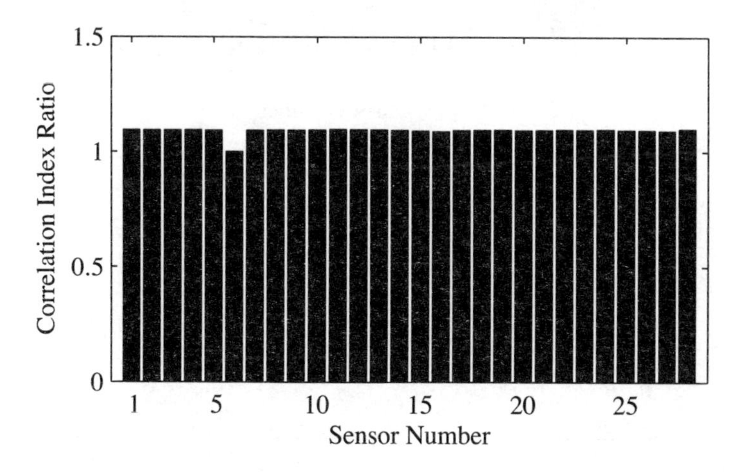

Figure 18. Correlation coefficient ratio under faulty and non-faulty conditions using modal residuals: Sensor 6 has an additive fault.

gives the results using modal residuals and clearly the fault has not been detected. Figure 19 shows the results of the parity space approach and shows that the technique is able to clearly isolate multiplicative faults. Furthermore it seems from these experiments that the neighbouring correlation ratios are not influenced by the faulty sensor(s). Figure 21 shows the results for a complete loss of sensor 8. Only the ratio of the corresponding sensor changes, while the others stay relatively unchanged. Finally the correlation ratio seems to be related to the amount of damage introduced to the sensors.

All of the above results are based on five modes and 28 sensor locations. As the number of modes used increases, the subspace in which the response lies increases, whereas the subspace in which the errors lie decreases. Thus the performance of the fault location scheme decreases. Similarly using more sensors improves the results, since by increasing the number of sensors the redundancy in the data is increased, which improves the detection and isolation of the faults. As an example a smaller number of sensors was used for the experiment, and more modes were included. Figure 22 shows the results using 10 sensors and nine modes, for the fault at sensor 8, and should be compared to Figure 21. Fault detection has now failed.

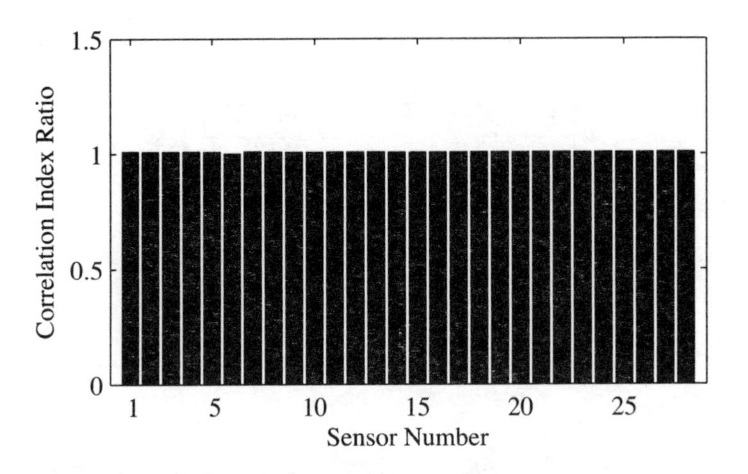

Figure 19. Correlation coefficient ratio under faulty and non-faulty conditions using modal residuals. Sensor 6 has a multiplicative fault.

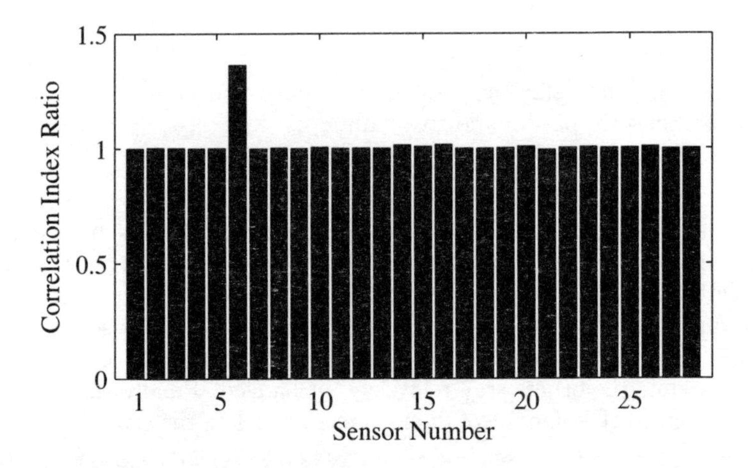

Figure 20. Correlation coefficient ratio under faulty and non-faulty conditions. Sensor 6 has a multiplicative fault.

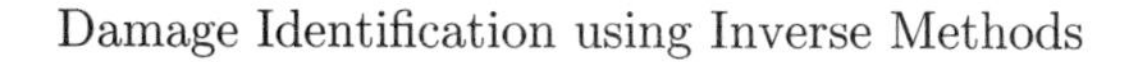

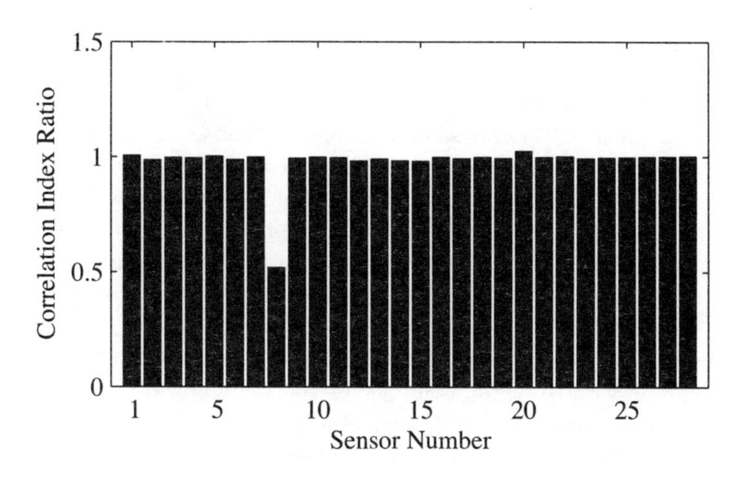

Figure 21. Correlation coefficient ratio under faulty and non-faulty conditions: complete loss of sensor 8.

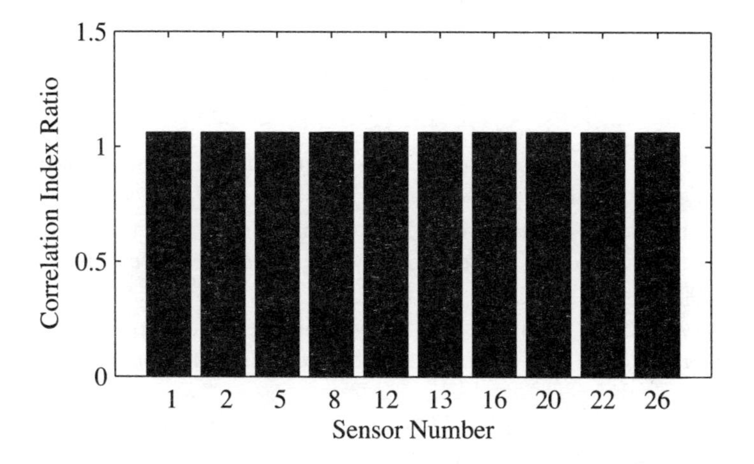

Figure 22. Correlation ratio under faulty and non-faulty conditions: Sensor 8 faulty. Nine modes used for the 10 sensors.

7 Conclusions

This chapter has given a brief introduction to the huge literature available on the approaches of damage identification based on inverse methods. The sensitivity based methods to identify physical parameters using subset selection for error localisation has been suggested as a viable approach. However, many difficulties remain to be fully solved, such as the modelling error between the model and the physical structure, and the influence of environmental factors. The most promising route is to include measurements of temperature, humidity and other environmental variables within the model, although this requires more stringent conditions on modelling error. At the very least these errors give a lower bound on the level of damage that can be detected and localised, and this can be formalised using statistics from the response of the undamaged structure in its normal operating environment. One scenario is that damage location using low frequency vibration is undertaken to identify those areas where more detailed local inspection should be concentrated. The application of robust damage detection and location algorithms based on monitoring the in-service response of a structure remains a challenge, although the availability of a model does open the way to more accurate prognosis and the estimation of the remaining life.

8 Acknowledgements

The author gratefully acknowledges the support of the Royal Society through a Royal Society-Wolfson Research Merit Award. The author also acknowledges the contribution of his co-workers, highlighted by reference to published papers throughout this chapter.

Bibliography

M. O. Abdalla, K. M. Grigoriadis, and D. C. Zimmerman. Enhanced structural damage detection using alternating projection methods. *AIAA Journal*, 36:1305–1311, 1998.

M. O. Abdalla, K. M. Grigoriadis, and D. C. Zimmerman. Structural damage detection using linear matrix inequality methods. *Journal of Vibration and Acoustics*, 122: 448–455, 2000.

M. Abdelghani and M. I. Friswell. A parity space approach to sensor validation. In *Proceedings of the 19th International Modal Analysis Conference*, pages 405–411, Orlando, Florida, 2001.

M. Abdelghani, C. T. Chou, and M. Verhaegen. Using subspace methods in the identification and modal analysis of structures. In *Proceedings of the 15th International Modal Analysis Conference*, pages 1392–1398, Orlando, Florida, 1997.

M. Abdelghani, M. Goursat, T. Biolchini, L. Hermans, and H. van der Auweraer. Performance of output-only identification algorithms for modal analysis of aircraft structures. In *Proceedings of the 17th International Modal Analysis Conference*, pages 224–230, Kissimmee, Florida, 1999.

D. E. Adams and M. Nataraju. A nonlinear dynamical systems framework for structural diagnosis and prognosis. *International Journal of Engineering Science*, 40:1919–1941, 2002.

S. Adhikari and M. I. Friswell. Eigenderivative analysis of asymmetric non-conservative systems. *International Journal for Numerical Methods in Engineering*, 51:709–733, 2001.

H. Ahmadian, J. E. Mottershead, and M. I. Friswell. Regularisation methods for finite element model updating. *Mechanical Systems and Signal Processing*, 12:47–64, 1998.

K. J. Atherton, C. A. Paget, and E. W. O'Brien. Structural health monitoring of metal aircraft structures with modified acoustic emission. In *Proceedings of SEM X International Conference on Experimental and Applied Mechanics*, California, USA, 2004.

E. Balmes. *Structural Dynamics Toolbox: For use with MATLAB*. Users Guide, Version 4, 2000.

T. Belytschko, Y. Y. Lu, and L. Gu. Crack propagation by element-free Galerkin methods. *Engineering Fracture Mechanics*, 51:295–315, 1995.

J. A. Brandon. Some insights into the dynamics of defective structures. *Proceedings of the Institution of Mechanical Engineers Part C: Journal of Mechanical Engineering Science*, 212:441–454, 1998.

S. H. S. Carneiro and D. J. Inman. Comments on the free vibrations of beams with a single-edge crack. *Journal of Sound and Vibration*, 244:729–736, 2001.

S. H. S. Carneiro and D. J. Inman. Continuous model for the transverse vibration of cracked Timoshenko beams. *Journal of Vibration and Acoustics*, 124:310–320, 2002.

P. Cawley and R. D. Adams. The location of defects in structures from measurements of natural frequencies. *Journal of Strain Analysis*, 14:49–57, 1979.

P. Cawley, R. D. Adams, C. J. Pye, and B. J. Stone. A vibration technique for non-destructively assessing the integrity of structures. *Journal of Mechanical Engineering Science*, 20:93–100, 1978.

S. Christides and A. D. S. Barr. One dimensional theory of cracked Bernoulli-Euler beams. *International Journal of Mechanical Science*, 26:639–648, 1984.

A. D. Dimarogonas. Vibration of cracked structures: a state of the art review. *Engineering Fracture Mechanics*, 55:831–857, 1996.

S. W. Doebling. Minimum-rank optimal update of elemental stiffness parameters for structural damage identification. *AIAA Journal*, 34:2615–2621, 1996.

S. W. Doebling, C. R. Farrar, and M. B. Prime. A summary review of vibration-based damage identification methods. *Shock and Vibration Digest*, 30:91–105, 1998.

J. V. A. dos Santos, C. M. M. Soares, C. A. M. Soares, and N. M. M. Maia. Structural damage identification in laminated structures using FRF data. *Comoposite Structures*, 67:239–249, 2005.

R. Dunia and S. J. Qin. Subspace approach to multidimensional fault identification and reconstruction. *AIChE Journal*, 44:1813–1831, 1998.

R. Dunia, S. J. Qin, T. F. Edgar, and T. J. McAvoy. Identification of faulty sensors using principal component analysis. *AIChE Journal*, 42:2797–2812, 1996.

C. R. Farrar and D. A. Jauregui. Comparative study of damage identification algorithms applied to a bridge: I. experiment. *Smart Materials and Structures*, 7:704–719, 1998a.

C. R. Farrar and D. A. Jauregui. Comparative study of damage identification algorithms applied to a bridge: II. numerical study. *Smart Materials and Structures*, 7:720–731, 1998b.

M. I. Friswell. Damage identification using inverse methods. *Special Issue of the Royal Society Philosophical Transactions on Structural Health Monitoring and Damage Prognosis*, 365:393–410, 2007.

M. I. Friswell and D. J. Inman. Sensor validation for smart structures. *Journal of Intelligent Material Systems and Structures*, 10:973–982, 1999.

M. I. Friswell and J. E. Mottershead. *Finite Element Model Updating in Structural Dynamics*. Kluwer Academic Publishers, 1995.

M. I. Friswell and J. E. T. Penny. Crack modelling for structural health monitoring. *Structural Health Monitoring: An International Journal*, 1:139–148, 2002.

M. I. Friswell, J. E. T. Penny, and D. A. Wilson. Using vibration data and statistical measures to locate damage in structures. *Modal Analysis: The International Journal of Analytical and Experimental Modal Analysis*, 9:239–254, 1994.

M. I. Friswell, J. E. T. Penny, and S. D. Garvey. Parameter subset selection in damage location. *Inverse Problems in Engineering*, 5:189–215, 1997.

M. I. Friswell, J. E. Mottershead, and H. Ahmadian. Finite element model updating using experimental test data: parameterisation and regularisation. *Transactions of the Royal Society of London, Series A*, 359:169–186, 2001.

C. P. Fritzen, D. Jennewein, and T. Kiefer. Damage detection based on model updating methods. *Mechanical Systems and Signal Processing*, 12:163–186, 1998.

G. M. L. Gladwell and H. Ahmadian. Generic element matrices suitable for finite element model updating. *Mechanical Systems and Signal Processing*, 9:601–614, 1995.

G.H. Golub and C. F. van Loan. *Matrix Computations*. The John Hopkins University Press, 1996.

P. C. Hansen. Analysis of discrete ill-posed problems by means of the L-curve. *SIAM Review*, 34:561–580, 1992.

P. C. Hansen. Regularisation tools: a MATLAB package for analysis and solution of discrete ill-posed problems. *Numerical Algorithms*, 6:1–35, 1994.

F. M. Hemez and C. Farhat. Bypassing the numerical difficulties associated with the updating of finite element matrices. *AIAA Journal*, 33:539–546, 1995.

K. M. Holford, A. W. Davies, R. Pullin, and D. C. Carter. Damage location in steel bridges by acoustic emission. *Journal of Intelligent Material Systems and Structures*, 12:567–576, 2001.

T. J. Johnson, R. L. Brown, D. E. Adams, and M. Schiefer. Distributed structural health monitoring with a smart sensor array. *Mechanical Systems and Signal Processing*, 18: 555–572, 2004.

A. M. Kabé. Stiffness matrix adjustment using modal data. *AIAA Journal*, 23:1431–1436, 1985.

T. Y. Kam and T. Y. Lee. Detection of cracks in structures using modal test datas. *Engineering Fracture Mechanics*, 42:381–387, 1992.

M. Kaouk and D. C. Zimmerman. Structural damage assessment using a generalised minimum rank perturbation theory. *AIAA Journal*, 32:836–842, 1994.

R. Kenigsbuch and Y. Halevi. Model updating in structural dynamics: A generalised reference basis approach. *Mechanical Systems and Signal Processing*, 12:75–90, 1998.

G. Kerschen, J. C. Golinval, and F. M. Hemez. Bayesian model screening for the identification of nonlinear mechanical structures. *Journal of Vibration and Acoustics*, 125: 389–397, 2003.

M. Kisa and J. A. Brandon. The effects of closure of cracks on the dynamics of a cracked cantilever beam. *Journal of Sound and Vibration*, 238:1–18, 2000.

A. V. Knyazev and M. E. Argentati. Principal angles between subspaces in an A-based scalar product: algorithms and perturbation estimates. *SIAM Journal on Scientific Computing*, 23:2008–2040, 2002.

S. S. Law, T. H. T. Chan, and D. Wu. Efficient numerical model for the damage detection of a large scale complex structure. *Engineering Structures*, 23:436–451, 2001.

Y.-S. Lee and M.-J. Chung. A study on crack detection using eigenfrequency test data. *Computers and Structures*, 77:327–342, 2001.

N. A. J. Lieven. Spatial correlation of mode shapes, the coordinate modal assurance criterion. In *Proceedings of the 6th IMAC*, pages 690–695, 1988.

M. Link and M. I. Friswell. Working group 1. generation of validated structural dynamic models - results of a benchmark study utilising the GARTEUR SM-AG19 testbed. *Mechanical Systems and Signal Processing, COST Action Special Issue*, 17:9–20, 2003.

H. Luo and S. Hanagud. Delamination detection using dynamic characteristics of composite plates. In *Proceedings of the AIAA/ASME/ASCE/AHS Structures, Structural Dynamics & Materials Conference,*, pages 129–139, California, USA, 1995.

P. M. Majumdar and S. Suryanarayan. Flexural vibrations of beams with delaminations. *Journal of Sound and Vibration*, 125:441–461, 1988.

I. W. Mayes and W. G. R. Davies. Analysis of the response of a multi-rotor-bearing system containing a transverse crack in a rotor. *Journal of Vibration, Acoustics, Stress and Reliability in Design*, 106:139–145, 1984.

S. Meyer and M. Link. Modelling and updating of local non-linearities using frequency response residuals. *Mechanical Systems and Signal Processing*, 17:219–226, 2003.

T. Mickens, M. Schulz, M. Sundaresan, A.Ghoshal, A. S. Naser, and R. Reichmeider. Structural health monitoring of an aircraft joint. *Mechanical Systems and Signal Processing*, 17:285–303, 2003.

A. J. Millar. *Subset Selection in Regression.* Monographs on Statistics and Applied Probability 40, Chapman and Hall, 1990.

J. E. Mottershead and M. I. Friswell. Model updating in structural dynamics: a survey. *Journal of Sound and Vibration*, 167:347–375, 1993.

J. E. Mottershead, M. I. Friswell, and C. Mares. A method for determining model-structure errors and for locating damage in vibrating systems. *Meccanica*, 34:153–166, 1999.

S. A. Neild, M. S. Williams, and P. D. McFadden. Nonlinear vibration characteristics of damaged concrete beams. *Journal of Structural Engineering*, 129:260–268, 2003.

J. M. Nichols, M. D. Todd, and J. R. Wait. Using state space predictive modeling with chaotic interrogation in detecting joint preload loss in a frame structure. *Smart Materials & Structures*, 12:580–601, 2003a.

J. M. Nichols, L. N. Virgin, M. D. Todd, and J. D. Nichols. On the use of attrator dimension as a feature in structural health monitoring. *Mechanical Systems and Signal Processing*, 17:1305–1320, 2003b.

J. C. O'Callahan, P. Avitabile, and R. Riemer. System equivalent reduction expansion process (SEREP). In *Proceedings of the 7th International Modal Analysis Conference, Las Vegas*, pages 29–37, 1989.

K. Ono. *Fatigue Crack Measurement: Techniques and Applications*, chapter Acoustic emission, pages 173–205. Engineering Materials Advisory Service Ltd., 1991.

W. Ostachowicz and M. Krawczuk. On modelling of structural stiffness loss due to damage. In *DAMAS 2001: 4th International Conference on Damage Assessment of Structures, Cardiff, UK*, pages 185–199, 2001.

A. K. Pandey and M. Biswas. Damage detection in structures using changes in flexibility. *Journal of Sound and Vibration*, 169:3–17, 1994.

A. K. Pandey, M. Biswas, and M. M. Samman. Damage detection from changes in curvature mode shapes. *Journal of Sound and Vibration*, 145:321–332, 1991.

A. Paolozzi and I. Peroni. Detection of debonding damage in a composite plate through natural frequency variations. *Journal of Reinforced Plastics and Composites*, 9:369–389, 1990.

G. Park, H. Cudney, and D. J. Inman. Impedance-based health monitoring of civil structural components. *ASCE Journal of Infrastructure Systems*, 6:153–160, 2000.

G. Park, H. Cudney, and D. J. Inman. Feasibility of using impedance-based damage assessment for pipeline systems. *Journal of Earthquake Engineering & Structural Dynamics*, 30:1463–1474, 2001.

E. Parloo, P. Guillaume, and M. van Overmeire. Damage assessment using mode shape sensitivities. *Mechanical Systems and Signal Processing*, 17:499–518, 2003.

B. Peeters and G. de Roeck. One-year monitoring of the Z24-bridge: enviromental effects versus damage events. *Earthquake Engineering & Structural Dynamics*, 30:149–171, 2001.

B. N. Rao and S. Rahman. A coupled meshless - finite element method for fracture analysis of cracks. *International Journal of Pressure Vessels and Piping*, 78:647–657, 2001.

P. F. Rizos, N. Aspragathos, and A. D. Dimarogonas. Identification of crack location and magnitude in a cantilever beam from the vibration modes. *Journal of Sound and Vibration*, 138:381–388, 1990.

L. M. Rogers. *Structural and Engineering Monitoring by Acoustic Emission Methods Fundamentals and Applications.* Technical Report, Lloyds Register of Shipping, London, England, 2001.

A. Rytter. *Vibration Based Inspection of Civil Engineering Structures.* PhD Dissertation, Aalborg University, Denmark, 1993.

M. J. Schulz, M. J. Pai, and D. J. Inman. Health monitoring and active control of composite structures using piezoceramic patches. *Composites Part B: Engineering*, 30:713–725, 1999.

C. B. Scruby and D. J. Buttle. *Fatigue Crack Measurement: Techniques and Applications*, chapter Quantitative fatigue crack measurement by acoustic emission, pages 207–287. Engineering Materials Advisory Service Ltd., 1991.

M.-H. H. Shen. *Structronic Systems: Smart Structures, Devices, And Systems, Vol. 1: Smart Materials and Structures*, chapter On-line structural damage detection, pages 271–332. World Scientific, 1998.

M.-H. H. Shen and C. Pierre. Free vibration of beams with a single-edge crack. *Journal of Sound and Vibration*, 170:237–259, 1994.

J. K. Sinha, M. I. Friswell, and S. Edwards. Simplified models for the location of cracks in beam structures using measured vibration data. *Journal of Sound and Vibration*, 251:13–38, 2002.

H. Sohn, M. Dzwonczyk, E. G. Straser, A.S. Kiremidjian, K. H. Law, and T. Meng. An experimental study of temperature effects on modal parameters of the Alamosa Canyon bridge. *Earthquake Engineering & Structural Dynamics*, 28:879–897, 1999.

H. Sohn, C. R. Farrar, N. F. Hunter, and K. Worden. Structural health monitoring using statistical pattern recognition techniques. *Journal of Dynamic Systems Measurement and Control*, 123:706–711, 2001.

H. Sohn, K. Worden, and C. R. Farrar. Statistical damage classification under changing environmental and operational conditions. *Journal of Intelligent Material Systems and Structures*, 13:561–574, 2002.

N. Stubbs, J.-T. Kim, and K. G. Topole. An efficient and robust algorithm for damage localization in offshore platforms. In *Proceedings of the ASCE 10th Structures Congress*, pages 543–546, 1992.

H. Tada, P. Paris, and G. Irwin. *The Stress Analysis of Cracks Handbook*. Del Research Corporation, 1973.

A. Teughels and G. De Roeck. Structural damage identification of the highway bridge Z24 by FE model updating. *Journal of Sound and Vibration*, 278:589–610, 2004.

A. Teughels, J. Maeck, and G. De Roeck. Damage assessment by FE model updating using damage functions. *Computers and Structures*, 80:1869–1879, 2002.

B. Titurus, M. I. Friswell, and L. Starek. Damage detection using generic elements: part I, model updating. *Computers and Structures*, 81:2273–2286, 2003a.

B. Titurus, M. I. Friswell, and L. Starek. Damage detection using generic elements: part II, damage detection. *Computers and Structures*, 81:2287–2299, 2003b.

J. J. Tracy and G. C. Pardoen. Effect of delamination on the natural frequencies of composite laminates. *Journal of Composite Materials*, 23:1200–1215, 1989.

I. Trendafilova and W. Heylen. Categorisation and pattern recognition methods for damage localisation from vibration measurements. *Mechanical Systems and Signal Processing*, 17:825–836, 2003.

D. Wang, M. I. Friswell, P.E. Nikravesh, and E. Y. Kuo. Damage detection in structural joints using generic joint elements. In *Proceedings of the 17th International Modal Analysis Conference (IMAC)*, pages 792–798, Orlando, Florida, 1999.

M. G. Wood. *Damage Analysis of Bridge Structures using Vibrational Techniques*. PhD Thesis, Aston University, UK, 1992.

K. Worden. Structural fault detection using a novelty measure. *Journal of Sound and Vibration*, 201:85–101, 1997.

K. Worden, G. Manson, and N. R. J. Fieller. Damage detection using outlier analysis. *Journal of Sound and Vibration*, 229:647–667, 2000.

L. M. Zhang, Q. Wu, and M. Link. A structural damage identification approach based on element modal strain energy. In *Proceedings of ISMA23*, pages 107–114, Leuven, Belgium, 1998.

Q. W. Zhang, L. C. Fan, and W. C. Yuan. Traffic induced variability in dynamic properties of cable-stayed bridge. *Earthquake Engineering & Structural Dynamics*, 31: 2015–2021, 2002.

D. C. Zimmerman and M. Kaouk. Structural damage detection using a minimum rank update theory. *Journal of Vibration and Acoustics*, 116:222–230, 1994.

D. C. Zimmerman, M. Kaouk, and T. Simmermacher. On the role of engineering insight and judgement in structural damage detection. In *Proceedings of the 13th International Modal Analysis Conference (IMAC)*, pages 414–420, Nashville, TN, 1995.

Y. Zou, L. Tong, and G. P. Steven. Vibration-based model-dependent damage (delamination) identification and health monitoring for composite structures – a review. *Journal of Sound and Vibration*, 230:357–378, 2000.

Time-Domain Identification of Structural Systems from Input-Output Measurements

Raimondo Betti[*]

[*] Department of Civil Engineering and Engineering Mechanics, Columbia University, 640 S.W. Mudd Bldg., New York, NY 10027. E-mail: betti@civil.columbia.edu

Abstract This paper presents a methodology that can be used for the identification of second-order models of structural systems using dynamic measurements of the input and of the structural response. This approach (and its variations) starts from an identified first order model of a structural system and obtain estimation of the structure's mass, damping and stiffness matrices. For these approaches, both the full instrumentation option and the partial instrumentation option are presented. For the case of partial/limited instrumentation, five different model types have been considered showing, for each models, the limitations imposed on the identification by the lack of available data. Once these dynamic characteristics have been determined, structural damage can be assessed by comparing the undamaged and damaged estimation of such parameters. A damage representative parameter is introduced: this parameter uses the identified models to detect the location and amount of structural damage. This methodology has been tested on simulated numerical results and its effectiveness in determining structural damage is evaluated.

1 Introduction

In engineering, the process of System Identification (SI) can be described as the identification of the properties of mathematical models that aspire to represent the dynamical behavior of systems using experimental data. The purpose and tools of system identification depend strongly on the applications for which the model is sought. In civil engineering, the goal of system identification is to obtain structural models of buildings and bridges that can be used for an accurate prediction of the structural response to future excitations as well as for damage detection purposes. In mechanical and aerospace applications, system identification provides control engineers with an invaluable tool for the design of proper control systems. Because of such a broad range of applications, it is reasonable to expect that there will be models that will be more suitable for certain applications than for others. As an example, in civil engineering applications, a structure could be modelled either as a first-order state-space model, for the purpose of predicting its response to future excitations or of designing a vibration control system, or as a physically meaningful second-order mass, damping and stiffness model for damage assessment purposes.

The identification of such models from the measured dynamic response has proven to be a tough challenge, and it is often referred to as the "inverse vibration problem". Some of the noteworthy efforts in the identification of linear structural systems include the works by Agbabian et al. (1991), Safak (1989a,b), Udwadia (1994), Beck and Katafygiotis (1998a,b), Luş et al.

(1999), Alvin and Park (1994); Alvin et al. (1995), Farhat and Hemez (1993), Doebling et al. (1993), Tseng et al. (1994) and DeAngelis et al. (2002). Although they have the common goal of identifying a mathematical/physical model of the system, all these works differ from the type of model they identify and from the methodology used in the identification process.

In the time-domain, one model that is commonly used in mechanical and aerospace engineering applications is the so-called "first-order" model in which the dynamics of the system is represented by a set of matrices that are not based on physical laws. Such models with no physical bases are often derived from realization theory and that be considered as a "black box": their only requirement is that they closely map the input-output data without any consideration given to the physical laws. In control theory, a realization is a reasonable model to use when the only variables to control are measured variables. An important advantage of this type of identification is that it is capable of identifying complex models even with just few input-output measurements, provided that the excitation is "rich" in frequency so to excite all vibrational modes. Although it is quite popular in control-type applications, this type of representation of a system, characterized by non-physical parameters, is rather inconvenient for damage identification purposes.

For damage assessment purposes, it is more convenient to work with models that have a physical foundation: among these, "second-order" models have the system's dynamics represented by a set of second-order differential equations whose coefficients are physical parameters, like mass, damping and stiffness. In structural applications, the most common tool is definitely represented by the Finite Element Method (FEM). However, upgrading FEM models of complex structures is quite cumbersome and requires solid experience by the engineer and substantial simplifications about the parameters to be identified. In addition, issues related to the ability of the FEM to uniquely identify the model parameters from data are still unresolved.

A recent trend among the most successful methodologies for the identification of physical characteristics of a structure (e.g. mass, damping and stiffness) is to start from an identified first-order representation of the system and, through proper transformations, to obtain the physical structural parameters (Luş et al. (1999), Alvin and Park (1994); Alvin et al. (1995), Tseng et al. (1994) and DeAngelis et al. (2002)). The drawback of this type of approaches is that these methodologies are strongly limited by the available number of input-output measurements. In fact, the common point among these methodologies is the assumption of having a full set of instrumentation (a sensor and/or an actuator on each degree of freedom). Evidently, this type of instrumentation setup requires a strong financial and computational effort and, unfortunately, rarely happens in real life applications, where complex structures are usually instrumented with only few sensors. Not having a complete set of measurements (either input and/or output) impairs the complete identification of structural properties (Luş et al. (2003)) and only limited information on the structural stiffness, damping and mass distribution can be retrieved. The size of such matrices is related to the number of input/output dynamic measurements available.

In this section, a new methodology for the identification of structural parameters such as the mass, damping and stiffness matrices of a system will be presented. This approach consists of 2 phases: 1) the identification of a first-order realization of the structural system using the input/output recorded time histories, and 2) the identification of a second-order models of the system, specifically mass, damping and stiffness matrix, by using a proper transformation from first-order to second-order model. The case of a full set versus an incomplete set of instrumentation will also be discussed.

2 Statement of the Problem

Consider an N degree-of-freedom viscously damped linear structural system, subjected to r external excitations. The equations of motion for such a system can be expressed as:

$$\mathcal{M}\ddot{\mathbf{q}}(t) + \mathcal{L}\dot{\mathbf{q}}(t) + \mathcal{K}\mathbf{q}(t) = \mathcal{B}\mathbf{u}(t) \tag{2.1}$$

where $\mathbf{q}(t)$ indicates the vector of the generalized nodal displacements, with $(\dot{\ })$ and $(\ddot{\ })$ representing respectively the first and second order derivatives with respect to time. The vector $\mathbf{u}(t)$, of dimension $r\times 1$, is the input vector containing the r external excitations acting on the system, with $\mathcal{B} \in \Re^{N\times r}$ being the input matrix that relates the inputs to the DOFs. The matrices $\mathcal{M} \in \Re^{N\times N}$, $\mathcal{L} \in \Re^{N\times N}$, and $\mathcal{K} \in \Re^{N\times N}$ are the symmetric positive definite mass, damping, and stiffness matrices, respectively. Let us assume that only m output time histories of the structural response are available, so that the measurement vector $\mathbf{y}(t)$, of dimensions $m \times 1$, can be written as:

$$\mathbf{y}(t) = \left[(\mathcal{C}_p\mathbf{q}(t))^T \quad (\mathcal{C}_v\dot{\mathbf{q}}(t))^T \quad (\mathcal{C}_a\ddot{\mathbf{q}}(t))^T\right]^T \tag{2.2}$$

where the matrices $\mathcal{C}_p$, $\mathcal{C}_v$, and $\mathcal{C}_a$ relate the measurements to positions, velocities, and accelerations, respectively, and the superscript $()^T$ denotes the transpose.

The objective of this study is to present a general methodology (Luş (2001), DeAngelis et al. (2002)) that, using only the input/output time histories and information about the input locations ($\mathcal{B}$ matrix), is capable of correctly identify the mass, damping and stiffness matrices of the structural model. In this analysis, an assumption is made that, at each degree-of-freedom of the system, there is either a sensor or an actuator with at least one degree-of-freedom having a co-located sensor-actuator pair. Later, this assumption will be removed and the methodology will be extended to the case of incomplete instrumentation setup.

First Order State Space Representation

In control theory, the most widely used model to represent the dynamics of a linear, time invariant system is the first order matrix differential equation form

$$\begin{aligned} \dot{\mathbf{x}}(t) &= \mathbf{A}_C\mathbf{x}(t) + \mathbf{B}_C\mathbf{u}(t) \\ \mathbf{y}(t) &= \mathbf{C}\mathbf{x}(t) + \mathbf{D}\mathbf{u}(t) \end{aligned} \tag{2.3}$$

with $\mathbf{x}(t) \in \Re^{n\times 1}$ representing the state vector (of order n), while $\mathbf{y}(t) \in \Re^{m\times 1}$ and $\mathbf{u}(t) \in \Re^{r\times 1}$ are the output and the input vectors, respectively.The continuous time system matrices are denoted by $\mathbf{A}_C \in \Re^{n\times n}$, $\mathbf{B}_C \in \Re^{n\times r}$, $\mathbf{C} \in \Re^{m\times n}$, and $\mathbf{D} \in \Re^{m\times r}$.

The discrete time representations of eq.(2.3) are dependent on the way the data is sampled (or rather, assumed to be sampled). For zero order hold sampling, which assumes that the input is constant in between two consecutive samples ($\mathbf{u}(t) = \mathbf{u}(k)$ for $k\times\Delta T \le t < k\times\Delta T + \Delta T$ with ΔT being the sampling time), the discrete time formulation of the first order state space representation can be written as

$$\begin{aligned} \mathbf{x}(k+1) &= \mathbf{\Phi}\mathbf{x}(k) + \mathbf{\Gamma}\mathbf{u}(k) \\ \mathbf{y}(k) &= \mathbf{C}\mathbf{x}(k) + \mathbf{D}\mathbf{u}(k) \end{aligned} \tag{2.4}$$

where k stands for $k{\times}\Delta T$ and $k+1$ for $k{\times}\Delta T+\Delta T$. It is noteworthy that the output matrices $\mathbf{C}$ and $\mathbf{D}$ of eq.(2.4) remain equal to their continuous time counterparts in eq.(2.3), while the discrete time matrices $\mathbf{\Phi}$ and $\mathbf{\Gamma}$ are related to $\mathbf{A}_C$ and $\mathbf{B}_C$. The solution of equations (2.4) is given by the following convolution sum (for $k \geq 1$);

$$\mathbf{x}(k) = \mathbf{\Phi}^k\mathbf{x}(0) + \sum_{j=0}^{k-1}\mathbf{\Phi}^{(k-1-j)}\mathbf{\Gamma}\mathbf{u}(j) \tag{2.5}$$

$$\mathbf{y}(k) = \mathbf{C}\mathbf{\Phi}^k\mathbf{x}(0) + \sum_{j=0}^{k-1}\mathbf{C}\mathbf{\Phi}^{(k-1-j)}\mathbf{\Gamma}\mathbf{u}(j) + \mathbf{D}\mathbf{u}(k) \tag{2.6}$$

The matrices $\mathbf{D}$ and $\mathbf{C}\mathbf{\Phi}^i\mathbf{\Gamma}$ $(\forall \quad i = 0,1,2,\ldots)$ are called the "Markov parameters" of the system. Here it is important to notice that a system has an infinite number of realizations, that is, given a minimal realization $(\mathbf{C},\mathbf{\Phi},\mathbf{\Gamma})$ and any non-singular transformation matrix $\boldsymbol{T}$, the triplet $(\boldsymbol{T}^{-1}\mathbf{\Phi}\boldsymbol{T},\boldsymbol{T}^{-1}\mathbf{\Gamma},\mathbf{C}\boldsymbol{T})$ is also a minimal realization and the eigenvalues are unchanged. On the other hand, the set of Markov parameters, representing the system's response to the unit pulse, is unique.

3 Determining the Markov Parameters of the System

In a time domain formulation, the identification of sampled impulse / pulse response, also known as Markov parameters, can be performed directly from the input-output data. In fact, by writing equations 2.6 for a sequence of "l" consecutive time steps and assuming zero initial conditions, it is possible to obtain:

$$\mathbf{Y}_{m\times l} = \mathbf{M}_{m\times rl}\mathbf{U}_{rl\times l} \tag{3.1}$$

where

$$\mathbf{Y} = [\mathbf{y}(0),\mathbf{y}(1),\mathbf{y}(2),\cdots,\mathbf{y}(l-1)] \tag{3.2}$$

$$\mathbf{M} = \left[\mathbf{D},\mathbf{C}\mathbf{\Gamma},\mathbf{C}\mathbf{\Phi}\mathbf{\Gamma},\cdots,\mathbf{C}\mathbf{\Phi}^{l-2}\mathbf{\Gamma}\right] \tag{3.3}$$

and

$$\mathbf{U} = \begin{bmatrix} \mathbf{u}(0) & \mathbf{u}(1) & \mathbf{u}(2) & \cdots & \mathbf{u}(l-1) \\ & \mathbf{u}(0) & \mathbf{u}(1) & \cdots & \mathbf{u}(l-2) \\ & & \ddots & & \vdots \\ & & & \mathbf{u}(0) & \mathbf{u}(1) \\ & & & & \mathbf{u}(0) \end{bmatrix} \tag{3.4}$$

The matrix $\mathbf{Y}$, of dimensions $m \times l$, is a matrix whose columns are the output vectors for the l time steps, while the matrix $\mathbf{U}$ contains the input vectors for different time steps arranged in an upper-triangular form. The matrix $\mathbf{M}$ contains the Markov parameters of the system (Phan et al. (1991)).

For an asymptotically stable system, let q be a sufficiently large integer such that $\mathbf{\Phi}^h \approx \mathbf{C}\mathbf{\Phi}^h\mathbf{\Gamma} \approx 0$ for any $h \geq q$. Then eq.(3.1) can be approximated as

$$\mathbf{Y}_{m\times l} \approx \mathbf{M}_{m\times r(q+1)}\mathbf{U}_{r(q+1)\times l} \tag{3.5}$$

where

$$\mathbf{Y} = \begin{bmatrix} \mathbf{y}(0) & \mathbf{y}(1) & \mathbf{y}(2) & \cdots & \mathbf{y}(q) & \cdots & \mathbf{y}(l-1) \end{bmatrix} \tag{3.6}$$

$$\mathbf{M} = \begin{bmatrix} \mathbf{D} & \mathbf{B}_C\mathbf{\Gamma} & \mathbf{B}_C\mathbf{\Phi}\mathbf{\Gamma} & \cdots & \mathbf{B}_C\mathbf{\Phi}^{q-1}\mathbf{\Gamma} \end{bmatrix} \tag{3.7}$$

and

$$\mathbf{U} = \begin{bmatrix} \mathbf{u}(0) & \mathbf{u}(1) & \mathbf{u}(2) & \cdots & \mathbf{u}(q) & \cdots & \mathbf{u}(l-1) \\ & \mathbf{u}(0) & \mathbf{u}(1) & \cdots & \mathbf{u}(q-1) & \cdots & \mathbf{u}(l-2) \\ & & \mathbf{u}(0) & \cdots & \mathbf{u}(q-2) & \cdots & \mathbf{u}(l-3) \\ & & & \ddots & \vdots & \cdots & \vdots \\ & & & & \mathbf{u}(0) & \cdots & \mathbf{u}(l-q-1) \end{bmatrix} \tag{3.8}$$

The matrices $\mathbf{M}$ and $\mathbf{U}$ (previously defined in eqs.(3.3) and (3.4)) have been truncated according to the value of q. If the data is to have a realization, then the Markov parameters approximately satisfy the equation $\mathbf{M} = \mathbf{Y}\mathbf{U}^{\dagger}$, where $\mathbf{U}^{\dagger}$ is the pseudo inverse of the input matrix $\mathbf{U}$, and the error due to such an approximation decreases as q increases. However, for lightly damped structures, the integer q, and therefore the size of $\mathbf{U}$, is practically too large to perform the inversion operation numerically. To improve the stability of the system, and thereby to make the problem better conditioned numerically, Phan et al. (1992, 1993) introduced an observer based identification concept (Observer Kalman filter IDentification, OKID) in which one first identifies an associated observer from which the system pulse responses are then recovered.

To understand the role of this observer, let us rewrite the system input-output relations in terms of a new set of state space equations which are obtained by adding and subtracting the term $\mathbf{R}\mathbf{y}(k)$ in eq. (2.4) as presented in Juang et al. (1993). This will lead to:

$$\begin{aligned} \mathbf{x}(k+1) &= \mathbf{\Phi}\mathbf{x}(k) + \mathbf{\Gamma}\mathbf{u}(k) + \mathbf{R}((\mathbf{C}\mathbf{x}(k) + \mathbf{D}\mathbf{u}(k)) - \mathbf{y}(k)) \\ &= (\mathbf{\Phi} + \mathbf{R}\mathbf{C})\mathbf{x}(k) + (\mathbf{\Gamma} + \mathbf{R}\mathbf{D})\mathbf{u}(k) - \mathbf{R}\mathbf{y}(k) \\ &= \bar{\mathbf{\Phi}}\mathbf{x}(k) + \bar{\mathbf{\Gamma}}\boldsymbol{\nu}(k) \\ \mathbf{y}(k) &= \mathbf{C}\mathbf{x}(k) + \mathbf{D}\mathbf{u}(k) \end{aligned} \tag{3.9}$$

where

$$\bar{\mathbf{\Phi}} = (\mathbf{\Phi} + \mathbf{R}\mathbf{C}) \tag{3.10}$$

$$\bar{\mathbf{\Gamma}} = [(\mathbf{\Gamma} + \mathbf{R}\mathbf{D}) \quad (-\mathbf{R})] \tag{3.11}$$

$$\boldsymbol{\nu}(k) = \begin{bmatrix} \mathbf{u}(k) \\ \mathbf{y}(k) \end{bmatrix} \tag{3.12}$$

The gain matrix $\mathbf{R}$ is chosen to make the system represented by eq.(3.9) as stable as desired. Although eqs.(2.4) and (3.9) are mathematically identical, eq.(3.9) can be considered as an observer equation and the Markov parameters of this new system, denoted as $\bar{\mathbf{M}}$, are called the observer's Markov parameters. If the matrix $\mathbf{R}$ is chosen in such a way that $\bar{\mathbf{\Phi}}$ is asymptotically stable, then $\mathbf{C}\bar{\mathbf{\Phi}}^h\bar{\mathbf{\Gamma}} \approx \mathbf{0}$ for $h \geq p$, and we can solve for $\bar{\mathbf{M}}$ from general input-output data using

$$\mathbf{Y}_{m\times l} \approx \bar{\mathbf{M}}_{m\times((r+m)p+r)}\mathbf{V}_{((r+m)p+r)\times l} \tag{3.13}$$

$$\bar{\mathbf{M}} = \mathbf{Y}\mathbf{V}^{\dagger} \tag{3.14}$$

where

$$\mathbf{Y} = \begin{bmatrix} \mathbf{y}(0) & \mathbf{y}(1) & \mathbf{y}(2) & \cdots & \mathbf{y}(p) & \cdots & \mathbf{y}(l-1) \end{bmatrix} \tag{3.15}$$

$$\bar{\mathbf{M}} = \begin{bmatrix} \mathbf{D} & \mathbf{C}\bar{\mathbf{\Gamma}} & \mathbf{C}\bar{\mathbf{\Phi}}\bar{\mathbf{\Gamma}} & \cdots & \mathbf{C}\bar{\mathbf{\Phi}}^{p-1}\bar{\mathbf{\Gamma}} \end{bmatrix} \tag{3.16}$$

and

$$\mathbf{V} = \begin{bmatrix} \mathbf{u}(0) & \mathbf{u}(1) & \mathbf{u}(2) & \cdots & \mathbf{u}(p) & \cdots & \mathbf{u}(l-1) \\ & \boldsymbol{\nu}(0) & \boldsymbol{\nu}(1) & \cdots & \boldsymbol{\nu}(p-1) & \cdots & \boldsymbol{\nu}(l-2) \\ & & \boldsymbol{\nu}(0) & \cdots & \boldsymbol{\nu}(p-2) & \cdots & \boldsymbol{\nu}(l-3) \\ & & & \ddots & \vdots & \cdots & \vdots \\ & & & & \boldsymbol{\nu}(0) & \cdots & \boldsymbol{\nu}(l-p-1) \end{bmatrix} \tag{3.17}$$

It is important to note that p is now much smaller than q, and so the numerical difficulties are overcome. This development can also be viewed as the introduction of an observer to the system, with $\mathbf{R}$ being the observer gain. For further details on the formulation and on how to account for the initial conditions, the reader is referred to the work by Phan et al. (1991), Phan et al. (1992) , Phan et al. (1993) , Juang et al. (1993) , Juang and Phan (1994a), and Phan et al. (1995).

Having identified the observer Markov parameters, the true system's Markov parameters can be retrieved using the recursive formula:

$$\mathbf{M}_k = \bar{\mathbf{M}}_k^{(1)} + \sum_{i=0}^{k-1} \bar{\mathbf{M}}_i^{(2)} \mathbf{M}_{k-i-1} + \bar{\mathbf{M}}_k^{(2)} \mathbf{D} \tag{3.18}$$

where

$$\bar{\mathbf{M}} = \begin{bmatrix} \bar{\mathbf{M}}_{-1} & \bar{\mathbf{M}}_0 & \bar{\mathbf{M}}_1 & \cdots & \bar{\mathbf{M}}_{p-1} \end{bmatrix} \tag{3.19}$$

$$\begin{aligned} \bar{\mathbf{M}}_k &= \mathbf{C}\bar{\mathbf{\Phi}}^k\bar{\mathbf{\Gamma}} \\ &= \left[\mathbf{C}(\mathbf{\Phi}+\mathbf{RC})^k(\mathbf{\Gamma}+\mathbf{RD}), \quad -\mathbf{C}(\mathbf{\Phi}+\mathbf{RC})^k\mathbf{R}\right] \\ &= \left[\bar{\mathbf{M}}_k^{(1)}, \ \bar{\mathbf{M}}_k^{(2)}\right]; \qquad k=1,2,3,... \end{aligned} \tag{3.20}$$

with $\bar{\mathbf{M}}_{-1} = \mathbf{D}$.

4 Identification of State-space Models from Markov Parameters

Once the Markov parameters of the system have been determined, they can be used in building an Hankel matrix whose singular value decomposition leads to the identification of a state-space model of the system. This method is called the Eigensystem Realization Algorithm (ERA) and it is one of the most widely used and studied algorithms in the mechanical / aerospace engineering arena. Different formulations of the ERA algorithm include: 1) ERA in the Frequency Domain (Juang and Suzuki (1988)), 2) ERA in a recursive form (Longman and Juang (1989)), 3) ERA with Data Correlation (ERA/DC) (Juang et al. (1988)). Other ERA-based algorithms include the works by Yang and Yeh (1990), Juang and Phan (1994b), Phan et al. (1995), Moonen et al. (1989), and Lim (1998).

To briefly present the fundamental theory of ERA, let us consider an NDOF system, represented by a first-order discrete time state space model as in equation 2.4, and assume that r unit pulse tests have been performed on a system with m outputs; i.e.

$$\mathbf{y}(k) = \left[\begin{array}{cccc} y_1(k) & y_2(k) & \cdots & y_m(k) \end{array} \right]^T.$$

Let us denote with $\mathbf{y}^j(k)$ a new vector, of dimension m, which represents the system's response at time $k(\Delta T)$ to a unit pulse at time zero applied at input $\mathbf{u}_j$ (same as Markov parameters). In this way, we can regroup the data as

$$\mathbf{Y}(k) = \left[\begin{array}{cccc} \mathbf{y}^1(k) & \mathbf{y}^2(k) & \cdots & \mathbf{y}^r(k) \end{array} \right], \quad \text{k=1,2,...} \tag{4.1}$$

and form the $ms \times rs$ Hankel data matrix

$$\mathcal{H}^s(k-1) = \left[\begin{array}{cccc} \mathbf{Y}(k) & \mathbf{Y}(k+1) & \cdots & \mathbf{Y}(k+s-1) \\ \mathbf{Y}(k+1) & \mathbf{Y}(k+2) & \cdots & \mathbf{Y}(k+s) \\ \vdots & \vdots & \ddots & \vdots \\ \mathbf{Y}(k+s-1) & \mathbf{Y}(k+s) & \cdots & \mathbf{Y}(k+2(s-1)) \end{array} \right] \tag{4.2}$$

where s is an arbitrary integer that determines the size of such a matrix. Looking at eqs. 2.6, it is possible to see that the first Markov parameter, i.e. $\mathbf{D}$, can be readily expressed as

$$\mathbf{D} = \mathbf{Y}(0) \tag{4.3}$$

If the recorded data permits a realization, then the full data sequence can be generated from the triplet $(\mathbf{C}, \mathbf{\Phi}, \mathbf{\Gamma})$ via the following equation

$$\mathbf{Y}(k) = \mathbf{C}\mathbf{\Phi}^{k-1}\mathbf{\Gamma}, \qquad k = 1,2,3,.. \tag{4.4}$$

which substituted into (4.2) leads to the following representation of the Hankel matrix

$$\mathcal{H}^s(i) = \left[\begin{array}{cccc} \mathbf{C}\mathbf{\Phi}^i\mathbf{\Gamma} & \mathbf{C}\mathbf{\Gamma}^{i+1}\mathbf{\Gamma} & \cdots & \mathbf{C}\mathbf{\Gamma}^{i+s-1}\mathbf{\Gamma} \\ \mathbf{C}\mathbf{\Gamma}^{i+1}\mathbf{\Gamma} & \mathbf{C}\mathbf{\Gamma}^{i+2}\mathbf{\Gamma} & \cdots & \mathbf{C}\mathbf{\Gamma}^{i+s}\mathbf{\Gamma} \\ \vdots & \vdots & \ddots & \vdots \\ \mathbf{C}\mathbf{\Gamma}^{i+s-1}\mathbf{\Gamma} & \mathbf{C}\mathbf{\Gamma}^{i+s}\mathbf{\Gamma} & \cdots & \mathbf{C}\mathbf{\Gamma}^{i+2(s-1)}\mathbf{\Gamma} \end{array} \right], \quad i\text{=0,1,....} \tag{4.5}$$

It can be shown that, if there exists a finite dimensional realization $(\mathbf{C}, \mathbf{\Phi}, \mathbf{\Gamma})$ of the data sequence $\mathbf{Y}(k)$, k=1,2,3,..., and if the dimension of a minimal realization is n, then

$$rank\ \mathcal{H}^s(i) = n, \quad \forall\ s \geq n, \ \text{ and } i\text{=0,1,2,...} \tag{4.6}$$

Once the system's Markov parameters have been determined and the corresponding Hankel matrix has been built, let the singular value decomposition of $\mathcal{H}^s(0)$ be denoted by

$$\mathcal{H}^s(0) = \mathcal{O}\mathcal{C} = \mathbf{U}\mathbf{\Sigma}\mathbf{V}^T = \left[\begin{array}{cc} \mathbf{U}_1 & \mathbf{U}_2 \end{array} \right] \left[\begin{array}{cc} \mathbf{S} & \mathbf{0} \\ \mathbf{0} & \mathbf{0} \end{array} \right] \left[\begin{array}{c} \mathbf{V}_1^T \\ \mathbf{V}_2^T \end{array} \right] = \mathbf{U}_1\mathbf{S}\mathbf{V}_1^T \tag{4.7}$$

where $\mathbf{U}_{ms \times ms}$ and $\mathbf{V}_{rs \times rs}$ are unitary matrices, and $\mathbf{S}$ is a square diagonal matrix (the non-zero partition of $\boldsymbol{\Sigma}_{ms \times rs}$) whose dimensions are equal to the rank of the $\boldsymbol{\mathcal{H}}^s(0)$ matrix. The basic theorem of the ERA realization states that, if the dimension of any minimal realization is n, then the following triplet is a minimal realization of the system for any $s \geq n$:

$$\boldsymbol{\Phi} = \mathbf{S}^{-\frac{1}{2}} \mathbf{U}_1{}^T \boldsymbol{\mathcal{H}}^s(1) \mathbf{V}_1 \mathbf{S}^{-\frac{1}{2}} \tag{4.8}$$

$$\boldsymbol{\Gamma} = \mathbf{S}^{\frac{1}{2}} \mathbf{V}_1{}^T \mathbf{E}_r \tag{4.9}$$

$$\mathbf{C} = \mathbf{E}_m^T \mathbf{U}_1 \mathbf{S}^{\frac{1}{2}} \tag{4.10}$$

where $\mathbf{E}_r = \begin{bmatrix} \mathbf{I}_{r \times r} & \mathbf{0}_{r \times r} & \mathbf{0}_{r \times r} & \cdots & \mathbf{0}_{r \times r} \end{bmatrix}^T_{r \times rs}$ (with $\mathbf{I}_{r \times r}$ denoting the $r \times r$ identity matrix and $\mathbf{0}_{r \times r}$ denoting an $r \times r$ matrix whose elements are all zeros), and $\mathbf{E}_m$ is defined analogously. Hence, the initially unknown matrices $\boldsymbol{\Phi}$, $\boldsymbol{\Gamma}$, $\mathbf{C}$ and $\mathbf{D}$, can be obtained using the ERA formulation once the system's Markov parameters have been identified.

By converting the realized discrete time system matrix $\boldsymbol{\Phi}$ to the continuous time equivalent one, $\mathbf{A}_C$, and considering its eigenvalues, it is then possible to extract the modal frequencies and damping factors of the identified structural system.

In order to reduce the bias due to noise in the data, an alternative formulation of the ERA can be used (Juang et al. (1988)). Such an identification algorithm, called the ERA/DC, combines the minimum order realization approach with insights from the Correlation Fit method, using auto correlations and cross correlations of output data instead of actual response data. In this formulation, the Hankel matrix $\boldsymbol{\mathcal{H}}$ is replaced by a block correlation matrix $\boldsymbol{\mathcal{U}}$ which is defined as:

$$\boldsymbol{\mathcal{U}}(q) = \begin{bmatrix} \boldsymbol{\mathcal{R}}(q) & \boldsymbol{\mathcal{R}}(q+\gamma) & \cdots & \boldsymbol{\mathcal{R}}(q+\beta\gamma) \\ \boldsymbol{\mathcal{R}}(q+\gamma) & \boldsymbol{\mathcal{R}}(q+2\gamma) & \cdots & \vdots \\ \vdots & \vdots & \ddots & \vdots \\ \boldsymbol{\mathcal{R}}(q+\alpha\gamma) & \cdots & \cdots & \boldsymbol{\mathcal{R}}(q+(\alpha+\beta)\gamma) \end{bmatrix} \tag{4.11}$$

where $\boldsymbol{\mathcal{R}}(q) = \boldsymbol{\mathcal{H}}(q)\boldsymbol{\mathcal{H}}^T(0)$. Consequently, the identified system matrices will be expressed as functions of $\boldsymbol{\mathcal{U}}(0)$ and $\boldsymbol{\mathcal{U}}(1)$. For further details, the reader is referred to the work by Juang et al. (1988).

5 Optimizing Identified Models

In system identification, the primary goal is to identify a model of the system whose outputs are as close as possible to the ones of the real system when an identical input is applied to both. Most of the system identification algorithms focus their attention in minimizing the equation error, either of the system or of an observer. However, what one is really interested in system identification is in creating a model that has the property that when inputs are applied to the model, it produces outputs that are as close as possible to the ones of the real system. Hence, attention should be directed in minimizing the output error.

Optimization of the output error is a minimization of a nonlinear function that involves many parameters. Usually, such a nonlinear function is represented by the sum of the squares of the output errors of the model in fitting the data over the entire data length. It is common practice to use nonlinear least-squares approaches to tackle this minimization. However, the direct use of nonlinear least-squares methods is not recommended because of the presence of many local minima. To avoid such a problem, iterative methods have been proposed that, starting from an initial model guess, converge to a model that minimize such output error. By using, as initial guess, reliable models identified with the previously mentioned techniques, the ability to produce "better" outputs is substantially improved. Following this direction, Juang and Longman (1999) developed a set of system identification algorithms that minimized output error for Multi-Input Multi-Output (MIMO) and Multi-Input Single-Output (MISO) systems. This is done with sequential quadratic programming iterations on the nonlinear least-squares problem, with an eigendecomposition to handle indefinite second partials. To avoid the problem of converging to local minima, they starts the iterations from the OKID algorithm results.

6 From State-Space to Mass, Damping and Stiffness Models

The identified state-space matrices $\mathbf{A}_C$, $\mathbf{B}_C$, $\mathbf{C}$ and $\mathbf{D}$ are referred to arbitrary state-space coordinates. The state vector $\mathbf{x}$ does not correspond to physical quantities (displacements and/or velocities). In order to identify a physical model of the structure, it is then necessary to relate the identified state-space parameters to physically meaningful coordinates.

Consider a second-order representation of the system as presented in equations (2.1). By defining a state vector $\mathbf{z}(t) = [\mathbf{q}(t)^T, \dot{\mathbf{q}}(t)^T]^T$, the equations of motion (2.1) can be conveniently rewritten as:

$$\begin{bmatrix} \mathcal{L} & \mathcal{M} \\ \mathcal{M} & 0 \end{bmatrix} \dot{\mathbf{z}}(t) \begin{bmatrix} \mathcal{K} & 0 \\ 0 & -\mathcal{M} \end{bmatrix} \mathbf{z}(t) = \begin{bmatrix} \mathcal{B} \\ 0 \end{bmatrix} \mathbf{u}(t) \tag{6.1}$$

The advantage of using this formulation of the equations of motion is that the associated eigenvalue problem is now symmetric:

$$\begin{bmatrix} \mathcal{L} & \mathcal{M} \\ \mathcal{M} & 0 \end{bmatrix} \begin{bmatrix} \psi \\ \psi\Lambda \end{bmatrix} \Lambda = \begin{bmatrix} -\mathcal{K} & 0 \\ 0 & \mathcal{M} \end{bmatrix} \begin{bmatrix} \psi \\ \psi\Lambda \end{bmatrix} \tag{6.2}$$

where $\psi = [\psi_1, \psi_2, \ldots\ldots, \psi_{2N}]$ is the matrix that contains the eigenvectors of the complex eigenvalue problem:

$$(\lambda_i^2 \mathcal{M} + \lambda_i \mathcal{L} + \mathcal{K})\psi_i = 0 \tag{6.3}$$

and $\mathbf{\Lambda}$ is a $2N \times 2N$ diagonal matrix that contains the complex eigenvalues λ_i ($i = 1, 2, ..., 2N$).

In general, these eigenvectors can be arbitrarily scaled: however, if the scaling is done such as (Sestieri and Ibrahim (1994), Balmes (1997)):

$$\begin{aligned} \begin{bmatrix} \psi \\ \psi\Lambda \end{bmatrix}^T \begin{bmatrix} \mathcal{L} & \mathcal{M} \\ \mathcal{M} & 0 \end{bmatrix} \begin{bmatrix} \psi \\ \psi\Lambda \end{bmatrix} &= \mathbf{I} \\ \begin{bmatrix} \psi \\ \psi\Lambda \end{bmatrix}^T \begin{bmatrix} \mathcal{K} & 0 \\ 0 & -\mathcal{M} \end{bmatrix} \begin{bmatrix} \psi \\ \psi\Lambda \end{bmatrix} &= -\mathbf{\Lambda} \end{aligned} \tag{6.4}$$

then, for proportionally damped system, the real and imaginary parts of the components of the complex eigenvectors will be equal in magnitude.

Using a coordinate transformation $\mathbf{z}(t) = [\psi^T (\psi\mathbf{\Lambda})^T]^T \boldsymbol{\zeta}(t)$ and taking advantage of the normalization conditions (6.4), the equations of motion (6.1) and the output equation can be rewritten as:

$$\begin{aligned} \dot{\boldsymbol{\zeta}}(t) &= \boldsymbol{\Lambda}\boldsymbol{\zeta}(t) + \boldsymbol{\psi}^T \boldsymbol{\mathcal{B}}\mathbf{u}(t) \\ \mathbf{y}(t) &= \boldsymbol{\mathcal{C}}_p \boldsymbol{\psi}\boldsymbol{\zeta}(t) \end{aligned} \tag{6.5}$$

In this case, for ease of presentation, it is assumed that the output measurements are in the form of structural displacement time histories. However, the methodology is quite general and valid for any type of measurement (velocity, acceleration).

A similar transformation can also be applied to the identified state-space model of the system. Hence, considering the transformation $\mathbf{x}(t) = \boldsymbol{\varphi}\boldsymbol{\theta}(t)$, the state-space representation of the equations of motion of the system, eqs.(2.3), and the output equation can be transformed in:

$$\begin{aligned} \dot{\boldsymbol{\theta}}(t) &= \boldsymbol{\Lambda}\boldsymbol{\theta}(t) + \boldsymbol{\varphi}^{-1}\mathbf{B}_C\mathbf{u}(t) \\ \mathbf{y}(t) &= \mathbf{C}_C \boldsymbol{\varphi}\boldsymbol{\theta}(t) \end{aligned} \tag{6.6}$$

where Λ contains the continuous time eigenvalues of the identified state space model while $\boldsymbol{\varphi}$ is a matrix of dimensions $2N \times 2N$ containing the eigenvectors of the associated eigenvalue problem:

$$\mathbf{A}_C \boldsymbol{\varphi}_i = \lambda_i \boldsymbol{\varphi}_i \qquad (i = 1, 2, ...2N) \tag{6.7}$$

Equations (6.5) and (6.6) are different model representations of the same system. Therefore, there must be a transformation matrix $\boldsymbol{\mathcal{T}}$ that relates these two representations. This condition will imply that:

$$\begin{gathered} \boldsymbol{\mathcal{T}}^{-1}\boldsymbol{\Lambda}\boldsymbol{\mathcal{T}} = \boldsymbol{\Lambda} \\ \boldsymbol{\mathcal{T}}^{-1}\boldsymbol{\varphi}^{-1}\mathbf{B}_C = \boldsymbol{\psi}^T\boldsymbol{\mathcal{B}} \\ \mathbf{C}_C\boldsymbol{\varphi}\boldsymbol{\mathcal{T}} = \boldsymbol{\mathcal{C}}_p\boldsymbol{\psi} \end{gathered} \tag{6.8}$$

By examining equations (6.8), it appears that the transformation matrix $\boldsymbol{\mathcal{T}}$ has a twofold effect: 1) it transforms the eigenvectors from those of a nonsymmetric eigenvalue problem to those of a symmetric eigenvalue problem, and 2) to properly scale such eigenvectors. The input and output matrices, ($\boldsymbol{\mathcal{B}}$ and $\boldsymbol{\mathcal{C}}_p$, respectively), of the second-order model are considered to be known: they are assumed to contain binary information, with 0's and 1's to indicate the lack or presence of a sensor or an actuator.

The transformation matrix $\boldsymbol{\mathcal{T}}$ and the properly scaled complex eigenvectors $\boldsymbol{\psi}$ can be easily evaluated if we assume that each degree-of-freedom contains either an actuator or a sensor, with one degree-of-freedom containing a co-located sensor-actuator pair. If this co-located sensor-actuator pair is placed on the $i-th$ degree-of-freedom, the co-location requirement (Balmes (1997)) imposes that:

$$\boldsymbol{\mathcal{C}}_p(i,:)\boldsymbol{\psi} = [\boldsymbol{\psi}^T\boldsymbol{\mathcal{B}}(:,i)]^T \tag{6.9}$$

which leads to:

$$\begin{gathered} \mathbf{C}_C(i,:)\boldsymbol{\varphi}\boldsymbol{\mathcal{T}} = (\boldsymbol{\mathcal{T}}^{-1}\boldsymbol{\varphi}^{-1}\mathbf{B}_C(:,i))^T; \\ \mathbf{C}_C(i,:)\boldsymbol{\varphi}\boldsymbol{\mathcal{T}}^2 = (\boldsymbol{\varphi}^{-1}\mathbf{B}_C(:,i))^T \end{gathered} \tag{6.10}$$

If there are no repeated roots, it can be shown that the transformation matrix $\boldsymbol{\mathcal{T}}$ is diagonal and its elements are in complex conjugate pairs. Such elements can be easily obtained by solving eqs.(6.10).

Once these scaling factors have been determined, the complex eigenvector matrix can be easily obtained by using information at the sensor and actuator locations. For a sensor at the kth DOF, then the kth row of the matrix ψ can be evaluated using the proper equation (6.8) related to the sensor location:

$$\psi(k,:) = \mathbf{C}_C(k,:)\varphi\mathcal{T} \tag{6.11}$$

Similarly, for an actuator located at the kth DOF, then the kth row of the matrix ψ can be obtained by considering the proper equation (6.8) for an actuator location:

$$\psi(k,:) = (\mathcal{T}^{-1}\varphi^{-1}\mathbf{B}_C(:,k))^T \tag{6.12}$$

Repeating this condition for all the degrees-of-freedom with either a sensor or an actuator allows us to evaluate all the rows of the ψ matrix. Once the properly scaled eigenvector matrix ψ has been properly evaluated, the physical parameters of the second-order model, namely the mass, damping and stiffness matrices, can be easily obtained using the orthogonality conditions from equations (6.4):

$$\mathcal{M} = (\psi\Lambda\psi^T)^{-1}, \quad \mathcal{C} = -\mathcal{M}\psi\Lambda^2\psi^T\mathcal{M}, \quad \mathcal{K} = -(\psi\Lambda^{-1}\psi^T)^{-1} \tag{6.13}$$

From the orthogonality conditions, it is also shown that the normalized complex eigenvectors will satisfy the condition:

$$\psi\psi^T = \mathbf{0}. \tag{6.14}$$

Such a condition will become quite important when looking at the case of incomplete instrumentation setup.

7 Identification of Second-Order Models from a Limited Number of Input/Output Measurements

In real life applications, the structural system is usually instrumented with a limited number of sensors and actuators, leaving some degrees of freedom deficient of either a sensor or an actuator. In this context, the necessary condition that made possible the complete identification of a second-order model of the structure, namely that $(m + r = N + 1)$, ceases to exist. This implies that "full order" model of the system cannot be fully identified $(m+r < N+1)$, raising the question of searching for a reliable "reduced order" model of the system.

The condition $m+r < N+1$ implies that not enough information is available at the various degrees of freedom to fully retrieve the complex eigenvector matrix. This will imply that, using the $m+r$ available input/output data with one co-located sensor-actuator pair, only $n = m+r-1$ rows of the complex eigenvector matrix ψ can be determined while the remaining $p = N - n$ rows remain unknown (DeAngelis et al. (2002) and Luş et al. (2003)). By rearranging the known (ψ^{kn}) and unknown (ψ^{un}) components, the eigenvector matrix ψ can be represented as:

$$\psi = \begin{bmatrix} \psi^{kn}_{1,1} & \psi^{kn}_{1,2} & \cdots & \psi^{kn}_{1,2N} \\ \cdots & \cdots & \cdots & \cdots \\ \psi^{kn}_{n,1} & \psi^{kn}_{n,2} & \cdots & \psi^{kn}_{n,2N} \\ \psi^{un}_{n+1,1} & \psi^{un}_{n+1,2} & \cdots & \psi^{un}_{n+1,2N} \\ \cdots & \cdots & \cdots & \cdots \\ \psi^{un}_{N,1} & \psi^{un}_{N,2} & \cdots & \psi^{un}_{N,2N} \end{bmatrix} = \begin{bmatrix} \psi^{kn}_{n\times 2N} \\ \psi^{un}_{p\times 2N} \end{bmatrix} = \begin{bmatrix} \psi_1 \\ \psi_2 \end{bmatrix} \tag{7.1}$$

where the complex eigenvector matrix ψ has been partitioned into a known submatrix $\psi_1 = \psi^{kn}$, of dimension $n \times 2N$, and an unknown submatrix $\psi_2 = \psi^{un}$, of dimension $p \times 2N$. The single element $\psi_{i,l}$ represents the complex component of the lth mode to the ith degree of the freedom ($l = 1, 2, ..., 2N$ and $i = 1, 2, ..., N$). The superscripts kn and un indicate whether, at that degree of freedom, a sensor or an actuator is present or not, implying that the corresponding modal quantity can be identified (known) or not (unknown).

Starting from the general expression of the mass, damping and stiffness matrices in terms of the complex eigenvectors ψ as shown in Eqn.(6.13) (DeAngelis et al. (2002)), we can express these matrices in partitioned forms as functions of the known and unknown partitions of the eigenvector matrix. For example, the full order stiffness matrix $\mathbf{K}$ can be expressed as:

$$\mathcal{K} = -(\psi \mathbf{\Lambda}^{-1} \psi^T)^{-1} = -\begin{bmatrix} \psi_1 \mathbf{\Lambda}^{-1} \psi_1^T & \psi_1 \mathbf{\Lambda}^{-1} \psi_2^T \\ \psi_2 \mathbf{\Lambda}^{-1} \psi_1^T & \psi_2 \mathbf{\Lambda}^{-1} \psi_2^T \end{bmatrix}^{-1} = \begin{bmatrix} \mathcal{K}_{11} & \mathcal{K}_{12} \\ \mathcal{K}_{21} & \mathcal{K}_{22} \end{bmatrix} \tag{7.2}$$

with each subpartition given by the following expressions:

$$\mathcal{K}_{11} = -[(\psi_1 \mathbf{\Lambda}^{-1} \psi_1^T) - (\psi_1 \mathbf{\Lambda}^{-1} \psi_2^T)(\psi_2 \mathbf{\Lambda}^{-1} \psi_2^T)^{-1}(\psi_2 \mathbf{\Lambda}^{-1} \psi_1^T)]^{-1} \tag{7.3}$$

$$\mathcal{K}_{22} = -[(\psi_2 \mathbf{\Lambda}^{-1} \psi_2^T) - (\psi_2 \mathbf{\Lambda}^{-1} \psi_1^T)(\psi_1 \mathbf{\Lambda}^{-1} \psi_1^T)^{-1}(\psi_1 \mathbf{\Lambda}^{-1} \psi_2^T)]^{-1} \tag{7.4}$$

$$\mathcal{K}_{12} = -\mathcal{K}_{11}^{-1}(\psi_1 \mathbf{\Lambda}^{-1} \psi_2^T)(\psi_2 \mathbf{\Lambda}^{-1} \psi_2^T)^{-1} \tag{7.5}$$

$$\mathcal{K}_{21} = -\mathcal{K}_{22}^{-1}(\psi_2 \mathbf{\Lambda}^{-1} \psi_1^T)(\psi_1 \mathbf{\Lambda}^{-1} \psi_1^T)^{-1} \tag{7.6}$$

In the case of limited sensor/actuator capabilities, only the known part (ψ_1) of the eigenvector matrix is available and this will impair the use of Eqns.(7.2) and (7.3-7.6). Using only the known components of the complex eigenvector matrix, a “reduced” form for the mass, stiffness and damping matrices of the structural system, of dimension $n \times n$, can be expressed as (Luş et al. (2003)):

$$\mathcal{M}_r = (\psi_1 \mathbf{\Lambda} \psi_1^T)^{-1} = \mathcal{M}_{11} - \mathcal{M}_{12} \mathcal{M}_{22}^{-1} \mathcal{M}_{21} \tag{7.7}$$

$$\mathcal{K}_r = -(\psi_1 \mathbf{\Lambda}^{-1} \psi_1^T)^{-1} = \mathcal{K}_{11} - \mathcal{K}_{12} \mathcal{K}_{22}^{-1} \mathcal{K}_{21} \tag{7.8}$$

$$\mathcal{L}_r = -\mathcal{M}_r \psi_1 \mathbf{\Lambda}^2 \psi_1^T \mathcal{M}_r \tag{7.9}$$

The use of such “reduced” models of the structural system is quite limited. In fact, they cannot be used to represent the dynamic characteristics of the structural system: natural frequencies obtained from the reduced model do not match the ones from the real system. In addition, such models cannot be used to predict the structural response at future excitation since they represent small order systems that have been obtained by minimizing the output error at few locations. However, they can provide indications on the locations where damage has occurred, although quantifying the damage with such reduced models is quite difficult, if not impossible.

To solve this problems, attempts have been made to expand "reduced" second order models of a system, obtained from a limited number of instrumentations, to "full" second order models which preserve most of the system information such as the input/output relations and the normalized system eigenvalues. An interesting approach has been presented by Yu (2004) where the unknown components of the complex eigenvector matrix, ψ_2, are obtained through an optimization algorithm. The objective of the optimization is to minimize the difference of the total weighted response between the measured and simulated data under a set of constraints imposed on the variables to update. These constraints depend on the normalization condition imposed on the eigenvectors (Eqns.(6.14))and on the type of structural assumptions imposed to the system.

8 From "Reduced" to "Full" Order Models

For a structural engineer, having a "full " order representation of the system, represented by the complete mass, damping and stiffness matrices, is quite useful. In fact, such a representation can be used for predicting the structural response to future excitations at all possible locations in the structure as well as for locating and quantifying possible structural damage (e.g. by comparing the structural stiffness). This is not possible using reduced order models of the structure: in fact, such models cannot be used for predicting the structural response to future excitation even at instrumented locations while locations and amount of structural damage can only be vaguely pinpointed (DeAngelis et al. (2003)). The possibility of expanding these "reduced" order models to "full" order models depends on the type of structural model considered.

"Reduced order" models of structural systems can be grouped into different categories on the basis of different structural constraints. According to the different forms of the mass, damping and stiffness matrices, *five* representative scenarios have been defined and will be discussed separately. However, for all these scenarios, a common set of constraints can be defined using condition (6.14) that can be expressed in terms of ψ_1 and ψ_2 as:

$$\psi\psi^T = \mathbf{0}_{N\times N} = \begin{bmatrix} \psi_1\psi_1^T & \psi_1\psi_2^T \\ \psi_2\psi_1^T & \psi_2\psi_2^T \end{bmatrix} = \begin{bmatrix} \mathbf{0}_{n\times n} & \mathbf{0}_{n\times p} \\ \mathbf{0}_{p\times n} & \mathbf{0}_{p\times p} \end{bmatrix} \tag{8.1}$$

While the above constraint is automatically satisfied for the known part ψ_1, equations (8.1) provide two additional sets of conditions that can be used for the determination of the unknown part ψ_2. Such conditions are represented by the following sets of linear and non-linear equations:

$$\begin{aligned} \psi_1\psi_2^T &= \mathbf{0}_{n\times p}; \\ \psi_2\psi_2^T &= \mathbf{0}_{p\times p}. \end{aligned} \tag{8.2}$$

8.1 General $\mathcal{M}$, $\mathcal{L}$ and $\mathcal{K}$ - Case I

If the system can only be modeled with general (non diagonal, or block-diagonal, etc.) mass, damping and stiffness matrices, then very little can be done in expanding the initial reduced order model. In this case, the only constraint conditions that can be imposed on ψ_2 come from the above normalization equations, Eqn.(8.1). Hence, in addition to the minimization of the output error, Eqn.(8.2) will provide $n \times p$ and $p \times p$ equality constraints respectively. However, due to symmetry, only the upper diagonal part of $\psi_2\psi_2^T$ provides independent conditions so that the total number of constraints reduces to $\frac{p}{2}(2N - p + 1)$. In this case, the dimension of the solution space (SS_{dof}) becomes $\frac{p}{2}(N + p - 1)$, as reported in Table 1. For this type of structural models, the missing components of the eigenvector matrix ψ_2 and, consequently, the real mass normalized eigenvectors and the associated natural frequencies cannot be exactly identified, with serious consequences on the accuracy of the identified model. The only subpartition matrices that can be determined are $\mathcal{M}_r$, $\mathcal{L}_r$ and $\mathcal{K}_r$, as defined in Eqs.7.7-7.8, which however are of limited use for identification and damage detection purposes. In addition, although they reproduce the recorded structural output at the sensor locations quite closely, these models cannot be used to predict the response of the system to future excitation, even at the instrumented locations. The reason is that the optimization identifies a model by minimizing the error between the measured output and the predicted output for a given input. However, since the identified model does not represent the dynamic characteristics of the real system, it cannot be expected to provide accurate prediction of the structural behavior to a different input excitation.

8.2 Block Diagonal $\mathcal{M}$ - Case II

If we assume that the structural system can be represented with a block diagonal mass matrix (an assumption quite widely used in civil and mechanical applications), then we can introduce additional $n \times p$ conditions of the type $\psi_1 \mathbf{\Lambda}^{-1} \psi_2^T$ expressing the block diagonal character of the mass matrix. In this case, the subpartitions of the mass and damping matrices related to the instrumented degrees-of-freedom, $\mathcal{M}_{11}$, $\mathcal{L}_{11}$ can be exactly identified as:

$$\mathcal{M}_{11} = -(\psi_1 \mathbf{\Lambda} \psi_1^T)^{-1} = \mathcal{M}_r \tag{8.3}$$

$$\mathcal{L}_{11} = -\mathcal{M}_{11} \psi_1 \mathbf{\Lambda}^2 \psi_1^T \mathcal{M}_{11} \tag{8.4}$$

The remaining subpartitions of the full mass and damping matrices as well as the complete stiffness matrix cannot be determined uniquely. Similarly to the previous case, the retrieved full order system cannot be used for either response prediction or for damage assessment.

8.3 Assumption of Classical Damping - Case III

Introducing the assumption that the structural damping can be expressed as a function of the mass and stiffness matrices leads to a further improvement in the identification. In fact, using proper normalization conditions for the eigenvectors, it is possible to show that, in this case, the real and imaginary part of the eigenvectors are equal. This provides a substantial reduction in the number of parameters to optimize. After some simple derivations, it is possible to show that not only $\mathcal{M}_{11}$ and $\mathcal{L}_{11}$ are exact but also the subpartition of the stiffness matrix, $\mathcal{K}_{11}$, can be determined exactly. The fact that the assumption of classical damping allows the exact determination of subpartitions of the structural matrices related to the instrumented degrees-of-freedom is important. In fact, if damage occurs at the instrumented locations, the identified model will allow us to exactly detect and quantify the structural damage. However, in terms of structural response prediction, these identified models are still not capable of accurately predicting the response to future excitation.

8.4 Diagonal Mass Matrix and Tri-diagonal Damping and Stiffness Matrices - Case IV

This type of model can be used to represent discrete mass systems where each mass is only connected to the two adjacent ones by spring/dashpot elements and where the two end masses are linked to rigid supports (see Figure 1). If the system can be modeled with a diagonal mass matrix $\mathcal{M}$ and with tri-diagonal damping matrix $\mathcal{L}$ and stiffness matrix $\mathcal{K}$, the identified full-order system shows some interesting characteristics. By imposing the additional constraints in the optimization process, it is possible to show that the unknown part of the identified complex eigenvector matrix contains, for each row, only one undetermined factor α_i, with $i = 1, 2, \cdots, p$.

This implies that the exact (unknown) and the identified (determined through optimization) components of the eigenvector matrices are related by the following relationships:

$$\widehat{\psi}_1 = \psi_1(1,:); \qquad \widehat{\psi}_2 = \begin{bmatrix} \alpha_1\psi_2(1,:) \\ \alpha_2\psi_2(2,:) \\ \cdots \\ \alpha_{N-1}\psi_2(N-1,:) \end{bmatrix} \tag{8.5}$$

where ψ_1, ψ_2 and $\widehat{\psi}_1$, $\widehat{\psi}_2$ are the exact and the identified $\widehat{()}$ parts of the eigenvector matrix while $\alpha_1, \alpha_2, ..., \alpha_{(N-1)}$ are the unknown proportionality constants. This means that, while all the components of the complex eigenvectors at the instrumented locations are correctly identified, the identified components of the eigenvectors at the non instrumented locations are proportional to the exact values through some unknown constants. At each degree-of-freedom, the value of such a constant is the same for all the modes.

For the other cases where $p < p_{max}$, not all "zero" elements are independent. There are only $p^2 - p$ independent new constraints because, among the total p^2 conditions, the first $p^2 - p$ ones imply that there is an unknown factor α_i $(i = 1, 2, \cdots, p)$ at each unmeasured coordinate while the remaining p conditions are $0 \times \alpha_i = 0$. Therefore:

$$SS_{dof} = p^2 - [(N-1) \times (N-2) - p] = p \tag{8.6}$$

which implies that the identified $\widehat{\psi}_1$ and $\widehat{\psi}_2$ are similar to the corresponding ones in Eqn.(8.5).

$$\widehat{\psi}_1 = \begin{bmatrix} \psi_1(1,:) \\ \psi_1(2,:) \\ \cdots \\ \psi_1(N-p,:) \end{bmatrix}; \qquad \widehat{\psi}_2 = \begin{bmatrix} \alpha_1\psi_2(1,:) \\ \alpha_2\psi_2(2,:) \\ \cdots \\ \alpha_p\psi_2(p,:) \end{bmatrix} \tag{8.7}$$

This result can be easily presented by looking at a numerical example. Let's consider a 4 DOFs structure, described as 4 discrete masses, connected sequentially and anchored on the top and at the bottom to fixed supports. It is assumed that the system is excited by only one actuator, located at the bottom floor while the sensors will be placed in 3 different configurations. Structural accelerations are measured at different floors for different scenarios: 1) sensors at first, second and third floor, 2) sensors at first and second floor, and 3) sensor only at the first floor. These instrumentation setups guarantee the uniqueness of the solution.

Here, the (transpose) matrix of the identified eigenvectors is presented for the case of a full set of sensors (8.8, exact identification) and for the cases of 1 sensor (8.9), 2 sensors (8.10) and 3 sensors (8.11) missing.

$$\boldsymbol{\psi}^T = 10^{-1} \times$$

$$\times \begin{bmatrix} 0.8824 + 0.8824j & 3.4188 + 3.4188j & 3.9066 + 3.9066j & 4.6339 + 4.6339j \\ 0.8824 - 0.8824j & 3.4188 - 3.4188j & 3.9066 - 3.9066j & 4.6339 - 4.6339j \\ -0.6395 - 0.6395j & -1.9375 - 1.9375j & -0.8032 - 0.8032j & 3.7689 + 3.7689j \\ -0.6395 + 0.6395j & -1.9375 + 1.9375j & -0.8032 + 0.8032j & 3.7689 - 3.7689j \\ -2.2889 - 2.2889j & -0.8508 - 0.8508j & 2.2010 + 2.2010j & -0.6226 - 0.6226j \\ -2.2889 + 2.2889j & -0.8508 + 0.8508j & 2.2010 - 2.2010j & -0.6226 + 0.6226j \\ -2.9021 - 2.9021j & 0.9934 + 0.9934j & -1.3027 - 1.3027j & 0.2942 + 0.2942j \\ -2.9021 + 2.9021j & 0.9934 - 0.9934j & -1.3027 + 1.3027j & 0.2942 - 0.2942j \end{bmatrix} \tag{8.8}$$

$$\widehat{\psi}^T_{(1)} = \left[\begin{array}{cc} \widehat{\psi}^T_{1(1)} & \widehat{\psi}^T_{2(1)} \end{array} \right] = 10^{-1} \times$$
$$\times \left[\begin{array}{ccc|c} 0.8824+0.8824j & 3.4188+3.4188j & 3.9066+3.9066j & 8.0047+8.0047j \\ 0.8824-0.8824j & 3.4188-3.4188j & 3.9066-3.9066j & 8.0047-8.0047j \\ -0.6395-0.6395j & -1.9375-1.9375j & -0.8032-0.8032j & 6.5105+6.5105j \\ -0.6395+0.6395j & -1.9375+1.9375j & -0.8032+0.8032j & 6.5105-6.5105j \\ -2.2889-2.2889j & -0.8508-0.8508j & 2.2010+2.2010j & -1.0754-1.0754j \\ -2.2889+2.2889j & -0.8508+0.8508j & 2.2010-2.2010j & -1.0754+1.0754j \\ -2.9021-2.9021j & 0.9934+0.9934j & -1.3027-1.3027j & 0.5082+0.5082j \\ -2.9021+2.9021j & 0.9934-0.9934j & -1.3027+1.3027j & 0.5082-0.5082j \end{array} \right]$$
$$= \left[\begin{array}{c|c} \psi^T_{1(1)} & \alpha_{1(1)}\psi^T_{2(1)} \end{array} \right] \tag{8.9}$$

$$\widehat{\psi}^T_{(2)} = \left[\begin{array}{cc} \widehat{\psi}^T_{1(2)} & \widehat{\psi}^T_{2(2)} \end{array} \right] = 10^{-1} \times$$
$$\times \left[\begin{array}{cc|cc} 0.8824+0.8824j & 3.4188+3.4188j & 3.6993+3.6993j & 13.145+13.145j \\ 0.8824-0.8824j & 3.4188-3.4188j & 3.6993-3.6993j & 13.145-13.145j \\ -0.6395-0.6395j & -1.9375-1.9375j & -0.7606-0.7606j & 10.692+10.692j \\ -0.6395+0.6395j & -1.9375+1.9375j & -0.7606+0.7606j & 10.692-10.692j \\ -2.2889-2.2889j & -0.8508-0.8508j & 2.0842+2.0842j & -1.7661-1.7661j \\ -2.2889+2.2889j & -0.8508+0.8508j & 2.0842-2.0842j & -1.7661+1.7661j \\ -2.9021-2.9021j & 0.9934+0.9934j & -1.2336-1.2336j & 0.8346+0.8346j \\ -2.9021+2.9021j & 0.9934-0.9934j & -1.2336+1.2336j & 0.8346-0.8346j \end{array} \right]$$
$$= \left[\begin{array}{c|cc} \psi^T_{1(2)} & \alpha_{1(2)}\psi^T_{21(2)} & \alpha_{2(2)}\psi^T_{22(2)} \end{array} \right] \tag{8.10}$$

$$\widehat{\psi}^T_{(3)} = \left[\begin{array}{cc} \widehat{\psi}^T_{1(3)} & \widehat{\psi}^T_{2(3)} \end{array} \right] = 10^{-1} \times$$
$$\times \left[\begin{array}{c|ccc} 0.8824+0.8824j & 14.874+14.874j & 25.253+25.253j & 58.679+58.679j \\ 0.8824-0.8824j & 14.874-14.874j & 25.253-25.253j & 58.679-58.679j \\ -0.6395-0.6395j & -8.4294-8.4294j & -5.1925-5.1925j & 47.725+47.725j \\ -0.6395+0.6395j & -8.4294+8.4294j & -5.1925+5.1925j & 47.725-47.725j \\ -2.2889-2.2889j & -3.7015-3.7015j & 14.227+14.227j & -7.8832-7.8832j \\ -2.2889+2.2889j & -3.7015+3.7015j & 14.227-14.227j & -7.8832+7.8832j \\ -2.9021-2.9021j & 4.3218+4.3218j & -8.4210-8.4210j & 3.7255+3.7255j \\ -2.9021+2.9021j & 4.3218-4.3218j & -8.4210+8.4210j & 3.7255-3.7255j \end{array} \right]$$
$$= \left[\begin{array}{c|ccc} \psi^T_{1(3)} & \alpha_{1(3)}\psi^T_{21(3)} & \alpha_{2(3)}\psi^T_{22(3)} & \alpha_{3(3)}\psi^T_{23(3)} \end{array} \right] \tag{8.11}$$

It is worthy to observe that each row of the identified complex eigenvector matrices $\widehat{\psi}_2$ is proportional to the exact values ψ_2, while the identified $\widehat{\psi}_1$ coincides with the exact ψ_1.

Consequently, the identified full order mass, damping and stiffness matrices show the subpartitions corresponding to the measured degrees of freedom exactly identified while the subpartitions corresponding to the unmeasured degrees of freedom are proportional to the exact ones through these α coefficients. For example, the stiffness matrix for an N degrees-of-freedom system with only $n = N - p$ instrumented ones can be represented as:

$$\widehat{\mathcal{K}} = -\left(\widehat{\psi}\mathbf{\Lambda}^{-1}\widehat{\psi}^T\right)^{-1} = \left[\begin{array}{cc} \widehat{\mathcal{K}}_{11} & \widehat{\mathcal{K}}_{12} \\ \widehat{\mathcal{K}}_{21} & \widehat{\mathcal{K}}_{22} \end{array} \right] = \tag{8.12}$$

$$\left[\begin{array}{c|ccc} \mathcal{K}_{11} & \alpha_1^{-1} k_{n+1,2} & \cdots & 0 \\ \hline \alpha_1^{-1} k_{2,n+1} & (\alpha_1 \alpha_1)^{-1} k_{n+1,n+1} & \cdots & 0 \\ \vdots & \vdots & \vdots & \vdots \\ 0 & 0 & \cdots & (\alpha_p \alpha_p)^{-1} k_{NN} \end{array}\right] \tag{8.13}$$

where $\widehat{\mathcal{K}}_{11} = \mathcal{K}_{11}$ is the $n \times n$ known partition of the stiffness matrix while $\widehat{\mathcal{K}}_{12}$ and $\widehat{\mathcal{K}}_{22}$ are the $n \times p$ and $p \times p$ unknown partitions. For the 4-DOF's system presented before, the identified stiffness matrices for the for the case of 2 sensors missing has the form:

$$\widehat{\mathcal{K}}_{(2)} = \begin{bmatrix} \widehat{\mathcal{K}}_{11(2)} & \widehat{\mathcal{K}}_{12(2)} \\ \widehat{\mathcal{K}}_{21(2)} & \widehat{\mathcal{K}}_{22(2)} \end{bmatrix} = \left[\begin{array}{cc|cc} 4.0000 & -1.0000 & 0.0000 & 0.0000 \\ -1.0000 & 4.0000 & -3.1680 & 0.0000 \\ \hline 0.0000 & -3.1680 & 4.4606 & -0.3723 \\ 0.0000 & 0.0000 & -0.3723 & 0.1243 \end{array}\right]$$
$$= \left[\begin{array}{cc|cc} k_{11} & k_{12} & 0 & 0 \\ k_{21} & k_{22} & \alpha_{1(2)}^{-1} k_{23} & 0 \\ \hline 0 & \alpha_{1(2)}^{-1} k_{32} & \alpha_{1(2)}^{-2} k_{33} & \alpha_{1(2)}^{-1} \alpha_{2(2)}^{-1} k_{34} \\ 0 & 0 & \alpha_{1(2)}^{-1} \alpha_{2(2)}^{-1} k_{43} & \alpha_{2(2)}^{-2} k_{44} \end{array}\right]; \tag{8.14}$$

where the upper left subpartition of the matrix is exactly identified while the remaining ones are just proportional to the exact ones. Using the identified mass and stiffness matrices it is now possible to obtain the natural frequencies of the structural system, frequencies that are identical to the exact ones.

The effects of these α coefficients is also shown in the estimation of the system response. At the measured locations, the predicted response will be equal to the exact one while those predicted at locations with neither an actuator or a sensor, are proportional, through the same coefficients, to the exact ones:

$$\widehat{\mathbf{y}} = \begin{bmatrix} \widehat{\mathbf{y}}_1 \\ \widehat{\mathbf{y}}_2 \end{bmatrix}; \qquad \widehat{\mathbf{y}}_1 = \mathbf{y}_1; \qquad \widehat{\mathbf{y}}_2 = \begin{bmatrix} \alpha_1 \mathbf{y}_2(1,:) \\ \alpha_2 \mathbf{y}_2(2,:) \\ \cdots \\ \alpha_p \mathbf{y}_2(p,:) \end{bmatrix} \tag{8.15}$$

This result can be clearly seen in Figures 2- 4 where the exact and the identified outputs of the 4 DOFs structure are plotted for the three different instrumentation setups (three, two and one DOF instrumented). In addition, the ratios between each predicted and exact output, $\hat{y}_i(t)$ and $y_i(t)$ respectively, are calculated and plotted for each case. It can be observed that these ratios stay constant during the entire time histories: for the degrees-of-freedom with instrumentations, the values of these ratios is equal to one, indicating that the identified time-histories are identical to the exact ones, while for the unmeasured coordinates, these ratios are still constant but different from one. The value of this constant is exactly the unknown α parameter. It is interesting to see that these values differ from case to case (4th floor case 1: α = 1.7274, case 2: α = 2.8368 while for case 3, α = 12.663). If the structure was modeled as a shear type structure, these α parameters could be uniquely determined, as shown later, and so the matching between all the predicted and exact outputs would be accomplished. One interesting conclusion that can be drawn out of these figures is that the natural frequencies of the identified system are identical to the one of the exact system: because of the particular shape of the mass and stiffness matrices for these systems, the unknown α parameters cancel out in the characteristic equation of the associated eigenvalue problem as so the eigenvalues are not affected. On the contrary, the normalized real eigenvectors will present a similar pattern as the complex identified eigenvectors, keeping one part proportional to the exact one through the α parameters. Both these conclusions can also be seen by Figure 2-4: the frequency content of the identified time histories is identical to that of the exact one while the amplitude is constantly different.

Although these α parameters remain unknown, it is still possible to use the identified structural stiffness and mass matrices to obtain correct assessment of the structural damage. In fact, as shown in Yu (2004), by using assuming that the structural mass does not change because of the damage, appropriately defined parameters can be defined as ratios of identified stiffnesses and masses. Because of the nature of the identified matrices, these coefficients are such that the α parameters cancel out. By comparing these newly defined parameters before and after damage has occurred, a correct assessment of the structural damage and its location within the structure can be performed.

In the work by Yu (2004), a 'K/M ratio 'was introduced as a tool for damage detection. For the measured DOFs $(i, j = 1, 2, \cdots, n)$, such a ratio results in:

$$\frac{\widehat{k}_{ij}^2}{\widehat{m}_{ii}\widehat{m}_{jj}} = \frac{k_{ij}^2}{m_{ii}mjj} \tag{8.16}$$

providing the same value for the identified system as well as for the initial exact one.

For the unmeasured DOFs $(i, j = 1, 2, \cdots, p)$, it follows:

$$\frac{\widehat{k}_{ij}^2}{\widehat{m}_{ii}\widehat{m}_{jj}} = \frac{k_{ij}^2(\alpha_{i-n}\alpha_{j-n})^2}{m_{ii}(\alpha_{i-n})^2 m_{jj}(\alpha_{j-n})^2} = \frac{k_{ij}^2}{m_{ii}m_{jj}} \tag{8.17}$$

showing a very important property of the "$\mathbf{K}/\mathbf{M}$ ratio", that is, the "$\mathbf{K}/\mathbf{M}$ ratio" is a variable which is independent from the unknown factors included in the identified matrices $\widehat{\mathcal{M}}$ and $\widehat{\mathcal{K}}$. This is a quite valuable property that could be significant in damage detection processes that use "reduced order" models. The same "$\mathbf{K}/\mathbf{M}$ ratio" can be obtained considering the identified mass and stiffness matrices for the damaged system ($\widehat{\mathcal{M}}^d$ and $\widehat{\mathcal{K}}^d$). Considering that the mass $m_{ii}(i = 1, 2, .., N)$ does not change before and after damage, i.e. $m_{ii} = m_{ii}^{(d)}$, we can obtain information on damage location and amount in the entire system by a new damage index, the square root of the ratio between the two "$\mathbf{K}/\mathbf{M}$ ratios" corresponding to the damage and undamaged configurations, as shown:

$$\frac{k_{ij}^{(d)}}{k_{ij}} = \sqrt{\frac{\left(k_{ij}^{(d)}\right)^2}{(k_{ij})^2}} = \sqrt{\frac{\left(k_{ij}^{(d)}\right)^2 / \left(m_{ii}^{(d)} m_{jj}^{(d)}\right)}{(k_{ij})^2 / (m_{ii}m_{jj})}} = \sqrt{\frac{\left(\widehat{k}_{ij}^{(d)}\right)^2 / \left(\widehat{m}_{ii}^{(d)} \widehat{m}_{jj}^{(d)}\right)}{\left(\widehat{k}_{ij}\right)^2 / (\widehat{m}_{ii}\widehat{m}_{jj})}} \tag{8.18}$$

being $i, j = 1, 2, \cdots, N$. This approach allows us to explore the changes also in the off-diagonal terms of the stiffness matrix, giving the opportunity of refining the damage identification process. For details, the reader is referred to the work by Yu (2004).

8.5 A Special Case IV: Shear Type Structural Models - Case V

The identification of the complete mass, damping and stiffness matrices for a structural system with limited input-output data can only be performed in the case that the structure can be modeled as a shear-type structure, represented by concentrated masses connected with the adjacent ones, with only one mass at one end connected to a support. This type of models, defined shear-type models, represents a special case of the previous tri-diagonal systems. For these models, the first issue that needs to be addressed is whether there is a unique solution of the identification problem or not. In fact, depending on the type of input-output information available, there could be multiple structural systems that have the same input-output relations at the measured degrees of freedom. Studies conducted by Franco et al. (2006) and by Yu (2004) show that, in order to have a unique identification, it is sufficient to have one of the following 3 test scenarios:

- one sensor and one actuator at the first floor,
- one sensor and one actuator at the top floor,

- one actuator and two sensors at any two consecutive intermediate floors.

If one of these test setup is used, then the available data is sufficient to retrieve the complete second-order representation (mass, damping and stiffness matrices) for a shear-type structural system. Looking at the particular shape of the damping and stiffness matrices, additional constraints can be imposed by considering the following geometric relations:

$$\begin{cases} k_{(i-1)i} + k_{i(i+1)} = -k_{ii}, & \text{i = 2,3,...,N-1;} \\ k_{(i-1)i} = -k_{ii}, & \text{i = N.} \end{cases} \tag{8.19}$$

$$\begin{cases} c_{c(i-1)i} + c_{ci(i+1)} = -c_{cii}, & \text{i = 2,3,...,N-1;} \\ c_{c(i-1)i} = -c_{cii}, & \text{i = N.} \end{cases} \tag{8.20}$$

In this case, the optimization algorithm will lead to the identified matrices that have the same structures as those identified in the previous case. Now, because of the typology of the structure, it is possible to uniquely determine the "unknown" α parameters and use these parameters to retrieve all the information regarding the "full" order model of the system. All the modal parameters (natural frequencies and mode shapes) can be obtained and the response of the system at any degree-of-freedom can be accurately reproduced. For example, for a 4 DOFs shear type system analogous to the one from previous Case IV (see Figure 5 and Table2), with one actuator at the first floor and three sensors at each of the first three floors, the identified stiffness matrix can be expressed as:

$$\widehat{\mathcal{K}}_{(1)} = \begin{bmatrix} \widehat{\mathcal{K}}_{11(1)} & \widehat{\mathcal{K}}_{12(1)} \\ \widehat{\mathcal{K}}_{21(1)} & \widehat{\mathcal{K}}_{22(1)} \end{bmatrix} = \left[\begin{array}{ccc|c} 4.0000 & -1.0000 & 0.0000 & 0.0000 \\ -1.0000 & 4.0000 & -3.0000 & 0.0000 \\ 0.0000 & -3.0000 & 4.0000 & -0.5789 \\ \hline 0.0000 & 0.0000 & -0.5789 & 0.3351 \end{array}\right] \tag{8.21}$$

$$= \left[\begin{array}{ccc|c} k_{11} & k_{12} & 0 & 0 \\ k_{21} & k_{22} & k_{23} & 0 \\ 0 & k_{32} & k_{33} & \alpha_{1(1)}^{-1} k_{34} \\ \hline 0 & 0 & \alpha_{1(1)}^{-1} k_{43} & \alpha_{1(1)}^{-2} k_{44} \end{array}\right]; \tag{8.22}$$

The only unknown α factor can be determined using the identified values of $\widehat{k}_{43}$ and $\widehat{k}_{44}$ as shown:

$$\overline{k}_4 = k_{44} = [\alpha_{1(1)}]^2 \widehat{k}_{44}; \tag{8.23}$$

$$-\overline{k}_4 = k_{43} = \alpha_{1(1)} \widehat{k}_{43} \tag{8.24}$$

$$\Longrightarrow \alpha_{1(1)} = -\widehat{k}_{43}/\widehat{k}_{44} = 0.5789/0.3351 = 1.7274 \tag{8.25}$$

Then, once the $\alpha_{1(1)}$ has been obtained, the final identified stiffness matrix $\widehat{\mathcal{K}}_{(1)}$, can be retrieved:

$$\widehat{\mathcal{K}}_{(1)} = \begin{bmatrix} 4.0000 & -1.0000 & 0.0000 & 0.0000 \\ -1.0000 & 4.0000 & -3.0000 & 0.0000 \\ 0.0000 & -3.0000 & 4.0000 & -1.0000 \\ 0.0000 & 0.0000 & -1.0000 & 1.0000 \end{bmatrix} = \mathcal{K} \tag{8.26}$$

Having the possibility of determining the correct structural matrices allows us to perform accurate damage assessment. By comparing the stiffness matrix before and after damage has occurred, it is possible to pinpoint the exact locations where structural damage has occurred and its amount.

9 Conclusions

In this paper, a brief overview of some methodologies for identifying structural parameters like mass, damping and stiffness of a structural system has been presented. Methods based on the transformation from a first-order representation of the system to a second-order model have been considered. These methods show great potential when a complete set of input/output data is available. For complete set, we intend that each degree-of-freedom has been instrumented with either a sensor or an actuator. When the instrumentation set is not complete, these methodologies can provide only limited information. A procedure to attempt to expand the 'reduced'models has also been discussed. If the structure can be modeled as a system with mass, damping and stiffness matrices of general form, only the 'condensed'form of the structural matrices can be obtained. The 'expanded'matrices that are obtained through the optimization process do not capture any dynamics of the system. If the structure can be modeled with a diagonal mass matrix and with tri-diagonal damping and stiffness matrices, the 'expanded'matrices show that the components related to the unmeasured DOFs are proportional to the real ones. Although these coefficients of proportionality cannot be computed, these matrices can still be used to provide accurate information about the structural damage. If the structure can be modeled as a shear-type building, the exact full-order matrices can be obtained.

10 Acknowledgements

The author would like to acknowledge the fundamental contributions of many colleagues and students who have been working with him over the years on this topic and in particular, Professor R.W. Longman (Columbia University), Professor H. Luş (Bogazici University), Professor M. DeAngelis (Universita' degli Studi di Roma "La Sapienza"), Professor M. Imbimbo (Universita' degli Studi, Cassino) and Dr. J.Yu (Thorthon-Tomassetti).

Bibliography

M.S. Agbabian, S.F. Masri, R.K. Miller, and T.K. Caughey. System identification approach to detection of structural changes. *ASCE Journal of Engineering Mechanics*, 117(2):370–390, 1991.

K.F. Alvin and K.C. Park. Second-order structural identification procedure via state-space - based system identification. *AIAA Journal*, 32(2):397–406, 1994.

K.F. Alvin, L.D. Peterson, and K.C. Park. Method for determining minimum-order mass and stiffness matrices from modal test datan. *AIAA Journal*, 33(1):128–135, 1995.

E. Balmes. New results on the identification of normal modes from experimental complex modes. *Mechanical System and Signal Processing*, 11(2):229–243, 1997.

J.L. Beck and L.S. Katafygiotis. Updating models and their uncertainties.i: Bayesian statistical framework. *Journal of Engineering Mechanics*, 124(4):455–461, 1998a.

J.L. Beck and L.S. Katafygiotis. Updating models and their uncertainties.ii: Modal identifiability. *Journal of Engineering Mechanics*, 124(4):463–467, 1998b.

M. DeAngelis, H. Luş, and R. Betti. A new approach for reduced order models of structural systems. *ASME Journal of Applied Mechanics*, 70, 2003.

M. DeAngelis, H. Luş, R. Betti, and R.W. Longman. Extracting physical parameters of mechanical models from identified state space representations. *ASME Journal of Applied Mechanics*, 69(5):617–625, 2002.

S.W. Doebling, F.M. Hemez, M.S. Barlow, L.D. Petersonorse, and C. Farhat. Damage detection in a suspended scale model truss via modal update. *Proceedings of the 11th International Modal Analysis Conference, Society for Experimental Mechanics*, pages 1083–1094, 1993.

C. Farhat and F.M. Hemez. Updating finite element dynamic models using an element-by-element sensitivity methodology. *AIAA Journal*, 31(9):1702–1711, 1993.

G. Franco, R. Betti, and R.W. Longman. On the uniqueness of solutions for the identification of linear structural systems. *ASME Journal of Applied Mechanics*, 73(1):153–162, 2006.

J.N. Juang, J.E. Cooper, and J.R. Wright. An eigensystem realization algorithm using data correlations (era/dc) for model parameter identification. *Control Theory and Advanced Technology*, 4(1):5–14, 1988.

J.N. Juang and R.W. Longman. Optimized system identification. *NASA Technical Memorandum*, NASA/TM-1999-209711, 1999.

J.N. Juang and M. Phan. Identification of system, observer, and controller from closed-loop experimental data. *Journal of Guidance, Control, and Dynamics*, 17(1):91–96, 1994a.

J.N. Juang and M. Phan. Linear system identification via a backward-time observer. *Journal of Guidance, Control, and Dynamics*, 17(3):505–512, 1994b.

J.N. Juang, M. Phan, L.G. Horta, and R.W. Longman. Identification of observer/kalman filter markov parameters: Theory and experiments. *Journal of Guidance, Control, and Dynamics*, 16:320–329, 1993.

J.N. Juang and H. Suzuki. An eigensystem realization algorithm in frequency domain for modal parameter identification. *Journal of Vibration, Acoustics, Stress and Reliability in Design*, 110(1), 1988.

R.K. Lim. *System Identification, Observer Identification, and Data-Based Controller Design*. PhD thesis, Columbia University, 1998.

R.W. Longman and J.N. Juang. Recursive form of the eigensystem realization algorithm for system identification. *Journal of Guidance, Control, and Dynamics*, 12(5):647–652, 1989.

H. Luş. *Control Theory Based System Identification*. PhD thesis, Columbia University, 2001.

H. Luş, R. Betti, and R.W. Longman. Identification of linear structural systems using earthquake-induced vibration data. *Earthquake Engineering and Structural Dynamics*, 28:1449–1467, 1999.

H. Luş, M. DeAngelis, and R. Betti. A new approach for reduced order modelling of mechanical system using vibration measurements. *ASME Journal of Engineering Mechanics*, 70(5):715–723, 2003.

M. Moonen, B. DeMoor, L. Vandenberghe, and J. Vandewalle. On- and off-line identification of linear state space models. *International Journal of Control*, 49(1):219–232, 1989.

M. Phan, L.G. Horta, J.N. Juang, and R.W. Longman. Linear system identification via an asymptotically stable observer. *Journal of Optimization Theory and Applications*, 79(1):59–86, 1993.

M. Phan, L.G. Horta, J.N. Juang, and R.W. Longman. mprovement of observer/kalman filter identification (okid) by residual whitening. *Journal of Vibrations and Acoustics*, 117:232–238, 1995.

M. Phan, J.N. Juang, and R.W. Longman. On markov parameters in system identification. *NASA Technical Memorandum*, 104156, 1991.

M. Phan, J.N. Juang, and R.W. Longman. Identification of linear multivariable systems by identification of observers with assigned real eigenvalues. *Journal of Astronautical Sciences*, 40(2):261–279, 1992.

E. Safak. Adaptive modelling, identification, and control of dynamic structural systems.i: Theory. *Journal of Engineering Mechanics*, 115(11):2386–2405, 1989a.

E. Safak. Adaptive modelling, identification, and control of dynamic structural systems.ii: Applications. *Journal of Engineering Mechanics*, 115(11):2406–2426, 1989b.

A. Sestieri and S.R. Ibrahim. Analysis of errors and approximations in the use of modal co-ordinates. *Journal of Sound and Vibration*, 177(2):145–157, 1994.

D.H. Tseng, R.W. Longman, and J.N. Juang. Identification of gyroscopic and nongyroscopic second order mechanical systems including repeated problems. *Advances in Astronautical Sciences*, 87:145–165, 1994.

F.E. Udwadia. Methodology for optimum sensor locations for parameter identification in dynamic systems. *Journal of Engineering Mechanics*, 120(2):368–390, 1994.

C.D. Yang and F.B. Yeh. Identification, reduction, and refinement of model parameters by the eigensystem realization algorithm. *Journal of Guidance, Control, and Dynamics*, 13(6):1051–1059, 1990.

J. Yu. *Identification of Structural Systems with Limited Set of Instrumentation*. PhD thesis, Columbia University, 2004.

Constraints\Cases	$\sharp$ of Unknowns	$\sharp$ of Constraints	$\sharp$ of Solution Space
Case I	$2pN$	$\frac{p}{2}(2N - p + 1)$	$\frac{p}{2}(2N + p - 1)$
Case II	$2pN$	$\frac{p}{2}(4N - 3p + 1)$	$\frac{p}{2}(3p - 1)$
Case III	$2pN$	$\frac{p}{2}(4N - 3p + 1)$	$\frac{p}{2}(3p - 1)$
Case IV	$2pN$	$p(2N - 1)$	p
Case V	$2pN$	$2pN$	0

Table 1. Number of Total Constraints and Dimension of Solution Space for the Different Representative Scenarios of Reduced Order Systems

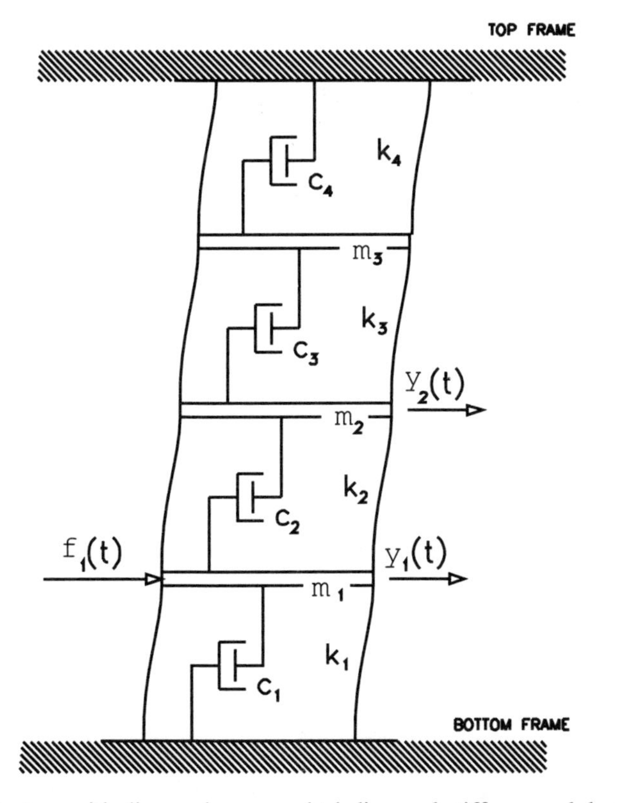

Figure 1. 3-DOF System with diagonal mass and tri-diagonal stiffness and damping matrices

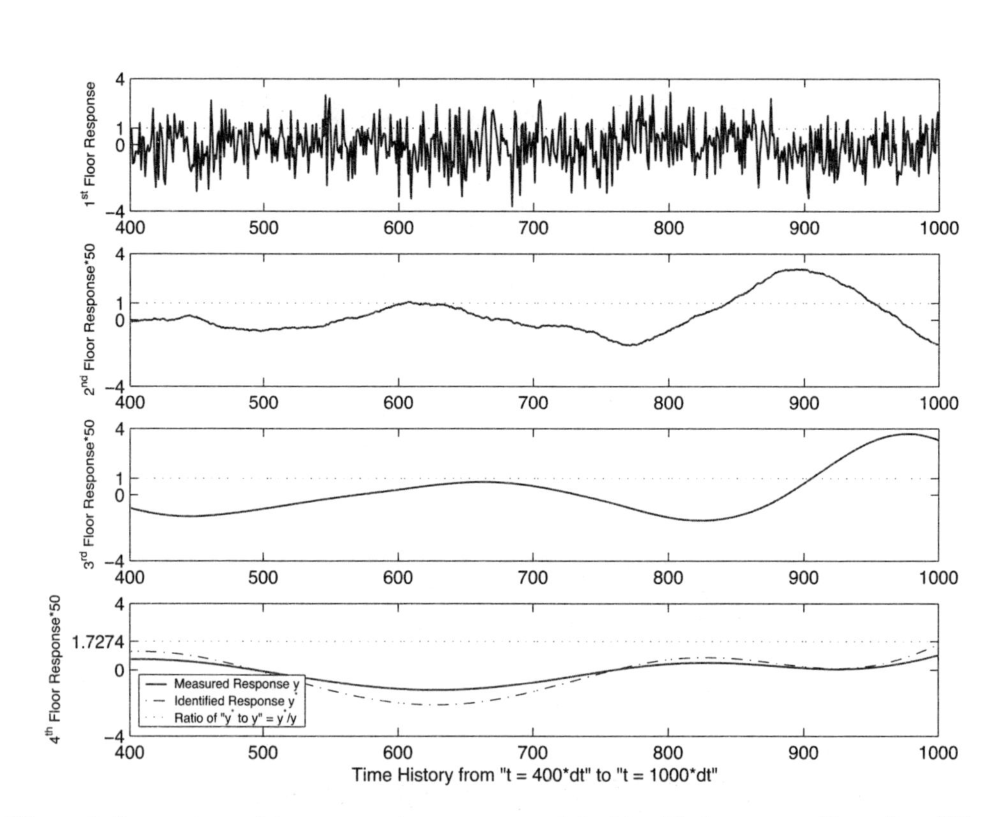

Figure 2. Comparison of the measured response $\mathbf{y}$ and the identified responses $\widehat{\mathbf{y}} = \mathbf{y}^*$ at different floors of the 4-DOF shear-type building for 1-DOF unmeasured case

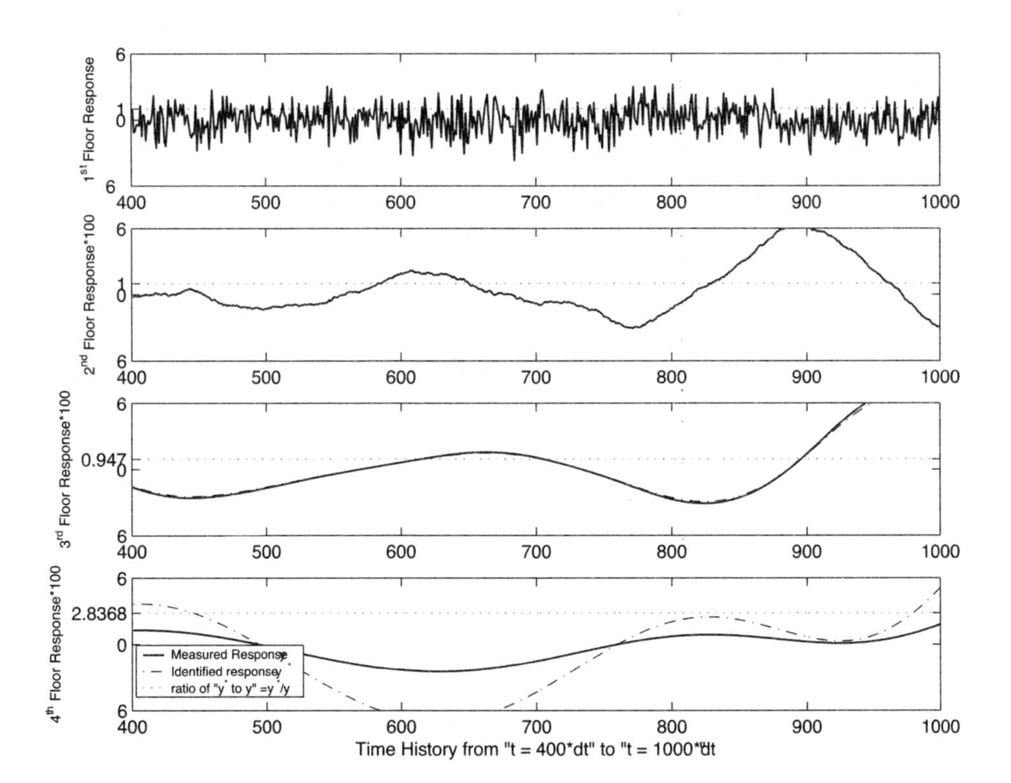

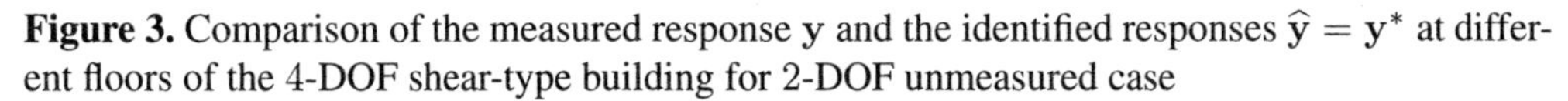

Figure 3. Comparison of the measured response $\mathbf{y}$ and the identified responses $\widehat{\mathbf{y}} = \mathbf{y}^*$ at different floors of the 4-DOF shear-type building for 2-DOF unmeasured case

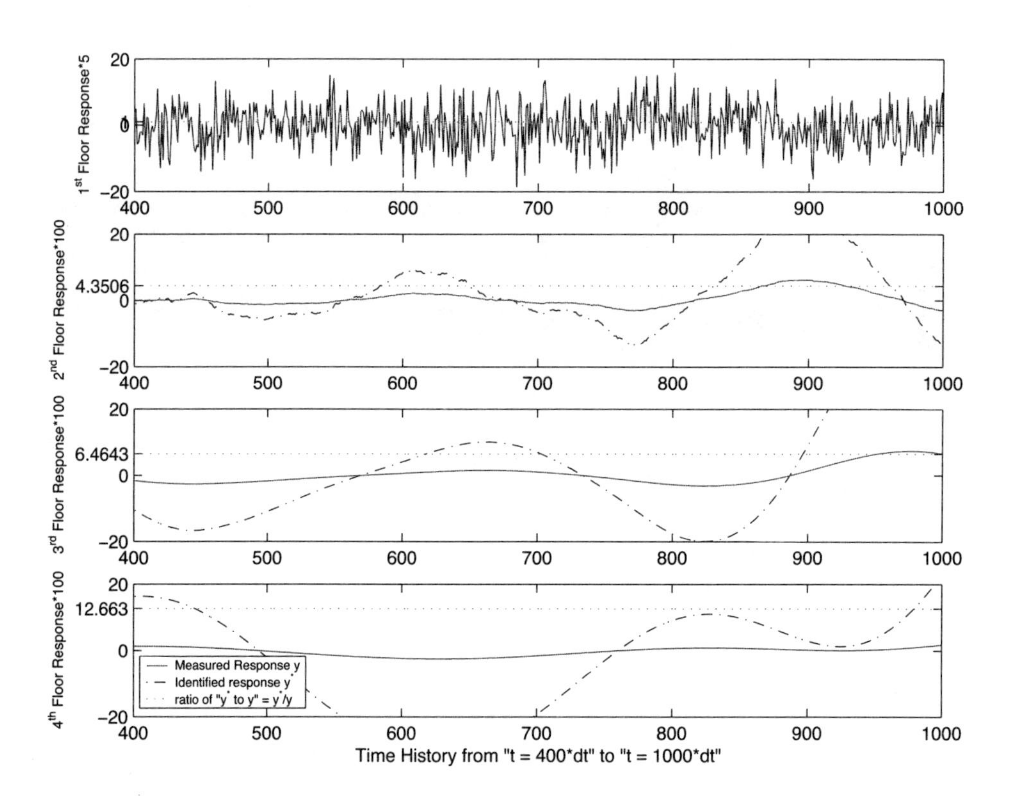

Figure 4. Comparison of the measured response $\mathbf{y}$ and the identified responses $\widehat{\mathbf{y}} = \mathbf{y}^*$ at different floors of the 4-DOF shear-type building for 3-DOF unmeasured case

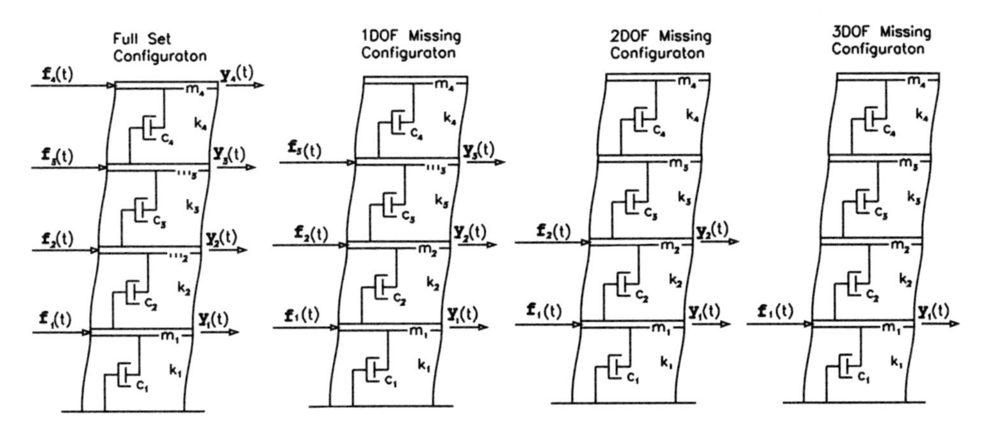

Figure 5. 4dof Shear-type Building System—Reduced-Order Scenarios

Exact System Information of the Test Model											
MASS M				DAMP C_c				STIFFNESS K			
0.8	0.0	0.0	0.0	0.48	-0.1	0.0	0.0	4.0	-1.0	0.0	0.0
0.0	2.0	0.0	0.0	-0.1	0.6	-0.3	0.0	-1.0	4.0	-3.0	0.0
0.0	0.0	1.2	0.0	0.0	-0.3	0.52	-0.1	0.0	-3.0	4.0	-1.0
0.0	0.0	0.0	1.0	0.0	0.0	-0.1	0.2	0.0	0.0	-1.0	1.0

Table 2. Mass, damping and stiffness matrices used for the system of Figure 5

Structural Identification Parametric Models and Idefem Code

Fabrizio Vestroni*

* Dipartimento di Ingegneria Strutturale e Geotecnica, University of Roma *La Sapienza*, Italy

Abstract This chapter describes structural identification techniques based on parametric models. Focus is given to physical models, where the inertial and mechanical characteristics are the goals of identification. These models are more attractive than modal models as they potentially yield local information. To reduce the computational effort of this complex problem, a refined algorithm is implemented which identifies the parameters of a structure's finite element model. Experimental data, usually represented by selected modal quantities, are used in the identification process.

1 Introduction

Great improvements have been made in modelling the dynamic behaviour of structural systems. However, the investigation of the response of complicated engineering structures cannot be reliably developed solely by means of a mathematical model. Simplifying assumptions are necessary and knowledge of the actual parameter values can only be approximate. Experimental data obtained from the real structure can be used to update the mathematical model, bearing in mind that it is important to verify the properties of the structure at the end of its construction and compare them to those expected. There is a growing interest in the experimental assessment of the mechanical characteristics of existing structures to better understand and predict their behavior and, more important, how their evolution may be used within a health monitoring strategy.

For large structural systems, frequently little information is available from experimental tests, as few points can be instrumented, and the number of conditions that can be investigated is limited. Moreover, the information that can be obtained mostly concerns general response quantities rather than specific characteristics of the structure. Hence, the best use of experimental data is achieved by referring to a system identification technique that allows for suitable correlation of a priori analytical information with test results (Natke, 1982; Mottershead and Friswell, 1993; Natke, 1991; Vestroni, 1991; Ghanem and Shinozuka, 1995; Capecchi and Vestroni, 1999; Ghanem and Sture Eds., 2000).

Structural identification is a process which determines a model M, within a fixed class $S(M)$, that best fits the structural experimental response, according to a judgment criterion. There will always be a difference between predicted (analytical) and measured (experimental) data, both due to modelling and measurement errors. Therefore, within $S(M)$, it is not possible to determine the true model, but the model that best fits the true one. In particular, dynamic identification is based on dynamic measurements.

Even restricting attention to elastic structures, many different techniques have been introduced concerning the class of interpretative models and the experimental results, the methods to evaluate the optimal estimate, and the procedures to solve the inverse problems, which are ill-conditioned and sometimes undetermined. Experimental data typically comprise response quantities such as displacement or acceleration components in time or frequencies and eigenvector components in the frequency domain.

Within parametric models, two important classes are modal models and physical models. In modal models the only assumption concerns the basic structure of the model; in physical models wider use is made of a priori information, and identification mainly consists of updating the model by adjusting the values of selected structural parameters. These parameters will effectively maintain physical meaning if the model is accurate in reproducing structural behaviour on the whole, otherwise they are merely instrumental quantities. The models considered here are held to be sufficiently complete for any uncertainty to be substantially restricted to the values of parameters.

Physical models are much more involved and due to their structure are more rigid with respect to modal models in fitting given experimental results. Thus parameter identification is much more difficult and onerous. This has lead to a more widespread use of modal models in practical engineering. However, physical models are more attractive, as they are more robust and refined, and typically used in direct problems. They are the only models potentially able to give local information, which is crucial to damage detection (Ewins, 2000; Friswell and Mottershead, 1995, 1998; Vestroni and Capecchi, 2003).

2 Modal Analysis and Structural Identification

The choice between physical or modal models thus depends on whether it is more reasonable to formulate a realistic analytical model with accurate values for the geometrical and mechanical characteristics or whether it is possible to develop a sufficiently wide experimental investigation where a high number of response quantities are measured.

In any case, even if the physical model of a structure is the final goal of the identification process, the first step is the experimental evaluation of modal parameters. However, in this case, only an incomplete set of modal parameters is determined and used as experimental data for identifying the physical model. The techniques to extract modal parameters from the measured time signals are presented in the first chapter.

Among the physical models reference is made to finite element models. Their response is governed by a set of coefficients that are strictly related to physical quantities, for example, the coefficients of mass or stiffness matrices. These physical quantities represent the parameters of the models. The estimation procedures of these quantities using experimental data is generally referred to as *structural identification* or *model updating.*

Identification based on a finite element (FE) model uses all a priori information on system behaviour derived from the theory of structures and the knowledge of the constitutive relationships, while guaranteeing a welldetailed description of the structural response. Uncertainties regarding system modelling are restricted to the values of some physical parameters governing the response of the model (Mottershead and Friswell, 1993). The optimal values of these are determined so as to minimize the difference between measured and predicted response quantities. The FE model assumed here is linear and the experimental data consist of the natural frequencies and normal mode components. Of course, this is not representative of every possible situation, but does cover quite a number. For a generic excitation, the model furnishes the response, which depends on its mechanical parameters. The characteristics of mass, stiffness, material constitutive moduli and geometric properties of a single element, or a group of elements, can be assumed as parameters.

3 Linear and Nonlinear Methods

In linear methods the best estimate of the parameters is determined by minimizing an input error, which is calculated as the difference between measured input quantities and input quantities evaluated by means of a finite element model, also using measured response. For instance, the input error in the equation of motion in the frequency domain is:

$$\mathbf{e}(\mathbf{x}) = \left[-\omega^2\mathbf{M}(\mathbf{x}) + i\omega\mathbf{C}(\mathbf{x}) + \mathbf{K}(\mathbf{x})\right]\overline{\mathbf{U}}(\omega) - \overline{\mathbf{F}}(\omega) \tag{3.1}$$

where $\mathbf{x}$ are the unknown parameters, $\overline{\mathbf{F}}(\omega)$ is the force vector, completely known, $\overline{\mathbf{U}}(\omega)$ is the measured displacement component vector (measured quantities).

In nonlinear methods, the best estimate of the parameters is determined by minimizing an output error, which is calculated as the difference between measured response quantities and response quantities evaluated by means of a finite element model. With reference to the equation of motion, as in the previous example, the output error is:

$$\mathbf{e}(\mathbf{x}) = \overline{\mathbf{U}}(\omega) - \overline{\mathbf{U}}(\omega, \mathbf{x}) = \overline{\mathbf{U}}(\omega) - \left[-\omega^2\mathbf{M}(\mathbf{x}) + i\omega\mathbf{C}(\mathbf{x}) + \mathbf{K}(\mathbf{x})\right]^{-1}\overline{\mathbf{F}}(\omega). \tag{3.2}$$

In linear methods, the error depends linearly on the model parameters. However, to evaluate a generic component of $\mathbf{F}$, the complete vector $\mathbf{U}$ is required, that is often unrealistic. In nonlinear methods, the error depends nonlinearly on the model parameters, but only the measured components of $\mathbf{U}$ are involved, while the vector $\mathbf{F}$ is always completely known.

4 Estimators

The relationship between observed quantities $\overline{\mathbf{z}}$ and the quantities provided by the analytical model $\mathbf{z}(\mathbf{x})$, which are functions of the parameters $\mathbf{x}$, contains an error $\mathbf{N}$, independent of $\overline{\mathbf{z}}$

$$\overline{\mathbf{z}} = \mathbf{z}(\mathbf{x}) + \mathbf{N}. \tag{4.1}$$

Because of the presence of errors, it is not possible to uniquely determine $\mathbf{x}$ from the previous relationship. However, we can choose an estimator of $\mathbf{x}$ which arbitrarily minimizes the effect of the errors. The methods of parametric estimation can be classified according to the criterion (probabilistic or deterministic) for parameter evaluation or according to the functional which synthesizes in a unique value the differences between observed and analytical quantities.

The aim of structural identification is to evaluate a reliable value for $\mathbf{x}$ given a known experimental value $\mathbf{z}$. This is done with the use of a statistical approach which takes into account the randomness of $\mathbf{N}$. One of the most popular and used approaches, adopted here, is the Bayesian approach (Collins et al., 1974; Beck and Katafygiotis, 1998). The optimal estimate $\hat{\mathbf{x}}$ of parameters takes into account information from experiments (affected by measurement errors) and from initial estimates (affected by some judgement uncertainty). Measurements and parameters are dealt with as random variables having their own probability distribution. A judgement criterion can be the least square error minimization

$$\min_{\hat{\mathbf{x}}} E\left[(\mathbf{x}-\hat{\mathbf{x}})^T(\mathbf{x}-\hat{\mathbf{x}})\right] = \min_{\hat{\mathbf{x}}} \int (\mathbf{x}-\hat{\mathbf{x}})^T(\mathbf{x}-\hat{\mathbf{x}})\, f(\mathbf{x}|\overline{\mathbf{z}})\, d\mathbf{x} \tag{4.2}$$

where $f(\mathbf{x}|\overline{\mathbf{z}})$ is the a posteriori probability density function.

Under the hypothesis that parameters and errors are normally distributed with mean values $(x_0, 0)$ and covariance matrices $(\mathbf{\Sigma}_x, \mathbf{\Sigma}_N)$ respectively, the estimator obtained by (4.2) is equivalent to the *maximum a posteriori probability* of the parameters

$$\hat{\mathbf{x}} = \max_x f(\mathbf{x}|\overline{\mathbf{z}}). \tag{4.3}$$

That is, the best estimate of parameters is the one which exhibits the greatest probability of occurrence given the measured quantities $\overline{\mathbf{z}}$. By making use of Bayes' formula, the a posteriori probability density function of the parameters can be expressed in terms of the a priori probability of the parameters and of the measurements

$$f(\mathbf{x}|\overline{\mathbf{z}}) = \frac{f(\overline{\mathbf{z}}|\mathbf{x})}{f(\overline{\mathbf{z}})} f(\mathbf{x}). \tag{4.4}$$

The probability $f(\overline{\mathbf{z}})$ is a known function. The probability density function $f(\overline{\mathbf{z}}|\mathbf{x})$ can be evaluated in terms of the probability density of N

$$f(\overline{\mathbf{z}}|\mathbf{x}) = f[\overline{\mathbf{z}} - \mathbf{z}(\mathbf{x})]. \tag{4.5}$$

Hence, the a posteriori probability density function of the parameters can be written as

$$f(\mathbf{x}|\overline{\mathbf{z}}) = \alpha \exp -\left\{[\overline{\mathbf{z}} - \mathbf{z}(\mathbf{x})]^T \mathbf{\Sigma}_N^{-1} [\overline{\mathbf{z}} - \mathbf{z}(\mathbf{x})] + [\mathbf{x} - \mathbf{x}_0]^T \mathbf{\Sigma}_{\mathbf{x}}^{-1} [\mathbf{x} - \mathbf{x}_0]\right\} \tag{4.6}$$

where α is a known constant. The values of the parameters which maximize this probability density function are those that minimize the objective function

$$l(\mathbf{x}) = \left\{[\overline{\mathbf{z}} - \mathbf{z}(\mathbf{x})]^T \mathbf{\Sigma}_N^{-1} [\overline{\mathbf{z}} - \mathbf{z}(\mathbf{x})] + [\mathbf{x} - \mathbf{x}_0]^T \mathbf{\Sigma}_{\mathbf{x}}^{-1} [\mathbf{x} - \mathbf{x}_0]\right\}. \tag{4.7}$$

The first term accounts for the disagreement between experimental and analytical values of the observed quantities. The second term accounts for the differences between the value of the parameters and their a priori estimate $\mathbf{x}_0$. Each term is normalized by the weighting matrix $\mathbf{\Sigma}_N^{-1}$ and $\mathbf{\Sigma}_{\mathbf{x}}^{-1}$, which take account of the reliability of the measurements and the a priori information on parameters.

In a Bayesian context it is possible to give an expression of the dispersion of the estimate $\mathbf{x}$ around the optimal solution $\hat{\mathbf{x}}$. This is described by the a posteriori covariance matrix $\mathbf{\Sigma}$. By accounting for the Gaussian hypothesis on the parameters and linearizing $\mathbf{z}(\mathbf{x})$

$$\mathbf{z}\left(\mathbf{x}\right) = \mathbf{z}\left(\hat{\mathbf{x}}\right) + \mathbf{H}\left(\mathbf{x} - \hat{\mathbf{x}}\right), \tag{4.8}$$

a useful and simple expression of the a posteriori covariance matrix $\mathbf{\Sigma}$ of the parameters is obtained:

$$\mathbf{\Sigma} = \left(\mathbf{H}^T \mathbf{\Sigma}_N^{-1} \mathbf{H} + \mathbf{\Sigma}_{\mathbf{x}}^{-1}\right)^{-1} \tag{4.9}$$

where $\mathbf{H}$ is the sensitivity matrix calculated in $\hat{\mathbf{x}}$, that is

$$H_{ij} = \left(\frac{\partial z_i}{\partial x_j}\right)_{\hat{x}}. \tag{4.10}$$

The a posteriori covariance matrix $\mathbf{\Sigma}$ of the parameters depends on the a priori covariance matrix $\mathbf{\Sigma}_{\mathbf{x}}$ modified by the first term which is related to the sensitivity of the observed quantities with respect to the assumed parameters. The choice of observed quantities and parameters governs the $\mathbf{\Sigma}$ value, and so this quantity can be used in the selection of $\mathbf{z}$ and $\mathbf{x}$.

5 Selection of Parameters and Measurements

The behaviour of a finite element model is described by many response quantities and depends on numerous parameters. Function $\mathbf{z}(\mathbf{x})$ relates p observable quantities to q parameters but, in practice, only $m < p$ values z_i are observed and it is meaningful to only consider $n < q$ as parameters x_i to be identified. It is thus necessary to use a criterion for optimal choice of the components of $\mathbf{z}$ and $\mathbf{x}$.

In the Bayesian identification, the problem can be approached by minimizing the dispersion of the estimated parameters, which is measured by the a posteriori covariance matrix $\mathbf{\Sigma}$. A good result of an estimation procedure is that which guarantees a small a posteriori covariance matrix $\mathbf{\Sigma}$ of parameters. When selecting the parameters it is convenient to consider only the part $\mathbf{A}$ of $\mathbf{\Sigma}$ (Fischer matrix) which directly depends on $\mathbf{H}$:

$$\mathbf{A} = \mathbf{H}^T \mathbf{\Sigma}_N^{-1} \mathbf{H}. \tag{5.1}$$

Matrix $\mathbf{A}$, as $\mathbf{\Sigma}$, depends on the parameters and measurements through $\mathbf{H}$. This is the sensitivity matrix of all the p observable quantities with respect to q parameters and depends on the experimental data and the model. In (Masri and Werner, 1985; Capecchi and Vestroni, 1993; Vestroni and Capecchi, 2003) a general procedure is described to choose the most relevant n of the q parameters. This often reduces to a choice of parameters associated to the n greater diagonal elements of $\mathbf{A}$. However, some parameters may not be independent. Hence, it is necessary to evaluate the eigenvalues of $\mathbf{A}$ to eliminate possible redundancy. The parameters can be ordered considering their eigenvalues, so that the most relevant n of q parameters are chosen.

Once the set of parameters are selected, the measurements can be chosen in order to determine the set which provides the most accurate estimate. An efficient estimator, such as the a posteriori maximum probability, is characterized by a minimum error covariance matrix $\mathbf{\Sigma}$, which is equal to the inverse of the Fischer matrix, disregarding $\mathbf{\Sigma}_{\hat{\mathbf{x}}}$.

A selection criterion for sensors consists of *minimizing the matrix* $\mathbf{A}^{-1}$, by finding the minimum trace of $\mathbf{A}^{-1}$ (trace = sum of diagonal terms) or the minimum of its determinant. An approximate criterion is that of maximizing the trace of $\mathbf{A}$ or its determinant. In this way, the contribution of each measurement is additive:

$$tr\left(\mathbf{A}\right) = \sum_{i=1}^{p} A_{ii} = \sum_{i=1}^{p}\left[\sum_{j=1}^{m}\frac{1}{\sigma_j^2}\mathbf{H}_{ji}^2\right] = \sum_{j=1}^{m}\left[\frac{1}{\sigma_j^2}\sum_{i=1}^{p}\mathbf{H}_{ji}^2\right] \tag{5.2}$$

and the optimal set of measures is straightforwardly determined by the m quantities with greatest values in the square brackets. The final selection must be validated in any case by means of the covariance matrix $\mathbf{\Sigma}$. It must be underlined that the selection performed is approximated because the Fischer matrix, or its inverse, are not calculated in the unknown value $\hat{\mathbf{x}}$, but in $x = x_0$.

6 Identifiability and Reliability of the Estimate

The full identifiability of the model depends on the possibility of uniquely determining the parameters $\mathbf{x}$ such that the objective function is at a minimum. Interest is limited to local identifiability, which is assured when the uniqueness exists in a subdomain of the parameter space. In this context, a necessary condition for identifiability is that the number of available data is at least equal to that of the unknown parameters. This is sufficient if the data items are independent of each other. Independence is difficult to ascertain if the measured quantities are eigenfrequencies and eigenmodes, particularly when dealing with a large number of experimental data.

When dealing with physical models and the parameters include the structure's mechanical and geometrical characteristics, it is difficult to establish a priori if the measured quantities are sufficient to identify the model. In practice it is convenient to analyse the definiteness of the Hessian matrix $\mathbf{H}_e$ of $l(\mathbf{x})$ in a subset B_x of $\mathbf{R}^n$, that ensures the uniqueness of the minimum of $l(\mathbf{x})$ in B_x, although this requires a time-consuming numerical evaluation of the matrix. Since $\mathbf{z}(\mathbf{x})$ is usually weakly nonlinear, $\mathbf{H}_e$ can be approximated by the linearized relation:

$$\mathbf{H}_e = \mathbf{H}^T\mathbf{\Sigma}_n^{-1}\mathbf{H} + \mathbf{\Sigma}_x^{-1} \tag{6.1}$$

and within this approximation it coincides with the inverse of the parameter covariance matrix $\mathbf{\Sigma}$; this implies that the estimated parameters have a finite variance if $\mathbf{H}_e$ is positive definite. When errors are present, the definiteness of $\mathbf{H}_e$, or $\mathbf{\Sigma}$, is no longer a sufficient condition for identifiability. More measurements than strictly necessary must be used to obtain a unique and/or acceptable solution.

The problem of identifiability is thus shifted to the evaluation of the accuracy of the parameter estimate. For this purpose reference must be made directly to the covariance matrix $\mathbf{\Sigma}$; the square root of each diagonal term furnishes the standard deviation of the parameters.

7 Estimate Algorithms

Iterative techniques must be employed in order to solve the nonlinear estimate problem. All these techniques have a similar structure: they need an initial estimate; starting from the initial estimate, they evaluate a search direction, then move in that direction in order to find the minimum. The solution at the i+1-th iteration is expressed as:

$$\mathbf{x}^{(i+1)} = \mathbf{x}^{(i)} + \lambda^{(i)}\mathbf{s}^{(i)} \tag{7.1}$$

where $\mathbf{s}$ indicates the search direction and λ the step length. There are different methods to calculate λ and $\mathbf{s}$, that employ linear or quadratic approximations of the response in the neighbourhood of the parameter estimate at the i-th step. At each iteration, a convergence test is applied: if the test is satisfied, the point $\mathbf{x}^{(i+1)}$ provides an updated solution, otherwise from this point a new search direction is evaluated.

Among the minimization techniques, trust-region methods are followed where the mimimum search is performed within a limited parameter domain, whose amplitude varies according to the iteration and depends on the objective function regularity. In particular, the minimization algorithm of Levenberg-Marquardt (Levenberg, 1944; Marquardt, 1963) is adopted

$$\min_{\mathbf{x}\in\Re^p} l\left(\mathbf{x}^{(i)}\right) \quad \text{with the constraint} \quad \left|\mathbf{s}^{(i)}\right| < \delta^{(i)} \tag{7.2}$$

where $\mathbf{s}^{(i)} = \mathbf{x}^{(i+1)} - \mathbf{x}^{(i)})$ represents the size of the trust-region. The objective function is approximated with a first order Taylor series expansion, where the last term accounts for the omitted terms

$$l\left(\mathbf{x}^{(i+1)}\right) = l\left(\mathbf{x}^{(i)}\right) + \left.\frac{\partial l\left(\mathbf{x}\right)}{\partial \mathbf{x}}\right|_{\mathbf{x}=\mathbf{x}^{(i)}} \left(\mathbf{x}^{(i+1)} - \mathbf{x}^{(i)}\right) + \lambda^{(i)}\left(\mathbf{x}^{(i+1)} - \mathbf{x}^{(i)}\right)^2 . \tag{7.3}$$

The minimum search operates within a region $\in \Re^p$ where the approximation assumed for $l\left(\mathbf{x}\right)$ is valid. The size of the region is modified at each iteration according to the real shape of the objective function. The minimization of $l\left(\mathbf{x}\right)$ provides the following iterative scheme:

$$\mathbf{x}^{(i+1)} = \mathbf{x}^{(i)} - \left[\lambda^{(i)}\mathbf{I} + \mathbf{\Sigma}_x^{-1} + \mathbf{H}^{(i)^T}\mathbf{\Sigma}_N^{-1}\mathbf{H}^{(i)}\right]^{-1}\left\{\mathbf{\Sigma}_x^{-1}\left(\mathbf{x}_0 - \mathbf{x}^{(i)}\right) + \right.$$
$$\left. + \mathbf{H}^{(i)^T}\mathbf{\Sigma}_N^{-1}\left[\overline{\mathbf{z}} - \mathbf{z}\left(\mathbf{x}^{(i)}\right)\right]\right\} \tag{7.4}$$

along with the constraint

$$\left\| \left[\lambda^{(i)} \mathbf{I} + \boldsymbol{\Sigma}_x^{-1} + \mathbf{H}^{(i)^T} \boldsymbol{\Sigma}_N^{-1} \mathbf{H}^{(i)} \right]^{-1} \left\{ \boldsymbol{\Sigma}_x^{-1} \left(\mathbf{x}_0 - \mathbf{x}^{(i)} \right) + \right. \right.$$
$$\left. \left. + \mathbf{H}^{(i)^T} \boldsymbol{\Sigma}_N^{-1} \left[\overline{\mathbf{z}} - \mathbf{z} \left(\mathbf{x}^{(i)} \right) \right] \right\} \right\| \leq \delta^{(i)} \quad . \tag{7.5}$$

At each iteration λ e $\mathbf{x}$ must be recalculated; if the inequality

$$\left\| \left[\boldsymbol{\Sigma}_x^{-1} + \mathbf{H}^{(i)^T} \boldsymbol{\Sigma}_N^{-1} \mathbf{H}^{(i)} \right]^{-1} \left\{ \boldsymbol{\Sigma}_x^{-1} \left(\mathbf{x}_0 - \mathbf{x}^{(i)} \right) + \mathbf{H}^{(i)^T} \boldsymbol{\Sigma}_N^{-1} \left[\overline{\mathbf{z}} - \mathbf{z} \left(\mathbf{x}^{(i)} \right) \right] \right\} \right\| \leq \delta^{(i)} \tag{7.6}$$

is satisfied, then $\lambda = 0$. Otherwise, λ is obtained from the previous constraint by imposing the equality. When λ and $\mathbf{s}$ are known, the solution $\mathbf{x}^{(i)}+\mathbf{s}^{(i)}$ is accepted as a new point $\mathbf{x}^{(i+1)}$ only if

$$l\left(\mathbf{x}^{(i)} + \mathbf{s}^{(i)}\right) < l\left(\mathbf{x}^{(i)}\right). \tag{7.7}$$

If the inequality is not satisfied, the minimum search is repeated within a smaller region. At each step of the iterative minimization process of $l(\mathbf{x})$, the sensitivity matrix of the observed quantities (for example, eigenvalues and eigenvectors or frequency response functions), must be calculated, that is done numerically.

In order to evaluate the error on the natural frequencies and on the mode components, which appears in the objective function $l(\mathbf{x})$, it is necessary to order the experimental and analytical modal quantities in the right sequence. In the correct order, the analytical and experimental eigenvectors have similar shapes. This correspondence is based on the similitude between eigenvectors, measured by their scalar product, known as the Modal Assurance Criterion (MAC). Two modes are considered similar if the MAC has a value near to unity; in this way, each analytical mode is associated with the most similar experimental mode.

At each step the following operations are performed:

1) the *Jacobian* of the *objective function* is calculated. This implies the evaluation of $l(\mathbf{x})$ in $2n$ points, in the parameter space. For each point:

a. the matrices $\mathbf{m}$ and $\mathbf{k}$ of the elements whose parameters are varied are calculated;

b. the global matrices $\mathbf{K}$ and $\mathbf{M}$ are updated accordingly and the observed quantities (for example, $\boldsymbol{\omega}$ and $\boldsymbol{\Phi}$) are calculated;

c. the varied modes are reordered and renormalized;

d. the components of the objective function are calculated.

2) The equation which provides $\mathbf{x}^{(i+1)}$ is solved. The trust-region size δ is calculated;

3) the objective function reduction is checked:

$$l\left(\mathbf{x}^{(i+1)}\right) < l\left(\mathbf{x}^{(i)}\right). \tag{7.8}$$

If this check is verified, the iterative process goes forward, otherwise, δ is reduced and the process goes back to 2);

4) the convergence of $\mathbf{x}$ is checked. If this is satisfied, the iterative process closes, otherwise, it goes back to 1).

8 IDEFEM: a procedure for the IDEntification of Finite Element Models

IDEFEM is a code which implements a procedure for the identification of linear finite element models (Capecchi et al., 1993; Antonacci et al., 1994; De Sortis et al., 2005). An important part of the program consists of a FE code utilized to evaluate function $l(\mathbf{x})$. Together with data necessary for the description of the FE model, information is requested regarding the definition of parameters, the observed response quantities and their experimental values. A Bayesian estimator is employed, thus the parameter and measurement covariance matrices $\mathbf{\Sigma}_x$ and $\mathbf{\Sigma}_N$ have to be defined. These are assumed to be diagonal with diagonal elements given by the variation coefficients of measurements and parameters v.

The code involves the following stages:

1. data of the base model; calculation of mass and stiffness matrices of each element and assemblying of the global matrix, which represents the whole structure; an assigned number of frequencies and eigenvectors is calculated;
2. parameter data and definition of the groups of elements associated with them;
3. experimental quantities definition and their values;
4. comparison between experimental and analytical modes by means of their scalar products;
5. objective function minimization according to an iterative process.

In the following analyses *natural frequencies* and *modal shapes* are assumed as the observed quantities. A certain number of frequencies n_f can be selected as measurements; moreover, only a few components n_m of modes can be used. The comparison between analytical and experimental response is made using the following objective function:

$$l(\mathbf{x}) = \frac{1}{2} \sum_{i=1}^{n_f+n_m} w_i e_i^2 \tag{8.1}$$

where $\mathrm{e}_i = \overline{z}_i - z_i$, is the error concerning the measurements. Each term of the objective function is normalized by weighting constants w_i:

$$\mathrm{w}_i = 1/\left(\mathrm{z}_i \nu_i\right)^2 \tag{8.2}$$

for the frequencies and:

$$\mathrm{w}_i = 1/\left(\mathrm{z}_{\max} \nu_i\right)^2 \tag{8.3}$$

for the i-th component of the j-th mode, where ν_i is the variation coefficient of z_i.

Due to the increasing size of the models and the number of parameters, the computational effort to solve the problem may become extremely large. The program structure has been optimized where calculation loops occur. Important steps are, for example, structural matrix updating and eigenvalues evaluation. The full re-assembly of the global matrices is avoided by adding only the matrix variation of elements which have varied stiffness or mass. Similarly, in the modal analysis care has been paid to save computations, providing storage for the eigenvectors of the base model. In turn, these are employed as an initial base for the subspace-iteration algorithm in updating eigenvalues and eigenvectors.

9 A Problem of Model Updating: Case-Study

The presented problem of model updating was developed by Elena Antonacci (DISAT, University of L'Aquila). The problem regards the identification of the finite element model of the lattice structure, represented in Figure 1, which best fits the measured dynamic response. The steel components are connected by welding. The structure is tested in free conditions under a random force applied at node 5. Vertical accelerations in nodes 1-5 are measured.

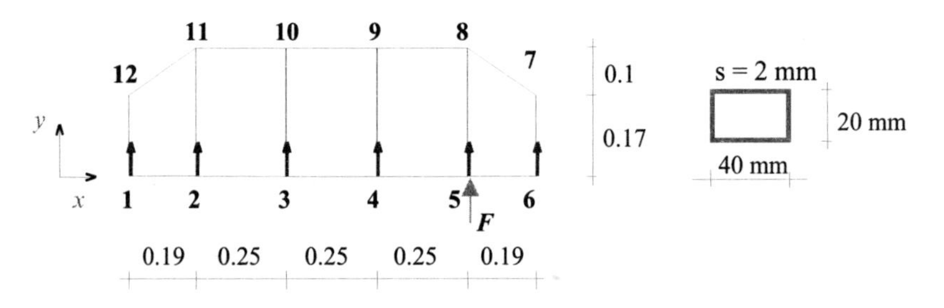

Figure 1. Scheme of the lattice structure.

Figure 2a reports the acceleration FRF; five peaks representative of the first five natural frequencies of the structure appear in the range 0-1000 Hz. Figures 2b-d show all the peaks in greater detail. By using a multi-mode algorithm (Beolchini and Vestroni, 1997; De Sortis et al., 2005), the experimental modal parameters are determined. The identified modes, shown in Figure 3, respect symmetry and skew-symmetry, with the exception of mode 3 that has a node at point 5, where the force is applied. This causes ill-conditioning and probably leads to incorrect results.

A comparison between experimental inertances and inertances reconstructed on the basis of the identified modal parameters exhibit good agreement (Figure 4), apart from the peak values, due to a low resolution of the experimental curve at peaks.

The first model (A) considered in the identification process is a finite element model with 12 nodes and 16 elements, a scheme is reported in Figure 5. Two cases are considered: a model with 4 parameters, which are the axial and flexural stiffness of stringers and uprights, and a model with 16 parameters, Young's modulus of every element.

The comparison between identified and FEM frequency response functions is shown in Figure 6. It can be seen that neither of the two models satisfactorily fits the experimental response in the neighbourhood of the 4th and 5th resonances. In particular, in a real structure, as in a finite element model, the sequence of natural frequencies is very rigid and largely depends on the modeling. Therefore, the model had to be refined.

The FEM B has 26 nodes and 30 elements, as shown in Figure 7. Once again, two cases are considered: a model with 4 parameters, which are the axial and flexural stiffness of stringers and uprights, and a model with 16 parameters, Young's modulus

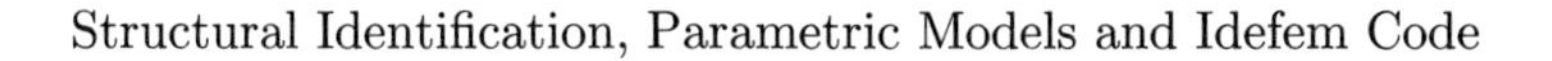

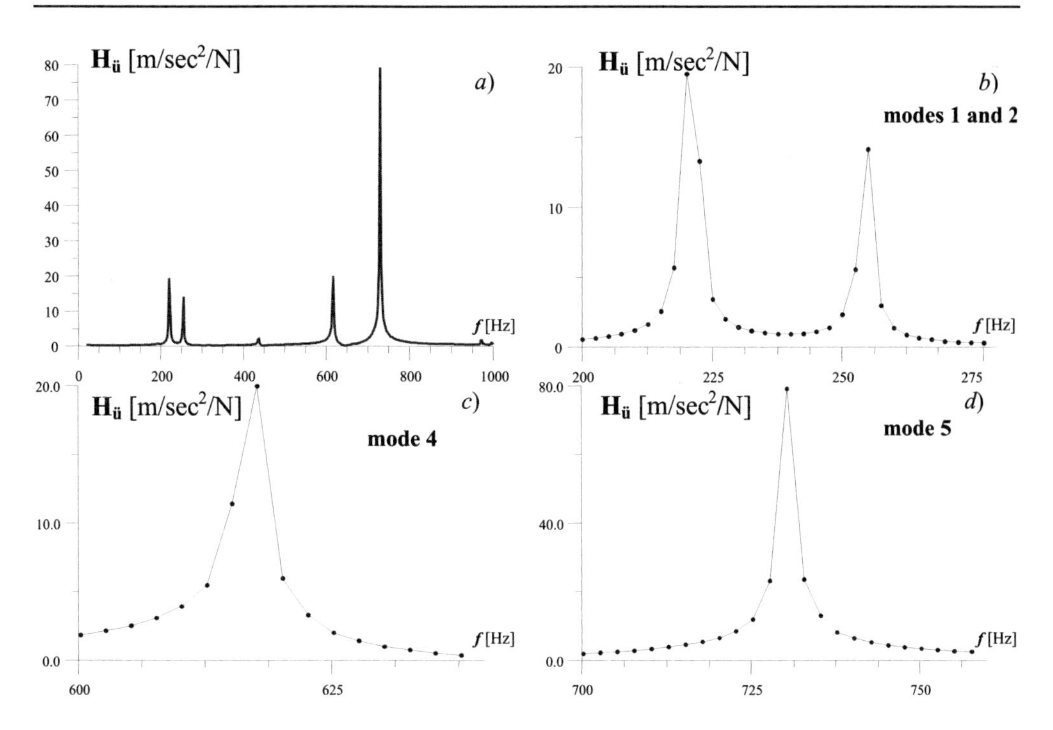

Figure 2. Experimental FRF of acceleration H_{45} at node 4 with force applied at node 5.

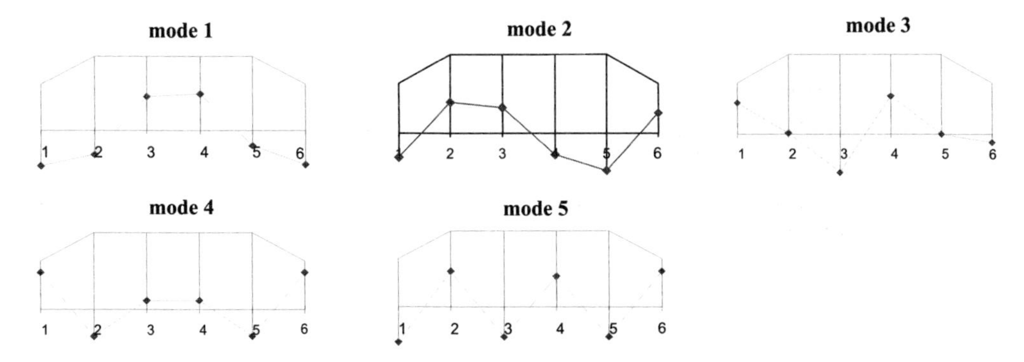

Figure 3. First five experimental mode shapes.

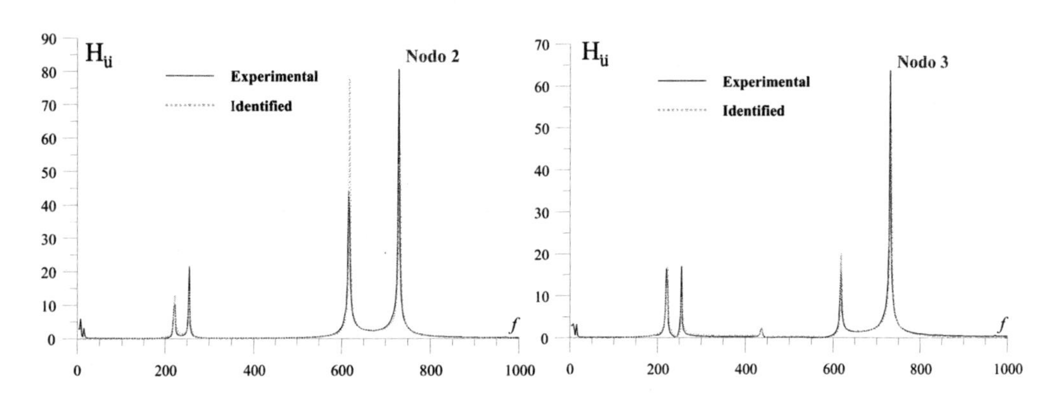

Figure 4. Comparison between experimental and identified model FRF.

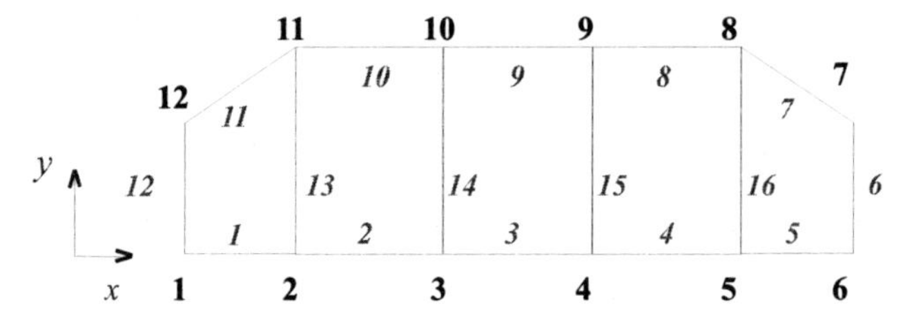

Figure 5. Finite element model A.

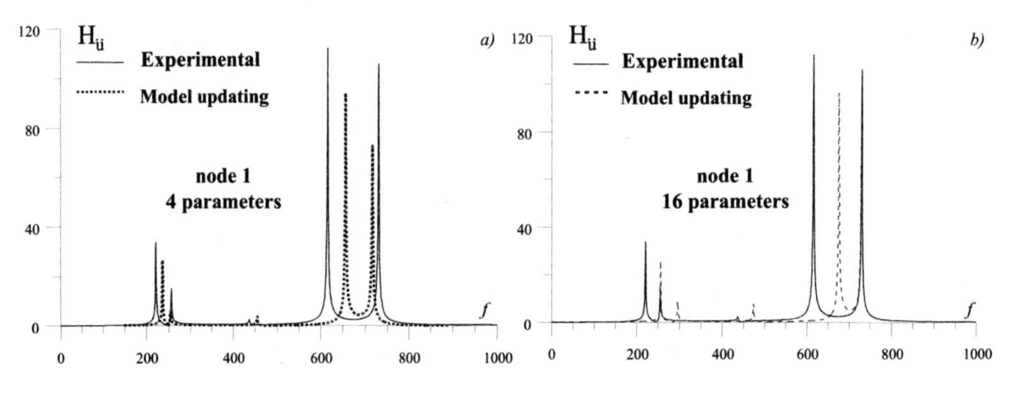

Figure 6. H_{45} comparison between experimental and updated model A.

of every principal element between two corners. The comparison between the FRF of the experimental and updated model, reported in Figure 8, shows that in both cases satisfactory results are obtained. However, the more refined model provides the right spacing between the 4th and the 5th frequencies.

In brief, this example shows that the modelling error can greatly hamper the identification of a finite element model of a structure.

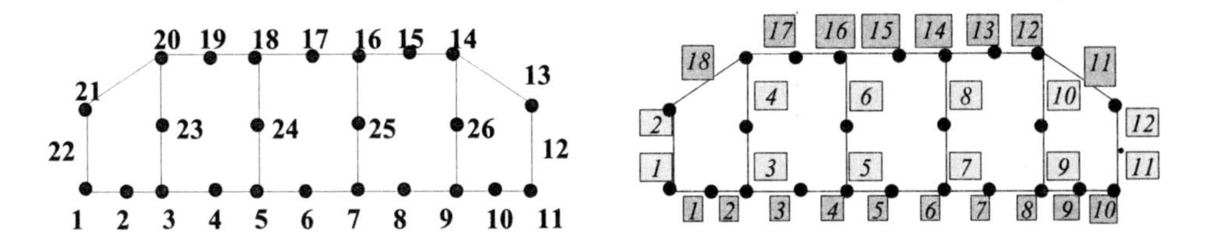

Figure 7. Refined finite element model B.

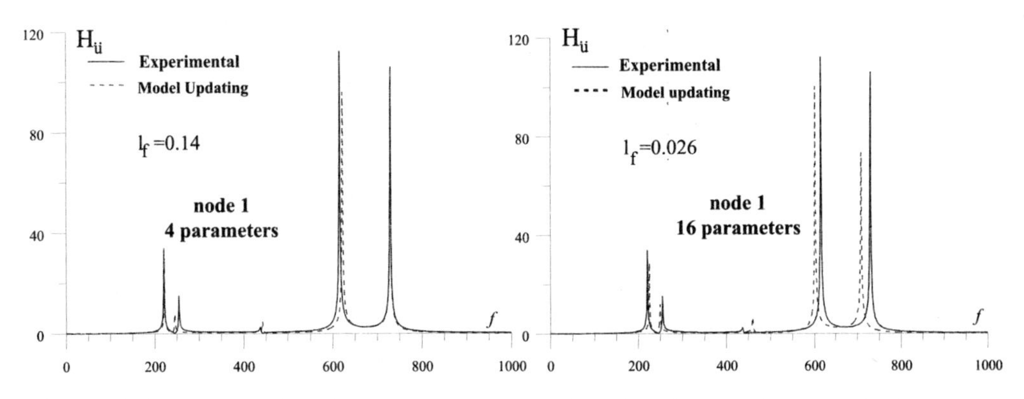

Figure 8. H_{45} comparison between experimental and updated refined model B.

10 Conclusions

Within the context of parametric model identification, the case of a finite element model of a structure has been dealt with. This is really an updating model, where the best estimate of selected physical quantities, assumed as parameters, is determined starting from given initial values and using the experimental data available from the real structure. The optimal values of parameters are determined by minimizing an objective function that accounts for the disagreement between experimental and analytical values of the observed quantities and for the differences between the value of the response parameters and their a priori estimate. A Bayesian approach has been used to combine a priori information on parameters and experimental information. Moreover, the expression of a posteriori covariance matrix of parameters, furnished by the probabilistic approach, has been adopted for the optimal selection of measurements and parameters.

The identification of model parameters is an onerous task. Thus, the procedure has been implemented in a code, characterized by two main parts, structural and minimization routines. A case study of a lattice structure with experimental results is presented. First, accurate modal quantities are determined, and then used as experimental data to identify the stiffness of selected elements of the model. It emerges that a finite element model has a very rigid structure evidenced, among others, by the frequency sequence. Hence, modeling assumptions are very important to pinpoint the main dynamic characteristics. This feature must be taken into account when the finite element model is used in damage detection.

Bibliography

E. Antonacci, D. Capecchi, and F. Vestroni. Updating of finite element models through nonlinear parametric estimation. In *Second International Symposium on Inverse Problems in Engineering Mechanics*, 1994.

J. L. Beck and L. S. Katafygiotis. Updating models and their uncertainties. *Journal of Engineering Mechanics*, 124(4):455–461, 1998.

G. C. Beolchini and F. Vestroni. Experimental and analytical study of dynamic behavior of a bridge. *Journal of Structural Engineering*, 123(11):1506–1511, 1997.

D. Capecchi and F. Vestroni. Identification of finite element models in structural dynamics. *Engineering Structures*, 15(1):21–30, 1993.

D. Capecchi and F. Vestroni. Monitoring of structural systems by using frequency data. *Earthquake Engineering and Structural Dynamics*, 28:447–461, 1999.

D. Capecchi, F. Vestroni, and E. Antonacci. Implementation of a procedure for identification of finite element models from experimental modal data. In *2nd Europ. Conf. on Structural Dynamics*, pages 593–600, 1993.

J. D. Collins, G. C. Hart, T. K. Hasselmann, and B. Kennedy. Statistical identification of structures. *AIAA Journal*, 12(2):185–190, 1974.

A. De Sortis, E. Antonacci, and F. Vestroni. Dynamic identification of a masonry building using forced vibration test. *Engineering Structures*, 27(2):155–165, 2005.

D. J. Ewins. *Modal Testing: theory, practice and application.* Reaserch Studies Press, 2000.

M. I. Friswell and J. D. Mottershead. *Finite Element Model Updating in Structural Dynamics.* Kluwer Academic Publishers, 1995.

M. I. Friswell and J. D. Mottershead. Identification in engineering systems, special issue. *Journal of Vibration and Control*, 4(1), 1998.

R. Ghanem and M. Shinozuka. Structural system identification. *Journal of Engineering Mechanics*, 121(2):255–273, 1995.

R. Ghanem and S. Sture Eds. *Journal of Engineering Mechanics*, 126(7):665–777, 2000.

K. Levenberg. A method for the solution of certain problems in least squares. *Quarterly Journal of Applied Mathematics*, 2:164–168, 1944.

D. W. Marquardt. An algorithm for least-squares estimation of nonlinear parameters. *SIAM Journal of Applied Mathematics*, 11:431–441, 1963.

S. F. Masri and S. D. Werner. An evaluation of a class of practical optimization techniques for structural dynamics applications. *Earthquake Engineering and Structural Dynamics*, 13(5):635–649, 1985.

J. E. Mottershead and M. I. Friswell. Model updating in structural dynamics: A survey. *Journal of Sound and Vibration*, 167(2):347–375, 1993.

H. G. Natke. *Identification of vibrating structures.* Springer–Verlag, 1982.

H. G. Natke. *Structural Dynamics.* W.B. Kratzing et al. Editors, Springer–Verlag, 1991.

F. Vestroni. *Nonlinear dynamics and structural identification.* World Scientific, 1991.

F. Vestroni and D. Capecchi. *Parametric identification and damage detection in structural dynamics*, pages 107–143. Research Signpost, Trivandrum, India, 2003. A. Luongo ed.

Structural Identification and Damage Detection

Fabrizio Vestroni*

* Dipartimento di Ingegneria Strutturale e Geotecnica, University of Roma *La Sapienza*, Italy

Abstract The use of parametric models is extended to damage detection, thereby exploiting their ability to describe local characteristics. Aspects closely related to parameter estimation are examined in detail here with reference to a specific case of a masonry building affected by *diffused* damage. Pseudo experimental and experimental data are considered in the identification procedure; this allowed different causes of ill-conditioning and other peculiarities to be investigated.

The changes induced by damage in the dynamic response are exploited to build a procedure for damage detection based on the variation of natural frequencies, both for continuous and discrete models of beams affected by *concentrated* damages. A new approach is proposed which considers the peculiar aspects of the damage identification problem. Some applications of the technique proposed are outlined with reference to numerical and experimental cases.

1 Introduction

In system identification problems, the system model is characterized by the structural parameters x, which are to be identified by the measured input I^* and output z^*. The system with generic parameters x, excited by the true input I^*, does not reproduce exactly the output z^*, but responds with generic output z. The challenge consists in identifying the correct values of the parameters x^*, so as to give the response values close to the measured ones.

When dealing with damaged structures, a modification of the response due to damage is observed. The same input I^* acting on an undamaged structure with parameters x^U produces an output z^U which is different from the output z^D of a damaged structure with parameters x^D. Variations in the behaviour of a structure can be associated with the decay of the system mechanical properties (Shen and Pierre, 1990; Morassi, 1993; Chondros et al., 1998). The response in a damaged state can be represented as a deviation from the undamaged state $z^U + \Delta z^D$ due to a variation of the parameters $x^U + \Delta x^D$.

Two possible approaches can be followed: in the first, the model parameters are identified in the two different states, the undamaged (x^U) and the damaged states (x^D); the variation in the parameters $(x^U - x^D)/x^U$ furnishes a degree of damage. In the second approach, using a given model of the structure, the modification $\Delta x^D/x^U$ are determined in a way to fit at best the experimental modification of the observed quantities Δz^D.

Both problems can be formulated as minimization problems of suitable functionals l, which in absence of errors, is equal to zero for the exact values x^* or (Δx^*) and are greater than zero for values x (or Δx) different from the exact ones:

$$l(x^*; z^*) = 0 \quad \text{and} \quad l(x; z^*) > 0 \quad \text{for} \quad x \neq x^* \tag{1.1}$$

$$l(\Delta x^*; \Delta z^*) = 0 \quad \text{and} \quad l(\Delta x; \Delta z^*) > 0 \quad \text{for} \quad \Delta x \neq \Delta x^* \tag{1.2}$$

2 Identification of Diffused Damage in a Masonry Building

An application to a real case is considered when damage is diffused over the structure. As a first step, the modal model is evaluated; then the experimentally determined modal quantities are used to identify the optimal parameters of a finite element model of the structure in the damaged and undamaged conditions. The masonry building under test is located in Italy, near L'Aquila. It is a two-storey structure with regular plan, built towards the end of the eighteenth century. The dynamic tests were performed in cooperation with ISMES (Bergamo), which provided the instruments and the technical support. A complete description of the dynamic tests can be found in (Capecchi and Vestroni, 1991; Vestroni et al., 1993).

The tests here discussed were performed on the portion of structure highlighted in Figure 1, which presents a square plan and is disconnected from the remainder of the building through a cut of the longitudinal walls and of the floor slab. The step-sine excitation technique was used to cover the frequency range of interest (1-11 Hz), with a frequency resolution of 0.02 Hz. A sinusoidal force, along the y-axis, was produced by a vibrodyne, with a maximum frequency of 24 Hz and a maximum amplitude of 100 kN.

The structural response was measured by 12 accelerometers placed on the walls and the slab and 4 seismometers at ground level (Figure 1). The transducer signals, after appropriate amplification and filtering, were acquired through an A/D converter connected to a computer. The experimental results considered are the inertance functions (acceleration/force) relative to the 12 accelerometer locations and to the force location.

2.1 Identification of the Modal Model

Figure 2 shows the FRFs obtained directly from the ratio between response and excitation amplitude with harmonic tests. The excitation intensity increases from tests 11 to 16 and the level is such to produce damage. Tests 10 and 17 were performed with a small amplitude oscillation and were used to analyse the irreversible effects of damage.

It is worth noticing that, by increasing the excitation intensity, the oscillation amplitude increases too but less than linearly; moreover, it was observed a decrease of resonant frequencies, a reduction of the FRF curve peaks at resonances and an increase of curve width in the neighbourhood of resonances. The last two phenomena are related to an expected increase in structural damping with damage. The decrease of resonant frequencies represents a reduction of stiffness which is reported in Figure 3 as a function of oscillation amplitude, together with a linear fitting of the experimental points. If results of test 10 and 17 are compared (Figure 2), a modification of shape is observed, which means that damage is not equally distributed, but it is greater in certain regions.

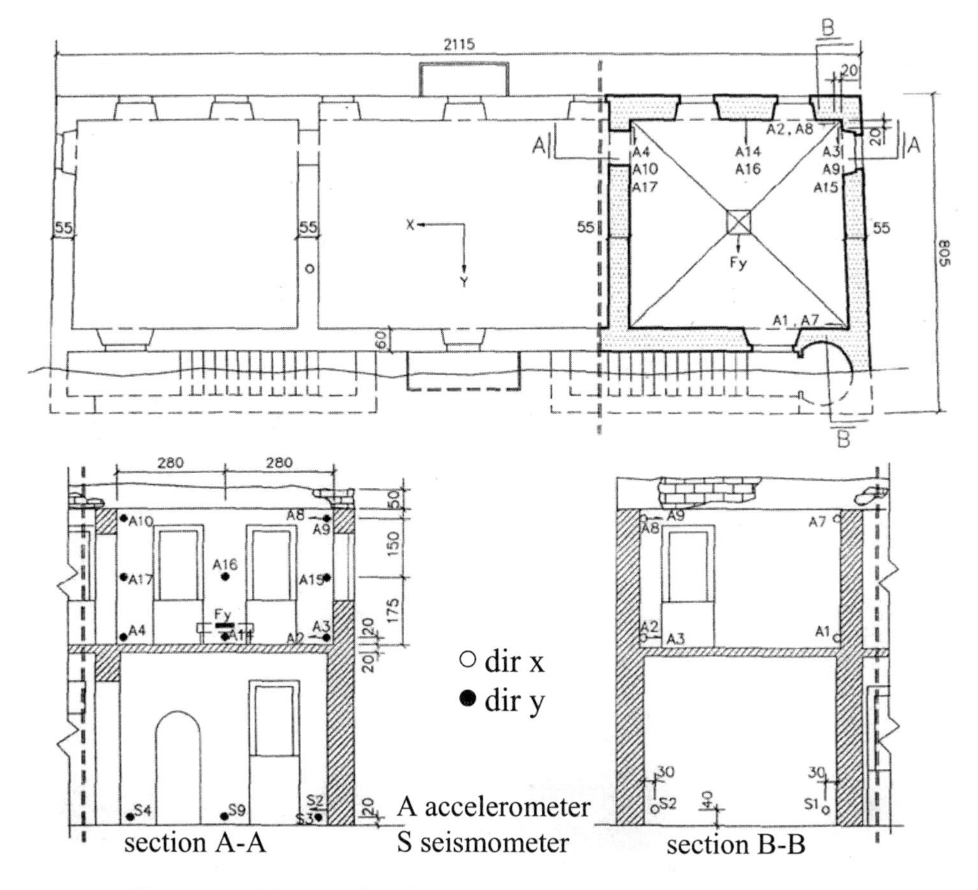

Figure 1. Masonry building: plan, sections and sensor locations.

The identification of modal parameters was performed on the small oscillation tests 10 and 17, which employed the same force intensity, before and after the damage. Figure 4 reports the results of the identification. The inspection of the Table, with reference to Test 10, indicates a strong modal coupling, due to a high number of peaks in a small frequency range; this is evident also from the mode shapes of Figure 4. Damping values are evenly shared among the different modes around an average value of about 1.7%. Examination of mode amplitudes and phases shows that some modes are complex, while others are practically real. In particular, the first mode is virtually real, the second, third, fourth and seventh mode present little phase differences among their components, while the remaining modes are complex. The complexity of higher order modes frequently observed (Pau and Vestroni, 2007) can be partly ascribed to the greater oscillation amplitudes, that may introduce nonlinear effects and then mode complexity.

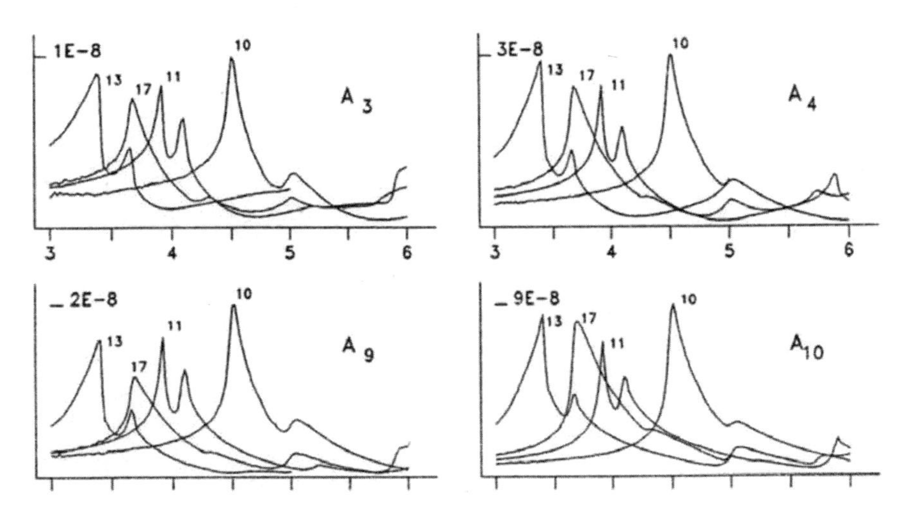

Figure 2. Experimental FRF for tests 10, 11, 13, 17 at sensors A3, 4, 9, 10.

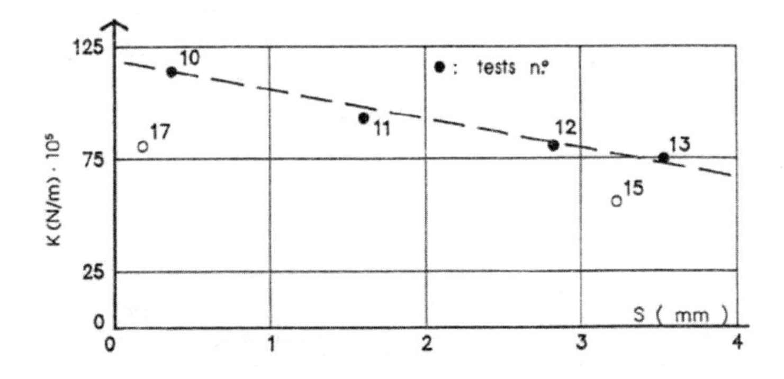

Figure 3. Generalized stiffness versus oscillation amplitude.

Figure 5 shows the experimental and curve-fitted inertances for four points on the first and second floor for T10 (Capecchi and D'Ambrogio, 1993). In most cases, the results are satisfactory. The largest deviations between experimental and fitted inertances occur for the response in the x-direction, i.e. normal to the excitation direction. Here, measurement errors are larger, due to a lower signal-to-noise ratio caused by lower response amplitudes, and nonlinear effects are important, since the motion in the x-direction is mainly caused by a torsion of the structure, which involves the connections among different structural elements, such as walls and floor slab. All the FRFs present a peak at about 11 Hz, which is the upper bound of the measured frequency range. In order to obtain a good fit, the contribution of this mode must be also considered. The inertance peaks related to modes 5 and 6, and to a lesser extent to mode 8, do not appear in some of the FRF curves, and could therefore be associated to local modes.

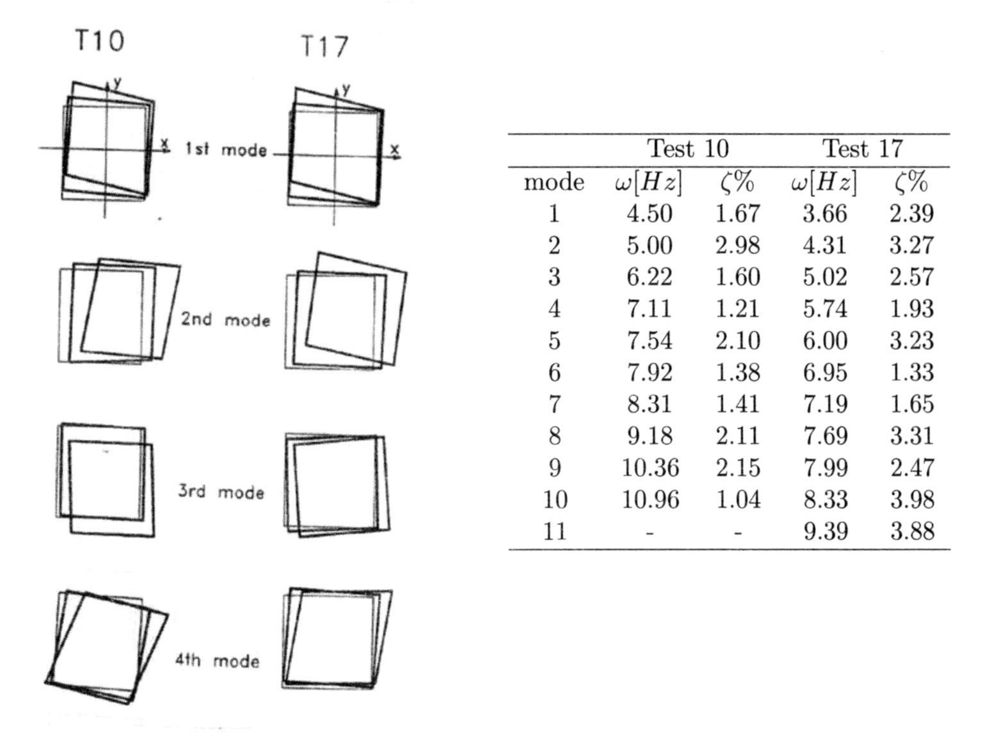

	Test 10		Test 17	
mode	$\omega[Hz]$	$\zeta\%$	$\omega[Hz]$	$\zeta\%$
1	4.50	1.67	3.66	2.39
2	5.00	2.98	4.31	3.27
3	6.22	1.60	5.02	2.57
4	7.11	1.21	5.74	1.93
5	7.54	2.10	6.00	3.23
6	7.92	1.38	6.95	1.33
7	8.31	1.41	7.19	1.65
8	9.18	2.11	7.69	3.31
9	10.36	2.15	7.99	2.47
10	10.96	1.04	8.33	3.98
11	-	-	9.39	3.88

Figure 4. Experimental modal parameters in the initial and damaged states.

The comparison between the experimental and identified FRF in the damaged state, presented in Figure 6, exhibits a good agreement too. The modification of dynamical characteristics is very evident in Figure 7; the use of variation of the identified parameters from T10 and T17 can furnish a better description and evaluation of structural damage.

2.2 Physical Model

In the identification procedure a FE model with shell elements is assumed (Figure 8) (Capecchi and Vestroni, 1991). The model has 160 nodes, 130 elements and five parameters: x_1, x_2, x_3, the elasticity modulus of walls 1, 2, 3 at 1st and 2nd floors, and x_4, x_5, the elasticity modulus at 1st and 2nd floors of wall 4, respectively. A simple model is employed for this complex structure, where many uncertainties, such as material and element connection behaviour, still remain.

As a first stage, the parameter identification is performed in the initial condition. Figure 9 reports the a priori values assumed for the parameters and their updated values at the end of the process. Large modifications in the parameter values are observed, notwithstanding the fact that the material should be the same. Notable decrease of the objective function is obtained, however, considerable errors between experimental

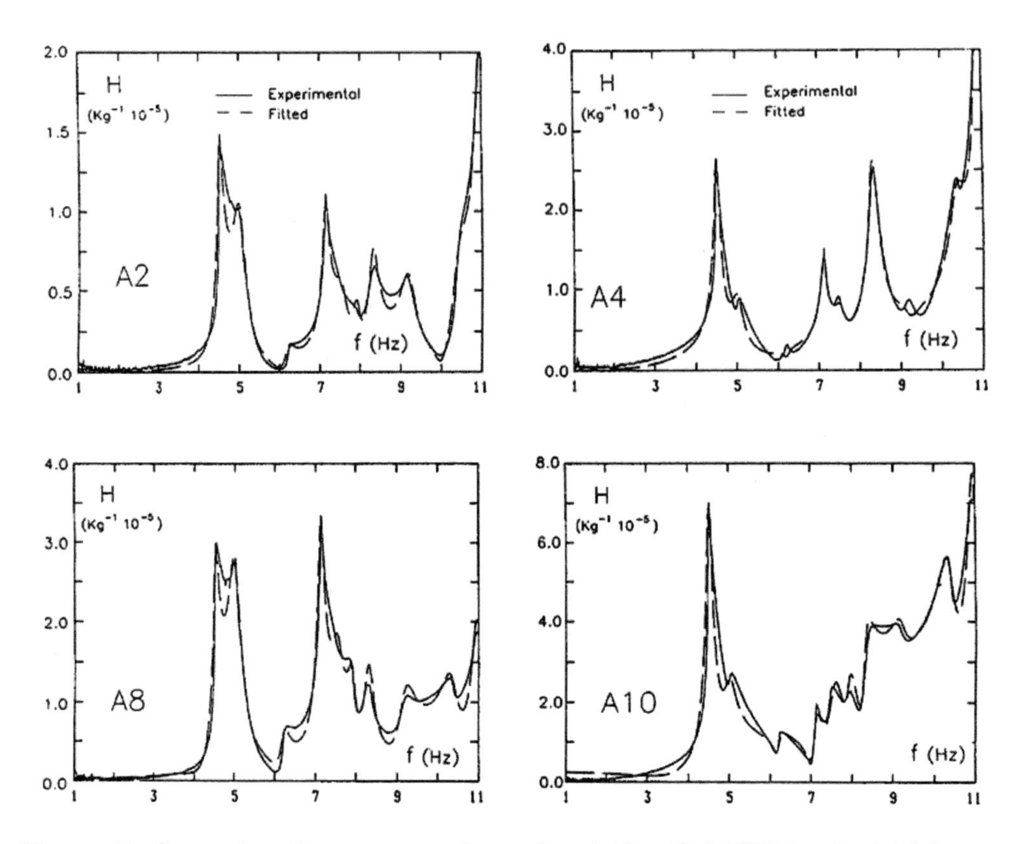

Figure 5. Comparison between experimental and identified FRF in the initial state (T10).

and identified modal parameters of the FE model remain, as denoted by the final errors on frequencies (e_f) and eigenvectors (e_u). Also the comparison between analytical and experimental mode shapes exhibit notable differences (Figure 9), especially as far as the second mode is concerned. It emerges an evident inaccuracy, substantially related to the complexity of the structural behaviour. Thus, before tackling the identification problem with experimental data, a preliminary analysis with pseudo-experimental data, that is using analytical values for the quantities assumed as measurements, is performed. This is useful to study cases where modelling and experimental errors are absent; numerical characteristics of the inverse problem and the minimum evaluation procedure can be seen in advance.

Identification cases, different for the number of parameters and the kind of measurements involved, are summarized in four series: A5 cases, characterized by all the five parameters and pseudo-experimental data; B5 cases, the same parameters as A5 but with experimental data; A2 cases, characterized by two parameters and pseudo-experimental data, and B2 cases, characterized by two parameters and experimental data.

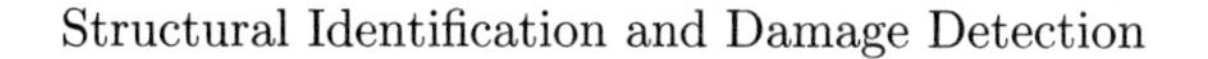

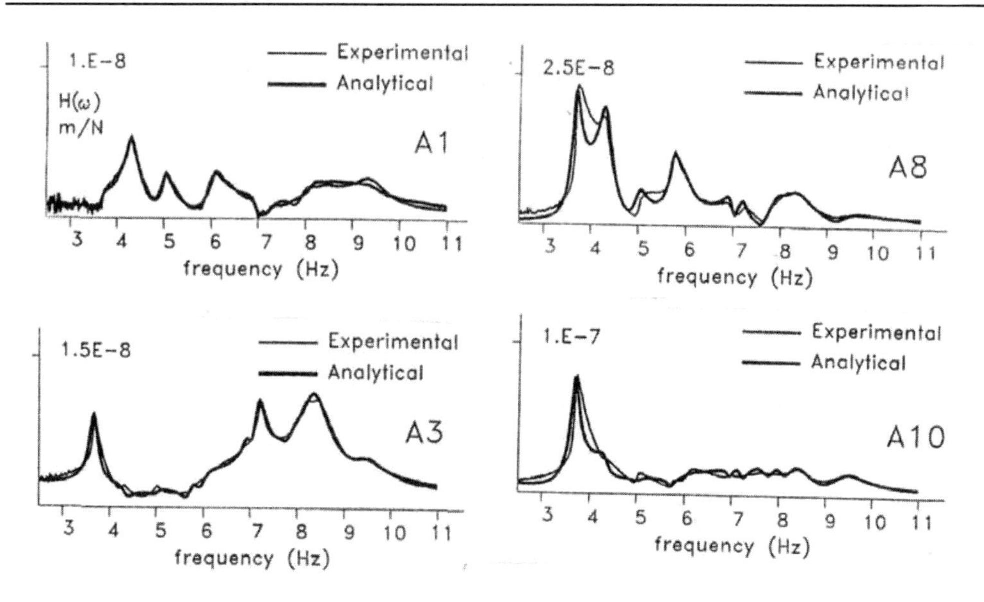

Figure 6. Comparison between experimental and identified FRFs in the damaged state (T17).

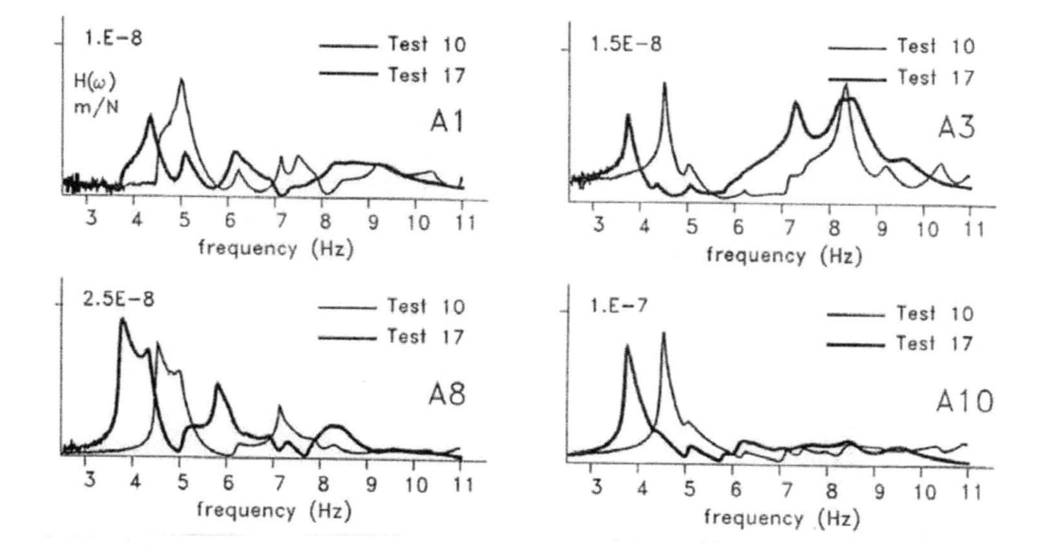

Figure 7. Comparison between the FRFs in the initial (T10) and damaged state (T17).

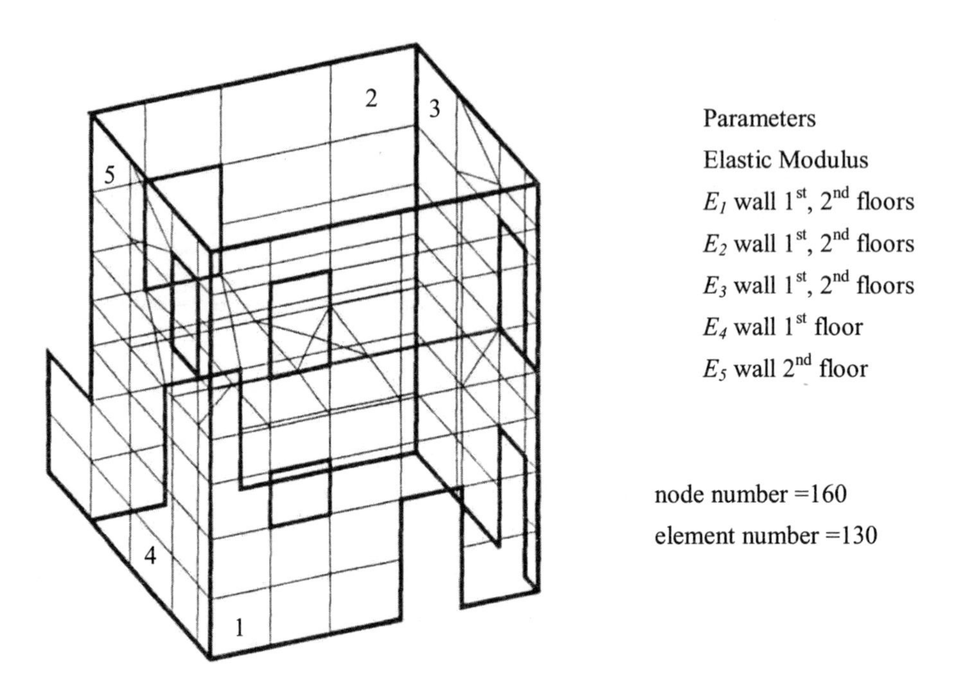

Figure 8. Finite element model of the tested masonry building.

Eighteen quantities are assumed as pseudo-experimental measurements in the A5 cases; they are the first two natural frequencies and eight components of the corresponding modes, furnished by the analytical model with the parameter value x_e = (5.46, 5.46, 6.89, 2.87, 1.76) 10^5 N/m^2, while as base value it has been assumed x_0 = (3.50, 3.50, 3.50, 3.50, 3.50) 10^5 N/m^2. Different cases corresponding to different starting points for the iterations procedure are reported in Table 1. Parameters are normalised with respect to the exact value, so it is easy to check the exactness of identified values to be compared with the unity. Among the four cases considered, one half (cases 2 and 3) has given good results, the other half (1 and 4) has not been successful, giving two different local minima.

With the assumed pseudo-experimental data, problems of identifiability arose. Actually, second and third frequencies are very close and the corresponding modes have very similar shape. In such cases, the identification procedure is not able to distinguish an eigenmode from another. For both unsuccessful cases ($A5_1$ and $A5_4$), at the end of identification, parameters obtained are associated to a model which has the second mode of the same kind as the third experimental eigenvector. Of course, these parameters cannot represent the exact solution.

To avoid identifiability problems, it is necessary to correctly order the analytical modes with respect to the experimental modes; when the system exhibits modes with similar shape, the comparison between analytical and experimental modes, by means of

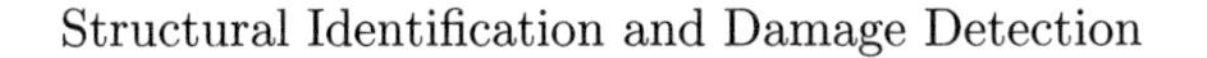

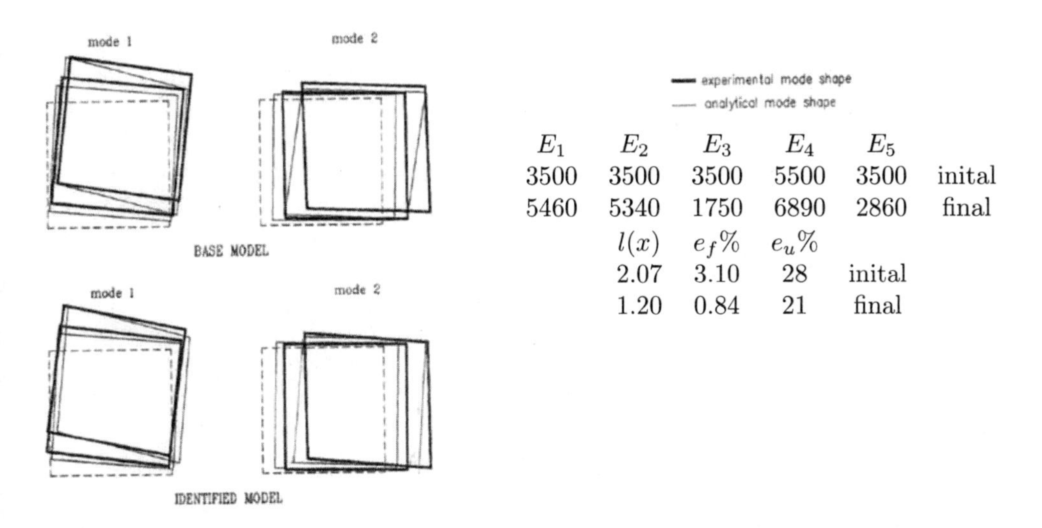

E_1	E_2	E_3	E_4	E_5	
3500	3500	3500	5500	3500	inital
5460	5340	1750	6890	2860	final

$l(x)$	$e_f\%$	$e_u\%$	
2.07	3.10	28	inital
1.20	0.84	21	final

Figure 9. Comparison between experimental and analytical mode shapes and estimated parameters in the initial state T10.

Table 1. Cases A5, initial and final identified parameters and objective function $l(x)$.

Case		x_1	x_2	x_3	x_4	x_5	$l(x)$
1	i	0.641	0.641	0.508	1.221	2.000	0.0886
	f	0.842	0.852	0.852	1.175	0.917	0.0083
2	i	1.007	0.916	1.016	0.977	0.971	0.0270
	f	1.000	1.000	1.000	1.000	1.000	$2*10^{-12}$
3	i	0.824	0.824	0.943	1.047	1.714	0.0387
	f	1.000	1.000	1.000	1.000	1.000	$2*10^{-9}$
4	i	1.099	0.916	1.161	0.698	0.857	0.1280
	f	0.840	0.839	0.785	1.258	0.906	0.0092

MAC, should involve a number of eigenvectors components greater than that actually used. In fact, when modes 1 and 4, which do not present similar characteristics, are used as experimental data, the exact solution is always found independently of the initial conditions.

Pseudo-experimental cases previously considered would suggest not to employ 2nd and 3rd modes, due to their similarity, but instead to prefer the 4th one. Actually, modal parameters of higher modes usually have less reliability than lower modes and so it was taken to use as experimental data the same 18 measurements (the first two frequencies and eight components of modes) already considered in A5 case.

Cases B5 use experimental data; results in Table 2, obtained for four different initial parameter values, evidence the existence of various minima, with comparable objective function values, but quite far from each other. The greatest differences concern parameters x_2 and x_3. A more in-depth local analysis, in the neighbourhood of the identified solutions, has shown a nearly flat objective function, with many small hollows. Thus, by varying the initial points the minimum procedures lead to results which may be very different regarding the parameter values against possibly similar values for the objective function values, which decrease only a little from the initial value.

To better understand the characteristics of the problem, cases with two parameters, which allow for visualization of the objective function, are investigated. The cases A2 refer to pseudo-experimental measurements. Parameters x_2 and x_3 are considered. The set of measurements formed by the eigenvectors z_1 e z_2 and the corresponding frequencies and the set of measurements formed by z_1 e z_4 and the corresponding frequencies have been considered. For the second set of measurements, the contour lines of $l(x_2, x_3)$ have regular shape, closed about the unique minimum. In such conditions, the success of identification is guaranteed, as proved by developed cases, not reported here.

The objective function for the first set of measurements, in a range of parameter variation between 30% of base values, is represented in Figure 10a. It shows a unique minimum, but a very irregular shape. As already shown for the 5 parameter case, the use of the second frequency and eigenvector as experimental measurements makes it necessary to perform a correct ordering of the modes. Due to similarity between the 2nd and 3rd modes, a recognition of modes is possible only if a high number of the analytical eigenvector components is used, allowing to reach the exact solution independently of the starting point.

Table 2. Cases B5, initial and identified parameters x and objective function $l(x)$.

Case		x_1	x_2	x_3	x_4	x_5	$l(x)$
1	i	1.000	1.000	1.000	1.000	1.000	0.138
	f	1.030	1.430	1.670	0.900	0.800	0.082
2	i	1.000	1.000	1.570	1.000	1.000	0.120
	f	1.070	1.140	1.920	0.760	0.750	0.098
3	i	1.560	1.560	1.968	0.819	0.500	0.119
	f	1.120	1.170	3.100	0.780	0.740	0.089
4	i	1.000	1.400	1.630	0.860	0.740	0.082
	f	1.050	1.860	3.730	0.840	0.560	0.065

In the B2 cases with experimental data, the shape of the contour lines, in Figure 10b, suggests a surface with small gradients, with a trend that, even though more irregular due to experimental errors, resembles that of A2 case. Five cases differing for the initial parameter values have been analyzed, shown in Table 3. Initial values for cases $B2_2$ e $B2_5$ fall in two zones interested by mode exchange and the corresponding minimum process leads to points external to the domain of parameter variation. In the other three cases examined, $B2_1$, $B2_3$, $B2_4$, the minimization of $l(x)$ leads to a unique point.

For the values of parameters identified the model has similar 2nd and 3rd mode; indeed $\text{MAC}(z_2,y_2) = 0.9994$, $f_2 = 5.032$ Hz; $\text{MAC}(z_3, y_3) = 0.9978$, $f_3 = 5.043$ Hz. Contribution

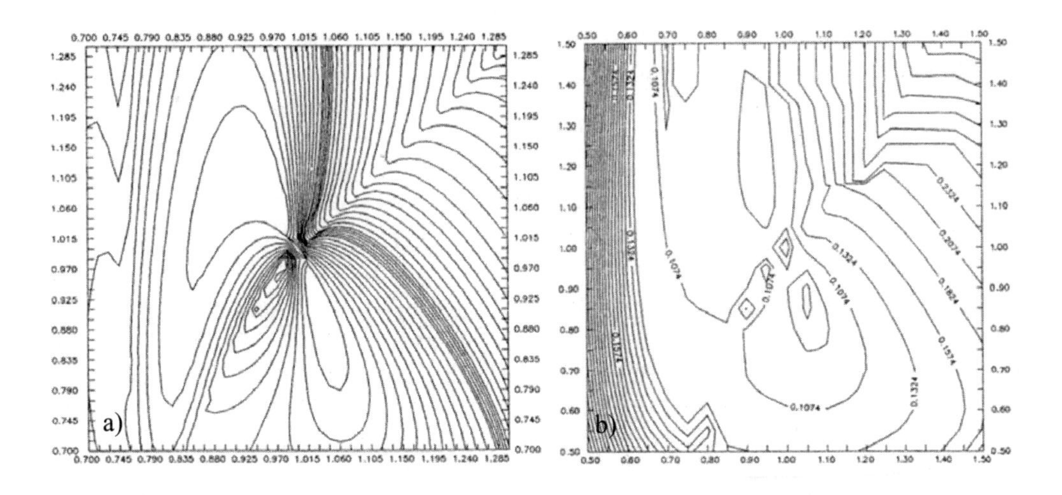

Figure 10. Contour lines of $l(x_2, x_3)$, modes 1 and 2. Pseudo-experimental (a) and experimental measurements (b).

to $l(x)$ of the terms corresponding to the 2nd experimental mode and corresponding frequency, evaluated by considering the 2nd or the 3rd analytical mode, are numerically very similar.

Obtained results show that while in the pseudo-experimental case a perfect recognition of the modes is always possible, despite their being very similar each other. In the real world, the presence of experimental and modelling errors, does not always allow this recognition which is a necessary condition to find an objective function which is sufficiently regular and with a unique minimum. In any case, in three cases ($B5_1$, $B5_3$, $B5_4$) the solution is attained at the same point in the parameter space and with the same value of the objective function.

Table 3. Cases B2, initial and final identified parameters and objective function $l(x)$.

Case	x_1^i	x_1^f	x_2^i	x_2^f	$l(x_i)$	$l(x_f)$
1	0.90	0.98	0.80	1.07	0.1100	0.0819
2	1.00	1.07	0.14	2.25	0.3000	0.2700
3	0.85	0.98	0.90	1.07	0.0990	0.0819
4	0.60	0.98	0.60	1.07	0.1580	0.0819
5	0.60	0.72	1.40	3.37	0.1580	0.0960

3 Damage Detection in Beam Structures

The case of identification of concentrated damages is dealt with in the following paragraphs. Damage modifies the dynamic response of a structure and, in turn, changes in the behavior of a structure can be associated with the decay of the system's mechanical properties. Based on these observations, a series of papers over the last two decades has examined the use of the measured variations of dynamic behavior for monitoring structural integrity (Cawley and Adams, 1979; Rizos et al., 1990; Shen and Pierre, 1990; Liang et al., 1991; Shen and Taylor, 1991; Morassi, 1993; Casas and Aparicio, 1994; Davini et al., 1995; Vestroni et al., 1996; Vestroni and Capecchi, 1996; Friswell et al., 1997; Morassi and Rovere, 1997; Salawu, 1997; Doebling et al., 1998; Vestroni and Capecchi, 2000; Cerri and Vestroni, 2000a,b; Vestroni and Capecchi, 2003; Cerri and Vestroni, 2003).

As in most inverse problems, ill-conditioning makes the search for a solution difficult because it largely depends upon the quantity and quality of experimental data. Focusing only on the basic aspects of the problem, the case of concentrated damage is considered, and linear behavior is assumed before and after the damage. Damage therefore is modelled as an open crack or the decay of the mechanical properties of a small beam element and is represented by a localized decrease of stiffness. According to the usual techniques of structural monitoring, a small amount of damage should be expected between two checks. Thus, in the detection problem, the unknown quantities are generally very limited in number, as are the location and stiffness reductions of damaged sections. Although this circumstance was known to a certain extent for many years (Cawley and Adams, 1979; Liang et al., 1991; Casas and Aparicio, 1994), a great number of papers on damage detection has been published practically ignoring the importance of this peculiarity of the problem. The desired solution should be such that the stiffness is known throughout to be equal to the undamaged value, except in the few damaged zones (Vestroni et al., 1996; Vestroni and Capecchi, 1996). A meaningful consequence is that a small number of measured data should be sufficient to obtain the solution. This is quite different from a reconstruction problem, in which the distribution of the stiffness parameter along the structure is completely unknown, and the solution calls for a quantity of data that is seldom available and some assumptions on the regularity of the solution are introduced.

The peculiarity of the damage detection problem with respect to the reconstruction problem has suggested an identification procedure, based on the analysis of all possible damage scenarios for a given number of damages. The procedure proposed is quite general and does not depend on the type of measured quantities. It is important to show that damage identification is not an underdetermined problem, as frequently thought. This avoids the addition of mathematical conditions, not physically related to the problem, or the use of regularization techniques that have not been felt suitable for damage detection where the solution is generally characterized by high discontinuities. The damage detection method is used here when frequencies are the observed quantities, as this renders some observations simpler. However, the proposed approach can be applied for any measured response quantities. Cases with pseudo experimental and experimental data are solved by means of the identification method proposed.

4 Frequency Modification Due to Damage: Direct and Inverse Problem

It is useful to discuss the peculiarities of damage detection, examining the direct and inverse problem of the frequency modification due to damage. First, reference is made to continuous models, for an easier and more explicit mathematical treatment. The results are then extended to discrete models, which are more important in practice.

4.1 Continuous Models

The simple case of concentrated damage in a supported beam is dealt with to discuss the problem. In this case, illustrated in Figure 11, a rotational spring can accurately model the dynamic behavior of the damaged beam (Liang et al., 1991; Vestroni et al., 1996). Damage is described by two parameters, location and magnitude; the stiffness value K of the spring is related to damage by means of different relationships, depending on the kind of damage. When crack characteristics are known, it is possible to determine K, whereas the inverse is not true, because the same value of K may correspond to different kinds of damage, equivalent in their dynamic behavior.

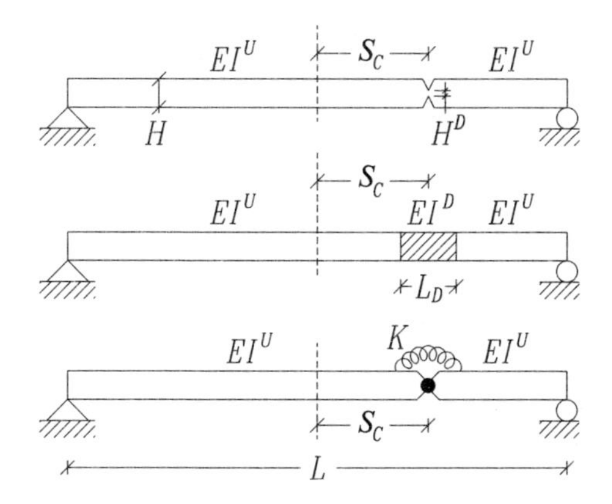

Figure 11. Model of a continuous beam with concentrated damage.

In the case of one crack only (Figure 11) the use of the characteristic equation provides an equation among λ, S_c and K, which is explicit with respect to K:

$$4k\sin(\underline{\lambda}L)\sinh(\underline{\lambda}L)+ \\ \underline{\lambda}L\{\sinh(\underline{\lambda}L)[\cos(\underline{\lambda}L)-\cos(s\underline{\lambda}L)]+\sin(\underline{\lambda}L)[\cosh(\underline{\lambda}L)-\cosh(s\underline{\lambda}L)]\}=0 \quad (4.1)$$

where the nondimensional quantities $s = S_c/(L/2)$ and $k = K/(EI/L)$ are introduced, and $\underline{\lambda} = \lambda(\rho A/EI)^{1/2}$ is the wave vector; the quantity k plays the role of the unknown x. Eq. 4.1 can be more synthetically written as:

$$kg_1(\underline{\lambda}) + g_2(\underline{\lambda}, s) = 0. \quad (4.2)$$

For given values of k and s, Eq. 4.2 is satisfied for each eigenvalue $\underline{\lambda} = \underline{\lambda}_r$. On the contrary, when $\underline{\lambda}_r$ is known, a 1-dimensional manifold $k_r(s)$ satisfies Eq. 4.1, i.e., for any location s there is a value of damage intensity k, for which the r-th eigenvalue is equal to the given $\underline{\lambda}_r$. Thus, two values of $\underline{\lambda}_r$ are, in principle, sufficient to determine s and k. To analyze the uniqueness problem, it is convenient to draw the function:

$$k_r(s) = -g_2(\underline{\lambda}_r)/g_1(\underline{\lambda}_r, s). \tag{4.3}$$

Curves obtained for different $\underline{\lambda}_r$ should cross at one s where damage is located. Once s is known, k is given by Eq. 4.3. For the damaged beam in Figure 11 with damage at $s = 0.15$ and $k = 33$, Figure 12 shows functions $k_r(s)$, with $r = 1, 3$. Not all couples define a unique solution; indeed k_1 and k_3 cross twice, and k_1 and k_2 and k_2 and k_3 cross once. However, it is clear that the addition of a new frequency is sufficient to eliminate the multiplicity; for example $\underline{\lambda}_1$, $\underline{\lambda}_2$, $\underline{\lambda}_3$ cross only at the exact s.

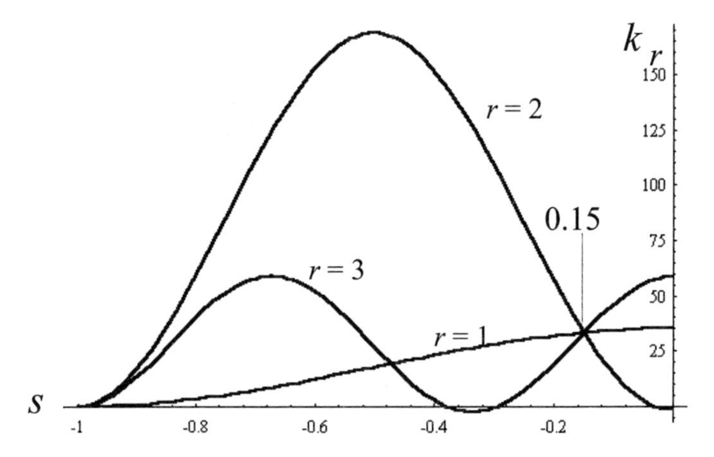

Figure 12. Functions $k_r(s)$ for the first three modes of a supported beam with concentrated damage at $s = 0.15$ and $k = 33$.

4.2 Discrete Models

The finding of an analytic expression for the eigenvalue problem becomes prohibitive when the number of cracks is high or when a beam with varying section is considered. Still greater difficulty arises when, instead of a single beam, a beam system is involved. In such cases, the use of a FE model is generally mandatory. Naturally, because a discrete model is substituted to a continuous one, some approximation is introduced. However, the discretization does not alter the characteristics of the problem (Cerri et al., 2005).

Here, reference is made to beams or frames but analogous considerations hold also for different kinds of structures. In beam structures, damage, which is actually more or less localized around a section, is assumed to be distributed throughout the element of the discretized beam to which the section belongs by considering a constant reduction in the stiffness of the whole element. If the elements are small enough, the distributed stiffness model reproduces well the frequency values of the lowest modes. Therefore, the damage of a single element is described by the nondimensional parameter:

$$x_i = \frac{\Delta k_i}{k_i^U} \tag{4.4}$$

where $\Delta k_i = k_i^U - k_i^D$, and k_i^U and k_i^D are the undamaged and damaged stiffnesses of the i-th element, respectively. The parameter value $x_i = 0$ is for the undamaged element and $x_i = 1$ for the completely damaged element.

The problem of locating and quantifying damage in a system discretized in n elements consists in the evaluation of the stiffness of $p \leq n$ elements or groups of elements, candidates to be damaged, on the basis of the knowledge of m response quantities, natural frequencies here, for the damaged and undamaged structure. The assumption of the stiffness of p elements or groups of elements as parameters of the problem, rather than that of all the n elements, is useful both to separate the discretization of the model from the location of damage and to use a smaller number of parameters. In this context, x_i refers either to an element or to a group of elements for which only one parameter is assumed to represent the decrease in their stiffness. This makes possible a very important solution strategy for large complex structures, in which the size of the zone where damage is sought is successively decreased. The procedure starts with parameters representing the stiffness of groups of elements and ends with parameters representing the stiffness of single elements only for those groups of elements where damage is found to be localized.

The general problem of the identifiability of beam model, stiffness and masses, has been studied and partially solved in (Gladwell, 1984; Gladwell et al., 1987). To obtain a complete and global solution to the problem, a very large number of data is needed. The damage problem is less general for two reasons. First of all, masses are assumed to be known. Secondly, and more important, only local identifiability is considered. The range of parameters in which the true solution is to be found is fixed on the basis of physical considerations; the values of the parameters should not diverge too widely from those of the undamaged structure and the stiffness of a single element of the beam is assumed not to be increasing ($x_i > 0$).

If no assumption is made on the number $r \leq p$ of expected damaged elements, the identification problem in absence of errors is equivalent to the algebraic problem

$$\boldsymbol{\lambda}_e - h(\mathbf{x}) = 0, \quad \mathbf{x} \in R^p, \ \boldsymbol{\lambda}_e \in R^m \tag{4.5}$$

where the m–dimensional vector $\boldsymbol{\lambda}_e$ collects the experimental eigenvalues in the damaged state and $h(\mathbf{x})$ is now the function that furnishes the corresponding analytical quantities through a finite element code, associated with the p–dimensional vector $\mathbf{x}$. The evaluation of $\mathbf{x}$ furnishes the variation of the stiffness with respect to the undamaged stiffnesses k_i^U, which are assumed to be known, although more realistically they, too, are the results of a preliminary identification procedure. When k_i^U is unknown, the procedure furnishes its reduction only, which is in any case a measure of damage.

Eq. 4.5 is assumed to possess at least one solution $\bar{\mathbf{x}}$. In such a circumstance, the implicit function theorem assures the generic existence of a $(p - m)$ dimensional submanifold of solutions to Eq. 4.5 passing through $\bar{\mathbf{x}}$. For the uniqueness of the solution around $\bar{\mathbf{x}}$, it is thus necessary that $m \geq p$; this condition is also sufficient when the Jacobian of $h(\mathbf{x})$ does not vanish at $\bar{\mathbf{x}}$. It is worth noting that, in general, no conditions can be stated for global uniqueness. Indeed, even if local uniqueness conditions are satisfied around $\bar{\mathbf{x}}$, it is possible that other isolated solutions $\hat{\mathbf{x}}$ to Eq. 4.5 exist.

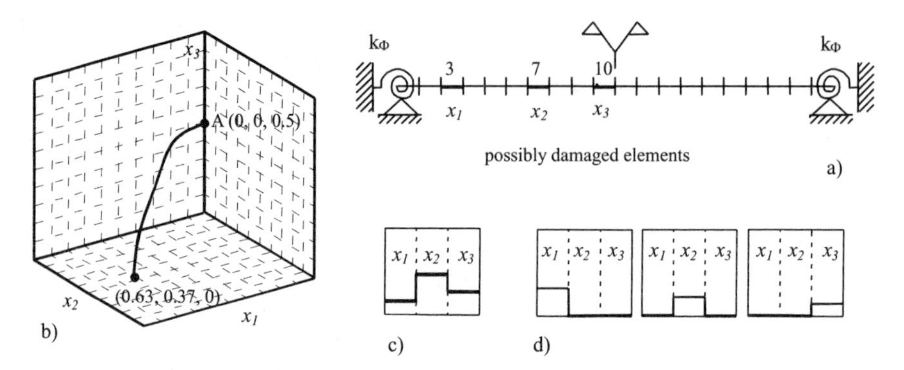

Figure 13. A discretized beam with 3 elements possibly damaged (a), manifold of solutions for 2 measured frequencies (b), in the case with three damaged elements (c) or one damaged element (d).

Figure 13 shows the situation in the case of a symmetric supported beam with elastic boundary restraints. To make possible a graphic representation, only three elements are candidates to be damaged ($p = 3$). This results in a model defined by three parameters, the stiffness of elements 3, 7, and 10 of Figure 13a. Two frequencies are assumed as data, in the virgin and damaged states, corresponding to the vector of parameters $\bar{\mathbf{x}} = (0.0, 0.0, 0.5)$, representing the damaged situation of a beam where only one element over the p candidates is damaged. The bold curve represents the $3 - 2 = 1$ dimensional manifold of solutions to Eq. 4.5, which of course contains $\bar{\mathbf{x}}$. The curve is drawn inside the unitary cube, which defines the range of parameters ($0 < x_i < 1$). It reaches the face $x_3 = 0$ of the cube, at the point $\tilde{\mathbf{x}} = (0.63, 0.37, 0.00)$. It is clear in Figure 13 that the submanifold of solutions contains only one point with a single damaged element (point $\bar{\mathbf{x}}$). Namely, if only one element of the beam is expected to be damaged and two measurements are used, the identification problem has one solution only and the solution has to be searched among those represented in Figure 13d and not that of Figure 13c.

This is a point of a certain importance. In a damage problem, a high number p of elements of the discretized structure are candidates for damage; in practice, only a much smaller number r of elements will actually be damaged. Thus, the proper identification problem would have the small dimension r and could easily be made over–determined. In particular, the example of Figure 13 shows that two frequencies are sufficient to evaluate damage located in one element without the knowledge of the damaged section location, but with the assumption that there is one damaged section only. This result seems straightforward enough, because only two unknowns, the position and the entity of the damage, need to be determined, as in the continuous beam problem. However, mainly when discrete models have been used, the problem is seen as a reconstruction problem and this has easily led to consider as unknown all the possible candidate elements.

It is possible to show (Vestroni and Capecchi, 1996, 2000) that $\mathbf{x}$ is a unique solution if $m \geq 2r$. In Figure 13, where $r = 1$, the solution must be sought on the axes x_i, where two frequencies are sufficient to guarantee a unique solution.

The result of such an approach is shown in Figure 14. A simply supported beam is discretized in $n = 14$ elements, whose stiffness is uniformly distributed (Figure 14a). If no hypothesis is made on the number of damaged elements and any element is considered as potentially hosting a stiffness reduction, as in the reconstruction problem, an irregular stiffness distribution in which every element is affected by a slight amount of damage can be obtained (Figure 14b). On the contrary, if r damages are expected, the reconstruction problem $\boldsymbol{\lambda}_e - h(\mathbf{x}) = 0$, where $\mathbf{x}$ is a n–dimensional vector, is replaced by an identification problem:

$$\boldsymbol{\lambda}_e - \boldsymbol{\lambda}_\alpha(\mathbf{x}, \boldsymbol{\alpha}) = 0 \quad \boldsymbol{\lambda} \in R^m,\ \mathbf{x} \in R^{r+},\ \|x\| \leq 1,\ \boldsymbol{\alpha} \in I^r \tag{4.6}$$

where $\boldsymbol{\alpha}$ is a vector which gathers the number of the candidate damaged elements.

Figure 14c and d show the results of the same problem conceived as an identification problem with $r = 2$ and $r = 1$: for each case, only two of the possible stiffness distributions, respectively in red and blue, are shown. When $r = 2$, a total of 91 problems with 2 unknowns only are obtained; when $r = 1$, 14 problems with 1 unknown only are obtained. The exact location of damage is the unique able to satisfy Eq. 4.6.

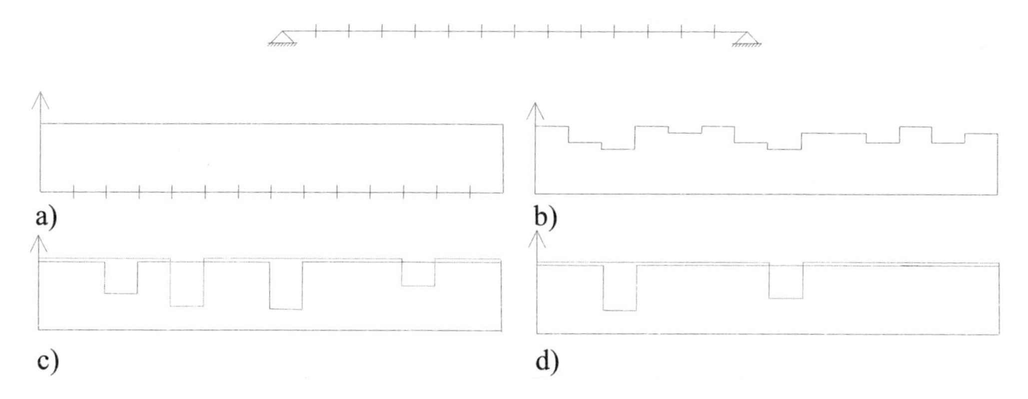

Figure 14. Identified stiffness when different numbers of elements are candidates to be damaged a) undamaged case, b) reconstruction problem, c) 2 elements d) 1 element.

5 Damage Detection

The direct problem is simple and well–posed; when the entity and location of damage are given, Eq. 4.1 furnishes the frequencies of the system for a continuous model and Eq. 4.5 for a discrete model. The inverse problem, which consists of determining the damage parameters when a certain number of λ_r is known, is usually more complex. From previous discussion, if r is the number of cracks, isolated solutions exist when $2r$ frequencies are known. This statement can be useful in selecting a specific approach for damage identification, different from the classical model identification approach.

The procedure of damage parameter identification, on the basis of known experimentally measured quantities, generally consists of two parts: (i) determination of the structural model in the undamaged conditions; and (ii) determination of the modified model when the structure is damaged. The comparison of the two models makes it possible to evaluate the damage parameters. Attention is focused here on the second phase, on the assumption that the representative model for the undamaged system has already been determined. In those cases in which the undamaged stiffness distribution cannot be determined with satisfactory accuracy, the damage evaluation is made by the second phase only, where the observed quantities are not the measured quantities, but their variations, and the identification object is the variation of the stiffness characteristics.

The problem is strongly ill-conditioned and therefore the success of the approach notably depends on the accuracy of the experimental data and on the interpretative model used in the solution technique. Below, the problem of a beam with length L, rectangular section width w and height d, with single or multiple cracks, as shown in Figure 15, is considered. It is assumed that the crack is characterised by a small thickness with respect to the beam length and by a constant depth d_c over the beam width. Moreover, the crack does not affect the mass of the beam.

When there is only one crack, the damage model is characterized by two parameters: location and degree. The damage due to multiple cracks can be analysed using models analogous to those for the single crack, considering a rotational spring for each crack when the cracks are independent, as they are localised at a distance larger than the conventional length of the damaged zone (Figure 15b). Alternatively, a model with a damaged zone with reduced stiffness where the cracks interact, as they are quite near, can be used (Figure 15c). In the latter case, the damage model is characterised by three parameters *location* $s_c = 2S_c/L$, *extension* of the damaged zone $b = L_D/L$ and *degree* $\beta = 1 - EI^D/EI$.

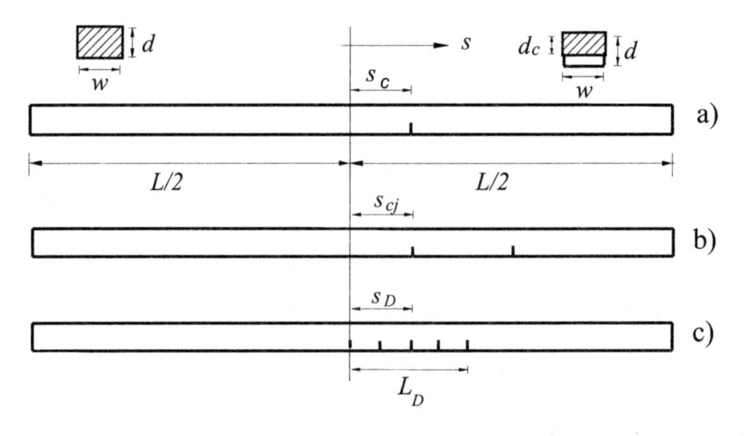

Figure 15. Beams with single a), multiple independent b) and interacting cracks c).

Cracks locally alter the stress distribution along an extension of the beam greater than the crack itself. Figure 16 shows the distribution of longitudinal stress in a zone of the beam around the crack, obtained by a finite element model. The stress distribution evidences where the beam undergoes a stiffness variation (Cerri and Vestroni, 2003).

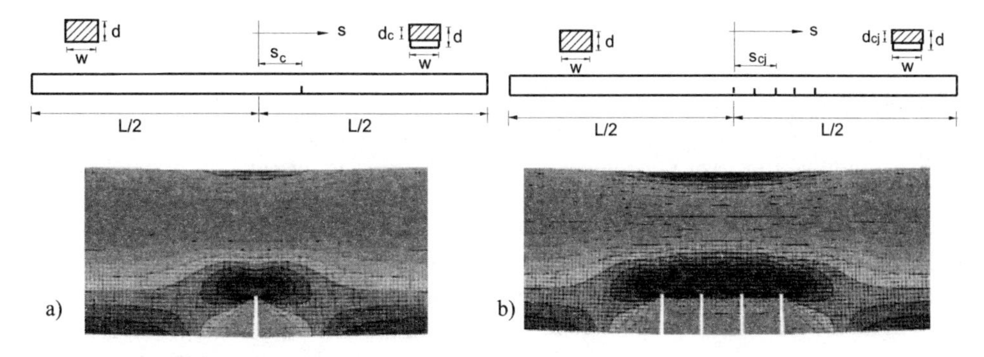

Figure 16. Stress distribution for single a) and interacting cracks b).

It has been shown that, for a given set of experimental frequencies, different equations of the eigenvalue problem are satisfied at the true solution:

a) the characteristic equation according to Eq. 4.1

$$g(\lambda_{e,r}, \mathbf{x}) = 0, \quad \forall \lambda_{e,r} \tag{5.1}$$

b) the classical comparison between experimental and analytical frequencies, as a function of damage parameter vector $\mathbf{x}$, according to Eq. 4.5:

$$\lambda_{e,r} - h_{e,r}(\mathbf{x}) = 0, \quad \forall \lambda_{e,r} \tag{5.2}$$

Based on the above expressions, two procedures for damage identification may be followed. The first is based on Eq. 5.1 and denoted as the modal equation procedure. The second is based on the direct output comparison (Eq. 5.2) and is denoted as the response quantity procedure. In the real world a noise term appears in Eqs. 5.1 and 5.2, due to experimental and modeling errors. Therefore, finding the correct values of $\mathbf{x}$ can be conveniently formulated as a minimum problem.

The modal equation approach is useful to more clearly illustrate the conditions for a unique solution and a minimum amount of data. However, the response quantities procedure based on the response comparison is more efficient in solving damage identification problems, mainly when finite element models are used. In this procedure it is convenient to directly introduce the objective function $l(\mathbf{x})$ which, for the case of a diffused damage characterized by three damage parameters, reads:

$$l(s, b, \beta) = \sum_r \left(\frac{\Delta\omega_{e,r}^U}{\omega_{e,r}^U} - \frac{\Delta\omega_r^D(s, b, \beta)}{\omega_r^U} \right)^2 \tag{5.3}$$

defined by the difference between experimental and analytical variations of the r–th frequency from the undamaged to the damaged state, normalized with respect to the undamaged frequency. The search for the minimum of function l can be divided in two phases and has an attractive mechanical meaning. For each scenario of damage location, defined by s, the objective function:

$$\tilde{l}\,(s) = \min_{b,\beta}\,(s, b, \beta) \tag{5.4}$$

as a function of parameter s only, is initially determined by the minimization of $l\,(s, b, \beta)$ (Eq. 5.4) with respect to parameters b and β. For each s, $\tilde{l}(s)$ gives the best value of b and β to minimize the error between experimentally and analytically observed quantities. The solution to the inverse problem is then given by the minimum of $\tilde{l}(s)$. If the problem exhibits only one solution $\bar{s}$, it is possible to determine one single global minimum of $\tilde{l}(s)$ for a number of frequencies $2r$, with r equal to the number of cracks.

The direct comparison procedure offers various advantages: it is possible and convenient to use any response quantities, frequencies, modal components or their combinations, which enables the selection of which are the most convenient to measure, and even though the model is discretized by a large number of elements, each of which is a possible candidate to be damaged, the evaluation of $\tilde{l}(s)$ requires the solution of a minimum problem with a small r dimension.

In Figure 17 and 18 some examples based on error free numerical investigations are outlined, where the observed quantities are the variation of frequencies Δf_r normalized with respect to their undamaged values. Figure 17 shows the result of the damage identification procedure based on the response comparison for the case of a beam with one crack. The objective function $\tilde{l}(s)$ presents a single minimum, therefore, both the unknowns are uniquely determined. In Figure 18, a case where multiple interacting cracks reduce the stiffness in a zone with length b is depicted. Also, in this case, the objective function built considering only three frequencies, exhibits a single minimum. It must be underlined that the solution of the identification problem involving three response quantities and three unknowns exists and is unique.

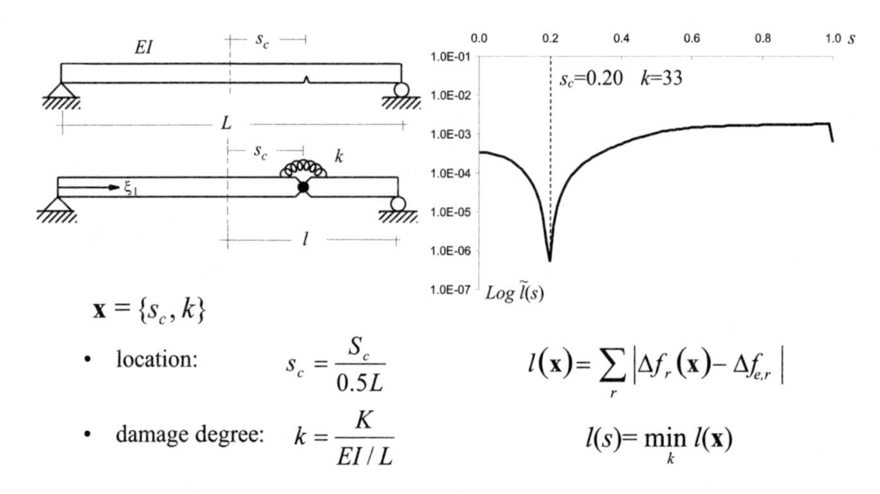

Figure 17. Damage identification in the case of one crack.

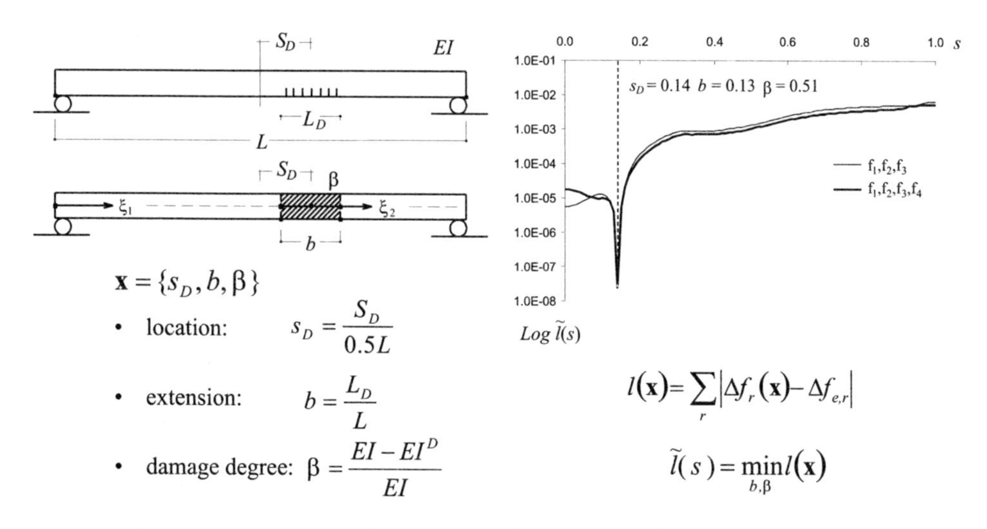

Figure 18. Damage identification in the case of interacting cracks.

The described procedures are applied to an experimental case. Two reinforced concrete beams were subjected to static and dynamic tests (Cerri and Vestroni, 2000a,b). The beams were simply supported in the static test configuration. For the dynamic tests, the beams were hung from very flexible springs that simulate free-free conditions. Seven load steps were performed: the first just beyond the cracking limit and then with regular increments up to yielding. After each load increment, the natural frequencies were measured. Figure 19 shows the moment-curvature diagram for the central region of the beam (a), which is the most stressed, and the curve of the applied load versus the midspan vertical displacement (b).

The flexural stiffness mean values of the central zone were evaluated from the moment-curvature diagram in the undamaged and most severely damaged conditions (lines a and b in Figure 19a). These are $EI^U \approx 0.9{*}10^{12}$ and $EI^D \approx 0.55{*}10^{12}$ N/mm^2, which represent a stiffness reduction of 40%, $\beta \approx 0.40$. The overall damage effect can be appreciated from the load-displacement curve of Figure 19b, where a continuous reduction of global stiffness versus the intensity of the applied load is clear, as well as the increase of residual displacements. The results are summarized in Table 4, where for each load step the ratio of maximum flexural bending with respect to the ultimate value M_{max}/M_y is reported, as well as the maximum displacement δ_{max}, the residual displacement at the unloading δ_{res} and the extension of the damaged zone L^D/L.

Dynamic tests have been performed in the undamaged initial conditions and at the various steps, after the application and removal of the load. For each step, experimental values of the frequencies are given in Table 5. Damage produces a non-negligible decrease in the first five frequencies observed, up to about 40%. The decrease of the first three frequencies versus the intensity of the load is quite regular for all frequencies and almost linear for the first mode. Notwithstanding the fact that the damage is in the midspan, the even mode frequencies also show a notable decrease, because the damage is not concentrated, but diffused over a zone that increases with the load intensity.

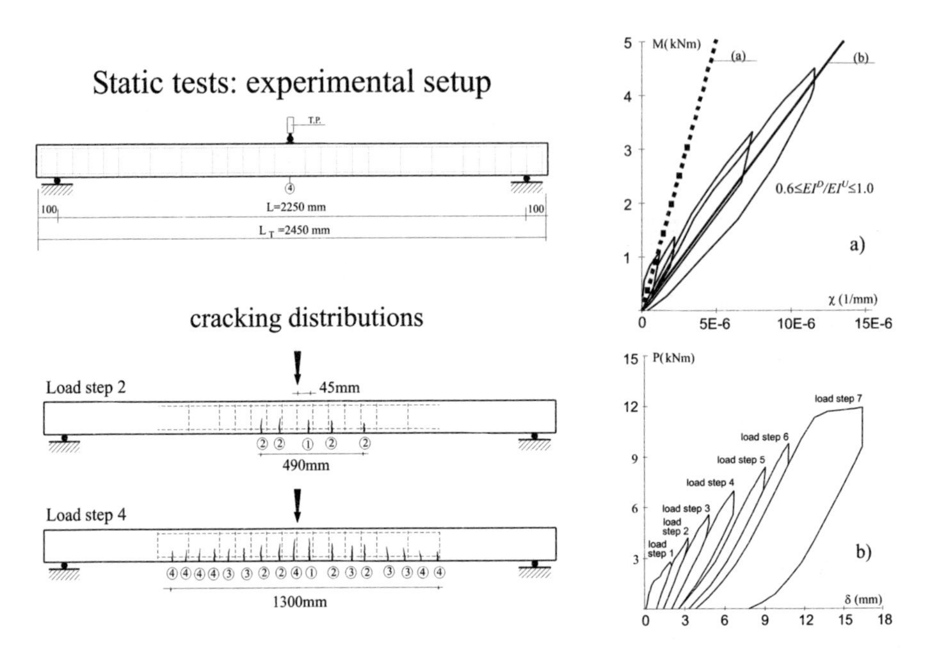

Figure 19. (a) Moment-curvature, (b) load-deflection for the seven load-steps and cracking distributions for load steps 2 and 4.

Table 4. Results of static tests.

Load step	P[kN]	M_{max}/M_y (%)	L^D/L(%)	δ_{max} [mm]	δ_{res} [mm]
1	2.8	24	0	1.9	0.8
2	4.2	35	20	3.2	1.4
3	5.6	47	29	4.9	2.0
4	7.0	59	53	6.7	2.6
5	8.4	71	60	9.1	3.3
6	9.8	82	67	10.9	3.8
7	11.9	100	67	16.4	7.5

Table 5. Dynamic tests: undamaged frequencies [Hz] and their variations.

	f_1 =89.50	f_2 =245.0	f_3 =467.5	f_4 =740.0	f_5 =1030.0
Load step	Δf_1(%)	Δf_2 (%)	Δf_3 (%)	Δf_4(%)	Δf_5(%)
1	9.2	3.3	4.5	4.4	3.4
2	14.0	7.3	6.7	7.1	5.1
3	17.0	11.0	10.7	9.8	7.5
4	24.9	22.0	19.6	-	14.6
5	30.4	26.9	23.2	18.5	-
6	34.4	30.2	27.7	-	20.2
7	39.1	33.2	31.8	28.4	23.5

With reference to the experimental frequencies of Table 5, the optimal value of damage parameters (s, b, β) is determined using three or five frequencies for each load step. The results obtained with the two sets of data do not always coincide, but they are very close. This implies that the problem is ill-conditioned, however, the different solutions determined are very close on the $b-\beta$ plane. The identified location suggests that damage in the intermediate steps is not symmetrical with respect to the midspan as it would be expected. In fact, non-symmetric damage was observed, especially in the initial phase. Figure 19 shows the experimentally observed crack pattern for steps 2 and 4. Clearly, at step 2 the eccentricity of the damaged zone with respect to the midspan is greater than that at step 4. With increasing damage extension the centerline of the damage tends to the beam midspan, as predicted by the identification procedure.

The identified values of b and β are reported in Table 6 for the load steps 3 and 5 only. For these two cases, the objective functions are also drawn in Figure 20. The curves are somewhat irregular and the minimum is not very accentuated, even though the absolute minimum is evident. A greater number of frequencies compromises the resolution of minimum points, as experimental errors are added with the new data. In any case, the parameters are very close, with a small difference on the location.

Table 6. Identified damage parameters s, b and β using 3 (a) and 5 frequencies (b).

Load step	(a)			(b)		
	s	b	β	s	b	β
3	0.19	0.40	0.38	0.19	0.41	0.38
5	0.13	0.59	0.53	0.01	0.57	0.53

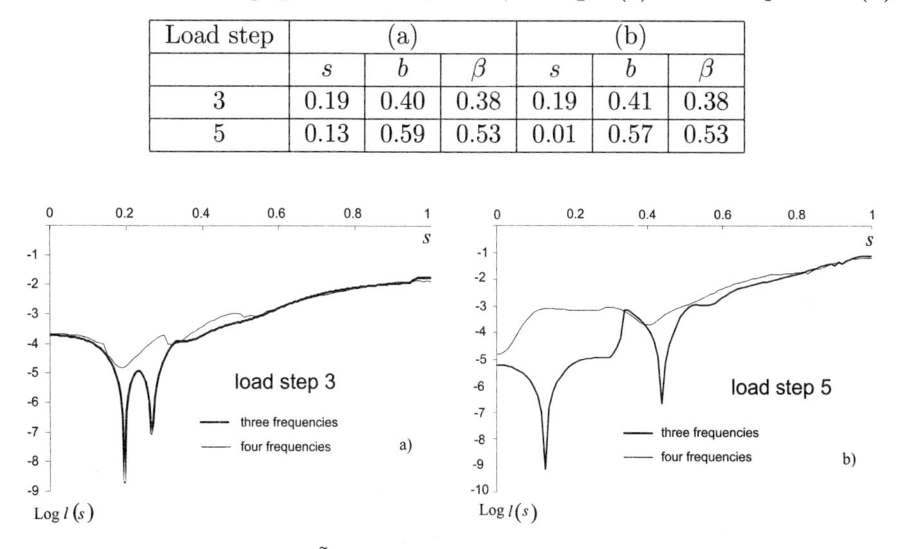

Figure 20. Objective function $\tilde{l}(s)$ for the load steps 3 and 5 with 3-4 frequencies.

6 Concluding Remarks

The parametric identification is used in damage detection; in particular the case of damage diffused over the structure is first dealt with making reference to an experimental investigation on a masonry building. Small oscillation amplitude tests were performed before and after damage; comparison of modal parameters in undamaged and damaged conditions reveals modification in the structural properties. Irreversible damage is detected, which is apparent by a strong modification of frequency response functions.

Notwithstanding the structure suffered a strong decay of its characteristics, very accurate modal models are obtained. On the contrary, difficulties are met in identification of a finite element model in undamaged and damaged conditions, due to the lack of modeling internal and external constraints; moreover, strong ill-conditioning is present, due to close and similar modes. Important role of preliminary analyses using pseudo-experimental data is remarked in order to evidence the ill-conditioning problem characteristics. The damage identification problem does not exhibit a unique solution; however, the results of the identification pointed out a different distribution of damage. Useful information on the walls which undergo most severe damage are obtained in agreement with inspections.

Parametric models and updating model techniques were successively used in the identification of concentrated damage. Within a classical monitoring condition, reference is made to scenarios where the number of damaged zones is small. Damage may be due to the decay of mechanical properties, to overstress in some zones, to fatigue and others causes. Under these assumptions, it is shown that damage identification differs from a problem of model reconstruction. In particular it is demonstrated that for a small number of damaged zones only a small quantity of data is necessary. A procedure which takes into account the peculiarity of the damage detection problem is presented and its effectiveness is demonstrated in different applications using experimental and pseudo-experimental data. To guarantee the unique identification of location and intensity of damage it suffices that the number of accurate measurements is comparable to the number of expected damaged zones. However, the problem is strongly ill-conditioned and only accurate experimental data and damage modelling can guarantee reliable results.

Bibliography

D. Capecchi and W. D'Ambrogio. Experimental modal analysis and damage detection on an ancient masonry building. *Meccanica*, 28(1):13–25, 1993.

D. Capecchi and F. Vestroni. Study of dynamic behaviour of an old masonry building. In *Proceedings of the 1st European Conference on Structural Dynamics*, 1991.

J. R. Casas and A. C. Aparicio. Structural damage identification from dynamic-test data. *Journal of Structural Engineering*, 120(8):2437–2450, 1994.

P. Cawley and R. D. Adams. The location of defects in structures from measurements of natural frequencies. *Journal of Strain Analysis*, 14(2):49–57, 1979.

M. N. Cerri and F. Vestroni. Detection of da,age in beams subhected to diffused cracking. *Journal of Sound and Vibration*, 234(2):259–276, 2000a.

M. N. Cerri and F. Vestroni. Use of frequency change for damage identification in reinforced concrete beams. *Journal of Vibration and Control*, 9(3-4):475–491, 2000b.

M. N. Cerri and F. Vestroni. Identification of damage due to open cracks by changes of measured frequencies. In *XVI AIMETA Congress of Theoretical and Applied Mechanics*, 2003.

M. N. Cerri, F. Vestroni, and S. Vidoli. Health monitoring of beam structures based on FE models and eigenfrequency test data. In *Proceedings of the XII International Congress on Sound and Vibration*, 2005.

T. G. Chondros, A. D. Dimarogonas, and J. Yao. A continuous cracked beam vibration theory. *Journal of Sound and Vibration*, 215(1):17–34, 1998.

C. Davini, A. Morassi, and N. Rovere. Modal analysis of notched bars: tests and comments on the sensitivity of an identification technique. *Journal of Sound and Vibration*, 179(3):513–527, 1995.

S. W. Doebling, C. R. Farrar, and M. B. Prime. A summary review of vibration-based damage identification methods. *Shock and Vibration Digest*, 30:91–105, 1998.

M. I. Friswell, J. I. T. Penny, and S. D. Garvey. Parameter subset selection in damage location. *Inverse Problems in Engineering*, 5:189–215, 1997.

G. M. L. Gladwell. The inverse problem for the vibrating beam. *Proceedings of the Royal Society of London Series A*, 393:277–295, 1984.

G. M. L. Gladwell, A. H. England, and D. Wang. Examples of reconstruction of an Euler-Bernoulli beam from spectral data. *Journal of Sound and Vibration*, 119:81–94, 1987.

R. Y. Liang, F. Choy, and J. Hu. Detection of cracks in beam structures using measurements of natural frequencies. *Journal of The Franklin Institute*, 328(4):505–518, 1991.

A. Morassi. Crack-induced changes in eigenparameters of beam structures. *Journal of Engineering Mechanics*, 119(9):1798–1803, 1993.

A. Morassi and N. Rovere. Localizing a notch in a steel frame from frequency measurements. *Journal of Engineering Mechanics*, 123(5):422–432, 1997.

A. Pau and F. Vestroni. Modal analysis of a beam with radiation damping: Numericl and experimenal results. *Journal of Vibration and Control*, 13(8):1109–1125, 2007.

P. F. Rizos, N. Asphagathos, and A. D. Dimarogonas. Identification of crack location and magnitude in a cantilever beam from the vibration modes. *Journal of Sound and Vibration*, 138(3):381–388, 1990.

O. S. Salawu. Detection of structural damage through changes in frequency: a review. *Engineering Structures*, 19(9):718–723, 1997.

M. H. H. Shen and J. E. Taylor. An identification problem for vibrating cracked beams. *Journal of Sound and Vibration*, 150(3):457–484, 1991.

S. H. Shen and C. Pierre. Natural modes of Bernoulli-Euler beams with symmetric cracks. *Journal of Sound and Vibration*, 138(1):115–134, 1990.

F. Vestroni, G. Beolchini, E. Antonacci, and C. Modena. Identification of dynamic characteristics of masonry buildings from forced vibration tests. In *Proceedings of the Eleventh World Conference on Earthquake Engineering*, 1993.

F. Vestroni and D. Capecchi. Damage evaluation in cracked vibrating beams using experimental frequencies and finite element models. *Journal of Vibration and Control*, 2:69–86, 1996.

F. Vestroni and D. Capecchi. Damage detection in beam structures based on frequency measurements. *Journal of Engineering Mechanics*, 126(7):761–768, 2000.

F. Vestroni and D. Capecchi. *Parametric identification and damage detection in structural dynamics*, pages 107–143. Research Signpost, Trivandrum, India, 2003. A. Luongo ed.

F. Vestroni, M. N. Cerri, and E. Antonacci. The problem of damage detection in vibrating beams. In *Proceedings Eurodyn '96 Conference*, pages 41–50, 1996.

Damage Detection in Vibrating Beams

Antonino Morassi
Department of Georesources and Territory, University of Udine, Udine, Italy

Abstract This paper deals with a class of inverse problems in vibration concerning the identification of damages in elastic beams by dynamic data. A review of some recent results is presented.

1 Introduction

Dynamic tests are commonly used as a diagnostic tool to detect structural modifications that may have occurred in mechanical systems during service. The final aim is to give an interpretation of the changes induced by a possible damage on the dynamic behavior of a system, using them to predict location and degree of magnitude of the degradation.

The goal of this paper is to present some recent approaches to damage identification in vibrating beams. In the paper there is no attempt to overview the vast literature on dynamic methods for damage detection in structures, see, for example, the articles by Friswell (2008) and Vestroni (2008), included in this volume, for an extensive presentation of the subject. Attention is focused, rather, to the presentation of specific lines of research developed by the author in the last years. Since, as noticed by Gladwell (2004) [Chapter 15, Par. 15.1] "the identification of damage in a vibrating structure from changes in vibratory behavior is an inverse problem in a loose interpretation of the term", some attempts to connect the specific application of diagnostic techniques with the general theory of inverse problems in vibration have been done here and there throughout the paper.

The structure of the paper is as follows. In Section 2 the damage is modelled as a reduction of the stiffness coefficient of the beam and it is shown that the frequency shifts caused by the damage contain information on some generalized Fourier coefficients of the unknown stiffness variation. Section 3 is devoted to the identification of localized damages, such as open notches or cracks, in beams from a minimal set of natural frequency measurements. In Section 4, the changes in the nodes of the mode shapes are used to identify localized damage in a beam. Extension of these results is briefly discussed in Section 5. All the diagnostic techniques presented in this paper have been tested on a basis of extensive series of dynamic tests carried out on real damaged beams. For the sake of conciseness, applications to experimental data are discussed only in few cases. The reader interested in details of the experiments is referred to the original papers collected in the references.

2 Damage Detection and Generalized Fourier Coefficients

2.1 Contents

This section deals with damage identification in a vibrating beam, either under axial or bending vibration, based on measurement of damage-induced changes in natural frequencies.

Assuming that the damaged configuration is a perturbation of the undamaged one and the linear mass density remains unchanged, the frequency shifts caused by the damage are correlated with some generalized Fourier coefficients of the unknown stiffness variation. This set of Fourier coefficients is determined on a suitable family of functions depending on the vibration modes of the undamaged system. When it is a priori known that the damage belongs to a half of the beam, the measurement of first M frequency shifts, roughly speaking, allows the first M generalized Fourier coefficients of the stiffness variation to be determined. A numerical procedure based on an iterative algorithm is proposed for solving the diagnostic problem. The idea of connecting the Fourier coefficients of the unknown coefficient with the frequency shifts is more deep and traces back to the fundamental contribution in inverse eigenvalue theory given by Borg (1946), see also Hald (1978) and Knobel and Lowe (1993) for more recent numerical applications. In the context of crack identification in elastic beams, Wu (1994) proposes a reconstruction method by eigenvalues shifts based on the determination of generalized Fourier coefficients of the stiffness variation induced by the damage. In particular, Wu (1994) considered an initially uniform pinned-pinned beam with a single symmetric crack at mid-span, see also Wu and Fricke (1990) for applications in acoustics to the identification of small blockages in a duct by eigenfrequency shifts.

The predictions of the theory and reliability of the proposed diagnostic technique were checked on the basis of results of several dynamic tests performed on free-free cracked steel beams, both under longitudinal and bending vibrations. It is found that the outcome of the damage analysis via Fourier coefficients depends on the accuracy of the analytical model that one uses for identification and on the severity of the damage. The technique provides a satisfactory identification of the damage, both for position and severity, when frequency shifts induced by the damage are bigger than modelling and measurement errors. For these cases, the results of the damage identification obtained via Fourier coefficient method have been compared with those obtained via a standard variational method based on frequency data. In all the cases considered, the comparison shows a good agreement. This leads to the conjecture that, at least for damage detection in simple beam models, updating the stiffness coefficient of the beam so that the distance between the first M measured and analytical frequencies is minimized, is equivalent to finding the first M generalized Fourier coefficients of the stiffness variation caused by the damage.

The plan of the section is as follows. The theoretical basis of the method is presented in Section 2.2 for a rod in longitudinal vibration. An iterative reconstruction procedure is shown in Section 2.3. Applications to real experimental data for damage identification in rods with localized damage are discussed in Section 2.4. The bending vibration case is studied in Section 2.5 and some extensions are discussed in Section 2.6. Finally, Section 2.7 is devoted to a comparison between the results obtained by the proposed diagnostic method and by a variational-type identification technique.

2.2 Theoretical Basis of the Method

The theoretical basis of the damage identification method is presented for a straight rod in longitudinal vibration.

Formulation of the eigenvalue problem It is assumed that the spatial variation of the infinitesimal free vibrations of an undamaged rod of length ℓ is governed by the differential equation

$$(au')' + \lambda \rho u = 0 \qquad \text{in } (0, \ell), \tag{2.1}$$

where $u = u(x)$ is the mode shape and $\sqrt{\lambda}$ is the associated natural frequency. The rod is assumed to have no material damping. The quantities $a = a(x) \equiv EA(x)$ and $\rho = \rho(x)$ denote the axial stiffness and the linear-mass density of the rod. E is Young's modulus of the material and $A(x)$ the cross-section area of the rod.

This analysis is concerned with rods for which $a = a(x)$ is a uniformly strictly positive and continuously differentiable function of x in $[0, \ell]$, namely

$$a \in C^1([0, \ell]), \quad a(x) \geq a_0 > 0 \quad \text{in } [0, \ell], \tag{2.2}$$

where a_0 is a given constant. The function $\rho = \rho(x)$ will be assumed to be a continuous and uniformly strictly positive function of x in $[0, \ell]$, that is

$$\rho \in C^0([0, \ell]), \quad \rho(x) \geq \rho_0 > 0 \quad \text{in } [0, \ell], \tag{2.3}$$

where ρ_0 is a given constant. Although the present analysis can be developed for general boundary conditions, to fix the ideas the beam is taken with free ends:

$$a(0)u'(0) = 0 = a(\ell)u'(\ell). \tag{2.4}$$

It is well known that for coefficients a and ρ satisfying (2.2), (2.3), respectively, and for end conditions (2.4), there is an infinite sequence $\{\lambda_m\}_{m=0}^{\infty}$ of real eigenvalues such that $0 = \lambda_0 < \lambda_1 < \lambda_2 < ...$, with $\lim_{m \to \infty} \lambda_m = \infty$, see Brezis (1986). Corresponding to every eigenvalue λ_m there exists a single eigenfunction $u_m = u_m(x)$, $m = 0, 1, 2, ...$, determined up to a multiplicative constant. In order to select uniquely the eigenfunctions, the following *normalization condition* will be used

$$\int_0^\ell \rho u_m^2 dx = 1, \qquad m = 0, 1, 2, ... \,. \tag{2.5}$$

The *fundamental mode* $u_0(x)$ of the free-free rod corresponds to $\lambda_0 = 0$ and $u_0(x) = \left(\int_0^\ell \rho dx \right)^{-\frac{1}{2}}$ in $[0, \ell]$.

Suppose that a damage appears on the rod. It is assumed that the presence of the damage can be described within the framework of the classical one-dimensional theory of rods and that it reflects on a reduction of the effective axial stiffness without altering the mass distribution, see, for example, Thomson (1949) and Petroski (1984). This assumption is rather common in damage detection studies and, in fact, a careful description of damage would be hardly worth doing, since it would require a detailed knowledge of degradation, which is not always available in advance in inverse analysis. More refined mechanical models of beams with localized damages have been presented, for example, by Christidies and Barr (1984) and Sinha et al. (2002). Therefore, in the present analysis, the axial stiffness of the damaged beam will be assumed as follows

$$a_\epsilon(x) = a(x) + b_\epsilon(x), \tag{2.6}$$

where the *stiffness variation* $b_\epsilon = b_\epsilon(x)$ introduced by the damage satisfies the conditions:

i) (regularity)

$$b_\epsilon \in C^1([0,\ell]); \tag{2.7}$$

ii) (uniform lower and upper bound of a_ϵ) there exists a constant $A_0 > 0$ such that

$$a_0 \leq a_\epsilon(x) \leq A_0 \quad \text{in } [0,\ell] \tag{2.8}$$

and

iii) (smallness)

$$\|b_\epsilon\|_{L^2} = \epsilon O(\|a\|_{L^2}), \tag{2.9}$$

for a real positive number ϵ, where $|O(\|a\|_{L^2})| < c\|a\|_{L^2}$ and c is a positive constant independent of ϵ.

In equation (2.9), the symbol $\|f\|_{L^2} \equiv \left(\int_0^\ell f^2 dx\right)^{\frac{1}{2}}$ denotes the norm of the Lebesgue space $L^2(0,\ell)$ of the square summable real-valued functions f on $(0,\ell)$.

A structural damage introduces a reduction of the axial stiffness of the rod, that is

$$b_\epsilon(x) \leq 0 \qquad \text{in } [0,\ell]. \tag{2.10}$$

The following analysis, however, holds even for a general variation $b_\epsilon(x)$ which takes positive and negative values in $[0,\ell]$.

Under the above assumptions, the damaged rod had a countable sequence of eigenpairs $\{(u_{m\epsilon}(x), \lambda_{m\epsilon})\}_{m=0}^{\infty}$, with $0 = \lambda_{0\epsilon} < \lambda_{1\epsilon} < \lambda_{2\epsilon} < \ldots$ and $\lim_{m\to\infty} \lambda_{m\epsilon} = \infty$. Moreover, under the condition (2.10), the variational formulation of the eigenvalue problem given, for example, in Weinberger (1965) shows that eigenvalues of the rod are decreasing functions of b_ϵ, that is

$$\lambda_{m\epsilon} < \lambda_m, \qquad m = 1, 2, \ldots . \tag{2.11}$$

Eigenfrequency sensitivity to damage The damaged rod is assumed to be a *perturbation* of the undamaged one, that is the number ϵ appearing in (2.9) is small:

$$\epsilon << 1. \tag{2.12}$$

The smallness of b_ϵ expressed by condition (2.12) allows in the analysis the inclusion of either *small* damages given on *large* portions of the interval $[0,\ell]$ (the so-called *diffuse damage*) or severe damages concentrated in small intervals of $[0,\ell]$ (*localized damages*).

For example, the coefficient $b_\epsilon(x) = -\epsilon a(x)$ in $[0,\ell]$ belongs to the first class; while $b_\epsilon(x) = \frac{a}{4}(1+\cos\frac{\pi(x-\ell/4)}{\epsilon\ell/4})$ in $[\frac{\ell}{4}(1-\epsilon), \frac{\ell}{4}(1+\epsilon)]$, with $\epsilon < 1/2$ and $b_\epsilon(x) = 0$ elsewhere in $[0,\ell]$, defines a severe damage localized near the cross-section of abscissa $\ell/4$.

Under assumption (2.12), an asymptotic eigenvalue expansion formula for $\epsilon \to 0$ holds true.

Let (u_m, λ_m), $m = 0,1,2,...$, be the mth normalized eigenpair of the problem (2.1), (2.4) corresponding to the undamaged rod, with a, ρ satisfying conditions (2.2) and (2.3), respectively. Denote by $(u_{m\epsilon}, \lambda_{m\epsilon})$, $m = 0,1,2,...$, the mth normalized eigenpair of the perturbed problem

$$(a_\epsilon u'_{m\epsilon})' + \lambda_{m\epsilon}\rho u_{m\epsilon} = 0 \qquad \text{in } (0,\ell), \tag{2.13}$$

$$a_\epsilon(0)u'_{m\epsilon}(0) = 0 = a_\epsilon(\ell)u'_{m\epsilon}(\ell), \tag{2.14}$$

where a_ϵ is defined by (2.6) and b_ϵ satisfies conditions (2.7)-(2.9), for a real positive number ϵ.

The following asymptotic eigenvalue expansion holds true:

$$\lambda_{m\epsilon} = \lambda_m + \int_0^\ell b_\epsilon (u'_m)^2 dx + r(\epsilon, m), \qquad m = 0,1,2,..., \tag{2.15}$$

where

$$\lim_{\epsilon\to 0} \frac{|r(\epsilon,m)|}{\|b_\epsilon\|_{L^2}} = 0. \tag{2.16}$$

Note that, since the fundamental mode is insensitive to changes of the stiffness coefficient, when $m = 0$ condition (2.15) reduces to the identity $\lambda_{0\epsilon} = \lambda_0$.

Formulas (2.15), (2.16) play an important role in this study and a proof of them is presented in Morassi (2007). The proof is based on two main results. The first one is represented by the following identity: for every $\epsilon > 0$ and for every integer number m, $m = 0,1,2,...$, one has

$$(\lambda_{m\epsilon} - \lambda_m) \int_0^\ell \rho u_m u_{m\epsilon} dx = \int_0^\ell b_\epsilon u'_m u'_{m\epsilon} dx. \tag{2.17}$$

Identity (2.17) can be obtained by multiplying (2.13) by u_m and (2.1) (with (u,λ) replaced by the mth eigenpair (u_m,λ_m)) by $u_{m\epsilon}$, and integrating by parts. The second result concerns with the asymptotic behavior of the eigensolutions $\{(u_{m\epsilon}, \lambda_{m\epsilon})\}_{m=0}^\infty$ as $\epsilon \to 0$, see Morassi (2007) for more details. One can prove that

$$u_{m\epsilon} \to u_m \quad \text{strongly in } H^1(0,\ell) \text{ as } \epsilon \to 0, \tag{2.18}$$

$$\lim_{\epsilon\to 0} \lambda_{m\epsilon} = \lambda_m, \quad m = 1,2,\dots . \tag{2.19}$$

Here, $H^1(0,\ell)$ is the Hilbert space formed by the measurable functions f, $f : (0,\ell) \to \mathbb{R}$, such that both f and its first derivative f' (in the sense of distributions) belong to the space $L^2(0,\ell)$ of the square summable real-valued functions on $(0,\ell)$.

One can notice that (2.15) can be read as a series Taylor expansion of the mth eigenvalue in terms of the variation b_ϵ. In fact, in an abstract context, the integral term in (2.15) is the *partial derivative* of $\lambda_{m\epsilon}$ with respect to the axial stiffness coefficient a_ϵ evaluated, at $\epsilon = 0$, on the *direction* b_ϵ. This partial derivative can be interpreted as the scalar product (in L^2-sense) between the *gradient* $\frac{\partial \lambda_{m\epsilon}}{\partial a_\epsilon(x)}|_{\epsilon=0} = (u'_m(x))^2$ and the direction b_ϵ, namely

$$< \frac{\partial \lambda_{m\epsilon}}{\partial a_\epsilon(x)}|_{\epsilon=0}, b_\epsilon >= \int_0^\ell b_\epsilon u_m'^2 dx. \tag{2.20}$$

The expression of the integral term in (2.15) shows that the so-called *sensitivity* of the mth eigenvalue to changes of the axial stiffness depends on the square of the first derivative of the corresponding mth vibration mode of the unperturbed system. When the perturbation b_ϵ is localized in a small interval centered in x_0, $x_0 \in (0,\ell)$, formula (2.15) indicates that the first order variation of the mth eigenvalue depends on the square of the longitudinal strain evaluated at x_0, see Section 3.2 for an analogous result in the extreme cases of cracks and notches modelled by translational elastic springs inserted at the damaged cross-sections. The explicit expression of the first derivative of an eigenvalue with respect to cracks or notches will be used in Sections 3.2 and 3.4 to identify localized damages in rods and beams by minimal frequency measurements. Analogous applications to discrete vibrating systems with a single localized damage are presented in Dilena and Morassi (2006).

The analysis has hitherto been related to rods under axial vibration with free ends. However, it is clear that, under analogous assumptions, the asymptotic eigenvalue expansion formula (2.15) holds true for rods with different boundary conditions, such as, for example, *supported* ($u(0) = 0 = u(\ell)$) or *cantilever* ($u(0) = 0$, $a(\ell)u'(\ell) = 0$).

2.3 A Reconstruction Procedure

The linearized problem Let the free vibrations of the reference rod and the perturbed rod be governed by the eigenvalue problems (2.1), (2.4) and (2.13), (2.14), respectively. The coefficients a and ρ are assumed to satisfy conditions (2.2) and (2.3), respectively. In this section, the inverse problem of determining the perturbation b_ϵ of the axial stiffness from measurements of the changes in the first M natural frequencies will be considered. The coefficient b_ϵ is assumed to satisfy (2.7)-(2.9) and, in addition, the *a priori* condition

$$\text{supp } b_\epsilon(x) \equiv \overline{\{x \in (0,\ell) \mid b_\epsilon(x) \neq 0\}} \subset \left(0, \frac{\ell}{2}\right). \tag{2.21}$$

The above condition plays an important role in the present study. It should be noticed that there are situations important for applications in which (2.21) appears as a rather natural assumption. For example, if the reference beam is symmetrical with respect to $x = \ell/2$, then the eigenvalues $\lambda_{m\epsilon}(b_{1\epsilon})$, $\lambda_{m\epsilon}(b_{2\epsilon})$ corresponding to two perturbations $b_{1\epsilon}(x)$, $b_{2\epsilon}(x)$ symmetrical with respect to $x = \ell/2$, e.g. $b_{1\epsilon}(\ell - x) = b_{2\epsilon}(x)$ in $[0,\ell]$, and such that supp $b_{1\epsilon} \subset (0,\ell/2)$, are exactly the same for every $m = 1, 2, \ldots$ Loosely speaking, one can say that the Neumann spectrum cannot distinguish left from the right. To avoid the indeterminacy due to the structural symmetry, condition (2.21) will be assumed to hold. In practical diagnostic applications, (2.21) is equivalent to a priori know that the damage is located on an half of the rod, see, for example, Davini et al.

(1993) and Davini et al. (1995) for applications via variational methods. It should be mentioned that diagnostic techniques based on mode shape measurements (see Yuen (1985) and Rizos et al. (1990)), node measurements (Section 4), simultaneous use of resonance and antiresonances (Section 5.1) have been recently proposed to avoid the non-uniqueness of the damage location problem in symmetric beam structures.

In order to illustrate the reconstruction procedure, the comparatively simple example of an initially uniform rod, with $a = const.$ and $\rho = const.$ in $[0, \ell]$, is first considered. The eigenpairs of the reference rod are given by

$$u_m(x) = \sqrt{\frac{2}{\rho \ell}} \cos \frac{m\pi x}{\ell}, \qquad \lambda_m = \frac{a}{\rho}\left(\frac{m\pi}{\ell}\right)^2, \quad m = 1, 2, \dots . \tag{2.22}$$

The rigid mode $u_0(x)$ is always insensitive to damage and, therefore, it will be omitted in the sequel. Putting the expressions of λ_m and $u_m(x)$ for $m \geq 1$ into equation (2.15) gives

$$\lambda_{m\epsilon} - \lambda_m = \left(\frac{m\pi}{\ell}\right)^2 \left(\frac{2}{\rho\ell}\right) \int_0^\ell b_\epsilon(x) \sin^2 \frac{m\pi x}{\ell} dx + r(\epsilon, m), \quad m = 1, 2, ..., \tag{2.23}$$

where $r(\epsilon, m)$ is a higher order term on ϵ, see condition (2.16).

The family $\{\Phi_m(x)\}_{m=1}^\infty$, with $\Phi_m(x) = \frac{(u_m'(x))^2}{\lambda_m} = \frac{2}{a\ell} \sin^2 \frac{m\pi x}{\ell}$ is a basis for the square summable functions defined on $(0, \ell/2)$. This means that any function f, with $f : [0, \ell/2] \to \mathbb{R}$ and f regular enough, can be expressed by the series

$$f(x) = \sum_{m=1}^\infty f_m \Phi_m(x), \tag{2.24}$$

where f_m is the mth *generalized Fourier coefficient* of the function f evaluated on the family $\{\Phi_m(x)\}_{m=1}^\infty$.

By neglecting, as a first approximation, the higher order term $r(\epsilon, m)$ in the asymptotic development of the mth eigenvalue and expressing b_ϵ in terms of the functions $\{\Phi_m(x)\}_{m=1}^\infty$, that is

$$b_\epsilon(x) = \sum_{k=1}^\infty \beta_{\epsilon k} \Phi_k(x), \tag{2.25}$$

one has

$$\delta\lambda_{m\epsilon} = \sum_{k=1}^\infty A_{mk} \beta_{\epsilon k}, \qquad m = 1, 2, ..., \tag{2.26}$$

where

$$\delta\lambda_{m\epsilon} \equiv \frac{\lambda_{m\epsilon} - \lambda_m}{\lambda_m}, \qquad m = 1, 2, ..., \tag{2.27}$$

$$A_{mk} \equiv \int_0^{\ell/2} \Phi_m \Phi_k dx = \frac{4}{a^2 \ell^2} \int_0^{\ell/2} \sin^2 \frac{m\pi x}{\ell} \sin^2 \frac{k\pi x}{\ell}, \quad k, m = 1, 2, \dots . \tag{2.28}$$

A direct calculation shows that

$$A_{mk} = \frac{2}{4a^2\ell} \text{ for } k \neq m, \quad A_{mk} = \frac{3}{4a^2\ell} \text{ for } k = m. \tag{2.29}$$

In real applications, only the eigenvalues of the first few vibrating modes are available. In fact, the number M typically ranges from $3-4$ to 10. Therefore, rather than studying the solution of the infinite linear system (2.26), the following analysis will be focussed on its M-*approximation*, that is the $M \times M$ linear system formed by

$$\delta\lambda_{m\epsilon} = \sum_{k=1}^{M} A_{mk}^{M} \beta_{\epsilon k}^{M}, \qquad m = 1, ..., M, \tag{2.30}$$

where $A_{mk}^{M} = A_{mk}$ for $k, m = 1, ..., M$, and $\{\beta_{\epsilon k}^{M}\}_{k=1}^{M}$ are the coefficients of the M-approximation of $b_{\epsilon}(x)$ evaluated on the family $\{\Phi_m(x)\}_{m=1}^{\infty}$.

A direct calculation shows that

$$\det A_{mk}^{M} = (2M+1)\left(\frac{1}{4a^2\ell}\right)^{M}, \tag{2.31}$$

$$(A_{mk}^{M})^{-1} = (4a^2\ell)\frac{2M-1}{2M+1} \quad \text{if } m = k, \quad (A_{mk}^{M})^{-1} = -(4a^2\ell)\frac{2}{2M+1} \quad \text{if } m \neq k, \tag{2.32}$$

$m, k = 1, ..., M$. Hence, the solution of (2.30) has the following explicit expression

$$\beta_{\epsilon k}^{M} = 4a^2\ell\left(\frac{2M-1}{2M+1}\delta\lambda_{k\epsilon} - \frac{2}{2M+1}\sum_{j=1, j\neq k}^{M} \delta\lambda_{j\epsilon}\right), \quad k = 1, ..., M, \tag{2.33}$$

and, going back to (2.25), the first order stiffness variation is given by

$$b_{\epsilon}(x) = 8a\sum_{k=1}^{M}\left(\frac{2M-1}{2M+1}\delta\lambda_{k\epsilon} - \frac{2}{2M+1}\sum_{j=1, j\neq k}^{M} \delta\lambda_{j\epsilon}\right)\sin^2\frac{k\pi x}{\ell}. \tag{2.34}$$

Expressions (2.33), (2.34) clarify how the relative eigenvalue shifts influence the various Fourier coefficients of the stiffness variation $b_{\epsilon}(x)$. Assuming that the relative eigenvalue shifts are, in average, all of the same order, it can be deduced from (2.33) that for relatively large values of M (starting from $M = 3-4$, for example), the kth Fourier coefficient $\beta_{\epsilon k}^{M}$ is mainly influenced by the variation of the corresponding kth eigenvalue. In fact, for a given k and, for example, for $M = 4$, the coefficient which multiplies $\delta\lambda_{k\epsilon}$ is equal to 0.78 about, whereas the coefficients of the remaining eigenvalue changes $\delta\lambda_j$, $j \neq k$, take the lower value 0.22. This difference becomes significant as M increases.

An iterative procedure and a numerical algorithm The above analysis is based on a linearization of Taylor's series expansion (2.23) for the eigenvalues of the perturbed rod. Therefore, the coefficient b_{ϵ} found by (2.34) does not satisfy identically equations (2.23). The estimation of b_{ϵ} can be improved by repeating the procedure shown above starting from the updated configuration $a^{(1)} = a + b_{\epsilon}$, with b_{ϵ} as calculated at the previous step.

This suggests the following iterative procedure for solving the inverse problem. The index ϵ will be omitted in this part to simplify the notation. Moreover, $\widetilde{\lambda}_m$ denotes the mth eigenvalue $\lambda_{m\epsilon}$ of the perturbed rod.

ITERATIVE PROCEDURE AND NUMERICAL ALGORITHM:
1) Let $a^{(0)}(x) = a(x)$, where $a(x)$ is the axial stiffness of the reference rod.
2) For $s = 0, 1, 2, ...$:
a) solve the linear system

$$\frac{\widetilde{\lambda}_m - \lambda_m^{(s)}}{\lambda_m^{(s)}} = \sum_{k=1}^{M} A_{mk}^M \beta_k^{M(s)}, \qquad m = 1, ..., M, \tag{2.35}$$

where $(\lambda_m^{(s)}, u_m^{(s)})$ is the mth normalized eigenpair of the problem

$$(a^{(s)}u')' + \lambda^{(s)}\rho u = 0 \qquad \text{in } (0, \ell), \tag{2.36}$$

$$a^{(s)}(0)u'(0) = 0 = a^{(s)}(\ell)u'(\ell). \tag{2.37}$$

The numbers $\{\beta_k^{M(s)}\}_{k=1}^M$ are the generalized Fourier coefficients of the unknown function $b^{(s)}(x)$, $b^{(s)}(x) = \sum_{k=1}^M \beta_k^{M(s)}\Phi_k^{(s)}$, and the matrix entries $A_{mk}^{M(s)}$ are given by

$$A_{mk}^{M(s)} = \int_0^{\ell/2} \Phi_m^{(s)}\Phi_k^{(s)} dx, \qquad m, k = 1, ..., M. \tag{2.38}$$

b) Update the coefficient $a(x)$:

$$a^{(s+1)}(x) = a^{(s)}(x) + b^{(s)}(x) \quad \text{in } [0, \ell/2]. \tag{2.39}$$

c) If the updated coefficient satisfies the condition

$$\frac{1}{M}\sum_{m=1}^{M}\left(\frac{\widetilde{\lambda}_m - \lambda_m^{(s+1)}}{\widetilde{\lambda}_m}\right)^2 < \gamma, \tag{2.40}$$

for a small given control parameter γ, then stop the iterations. Otherwise, go to step 2) and repeat the procedure.

With the exception of simple cases corresponding to special stiffness coefficients, e.g. $a(x) = const.$ in $[0, \ell]$, the eigenvalue problem (2.36)-(2.37) does not admits closed form eigensolutions. Therefore, for the practical implementation of the identification algorithm resort to numerical analysis in order. The procedure herein adopted is based on a finite element model of the rod with uniform mesh and linear displacement shape functions. The stiffness and mass coefficients are approximated by step functions, that is $a(x) = a_e = const.$, $\rho(x) = \rho_e = const.$ within the eth finite element. The local mass and stiffness matrices are given by

$$M_e = \rho_e\Delta\begin{pmatrix} 1/3 & 1/6 \\ 1/6 & 1/3 \end{pmatrix}, \quad K_e = a_e\Delta^{-1}\begin{pmatrix} 1 & -1 \\ -1 & 1 \end{pmatrix}, \tag{2.41}$$

where Δ is the element length. The discrete approximation of the eigenvalue problem (2.36)-(2.37) was solved by the Stodola-Vianello method. The derivative of the eigenfunctions was evaluated by using a finite difference scheme and the numerical integration was developed with a trapezoidal method.

In solving the linear system (2.35), the determination of the inverse of the matrix $A_{mk}^{M(s)}$ at each step s, $s = 1, 2, ...,$ is needed. If $s = 0$, then $\det A_{mk}^{M(0)} = (2M+1)(4a^2\ell)^{-M}$ by (2.31) and the inverse of the matrix $A_{mk}^{M(s)}$ exists. At the first step of the iteration scheme, $s = 1$, by (2.38) and recalling that

$$u_{m\epsilon}^{(1)} = u_m^{(0)} + \delta u_{m\epsilon}, \tag{2.42}$$

where $\delta u_{m\epsilon}$ is a *small* perturbation term such that $\|\delta u_{m\epsilon}\|_{H^1} \to 0$ as $\epsilon \to 0$, it turns out that

$$A_{mk}^{M(1)} = A_{mk}^{M(0)} + \delta A_{mk,\epsilon}, \tag{2.43}$$

where $\delta A_{mk,\epsilon} \to 0$ as $\epsilon \to 0$. Therefore, one can conclude that

$$\det A_{mk}^{M(1)} = \det A_{mk}^{M(0)} + \text{ small terms as } \epsilon \to 0, \tag{2.44}$$

and the inverse of the matrix $A_{mk}^{M(1)}$ is well defined. By proceeding step by step and within the assumption that the unknown stiffness coefficient is a perturbation of the initial one, the inverse of the matrix $A_{mk}^{M(s)}$ is well defined.

If, during the iterative procedure, the coefficient $a^{(s+1)}$ violates the ellipticity condition (2.8), then the perturbation $b_\epsilon^{(s)}$ is multiplied by a suitable step size $\alpha^{(s)}$, typically $\alpha^{(s)} = 1/2$, to obtain an updated coefficient satisfying (2.8) with $a_0 = \frac{1}{100} \min_{x\in[0,\ell]} a^{(0)}(x)$. This procedure is repeated at most five times during each step of the iterative process. After that, the iterations are stopped and the current stiffness distribution is taken as solution of the reconstruction procedure. Analogous considerations hold concerning the upper bound (2.8) with $A_0 = 2 \max_{x\in[0,\ell]} a^{(0)}(x)$.

The small parameter of the convergence criterion (2.40) is taken as $\gamma = 1.0 \cdot 10^{-12}$ and an upper bound of 50 iterations was introduced.

2.4 Applications

The reconstruction procedure presented in the previous section is applied to identify stiffness variations caused by localized damages in longitudinally vibrating beams. The principal results of identification are summarized in the sequel.

The experimental models are the two bars under free-free boundary conditions shown in Figure 1. Every specimen was damaged by saw-cutting the transversal cross-section. The width of each notch was approximately equal to 1.5 mm and, because of the small level of the excitation, during the dynamic tests each notch remains always open.

In the first experiment, the steel rod of series $HE100B$ (rod 1) shown in Figure 1(a) was considered, see Morassi (1997) for more details on dynamic testing. By using an impulsive dynamic technique, the first nine natural frequencies of the undamaged bar and of the bar under a series of two damage configurations ($D1$ and $D2$) were determined. The rod was suspended by two steel wire ropes to simulate free-free boundary

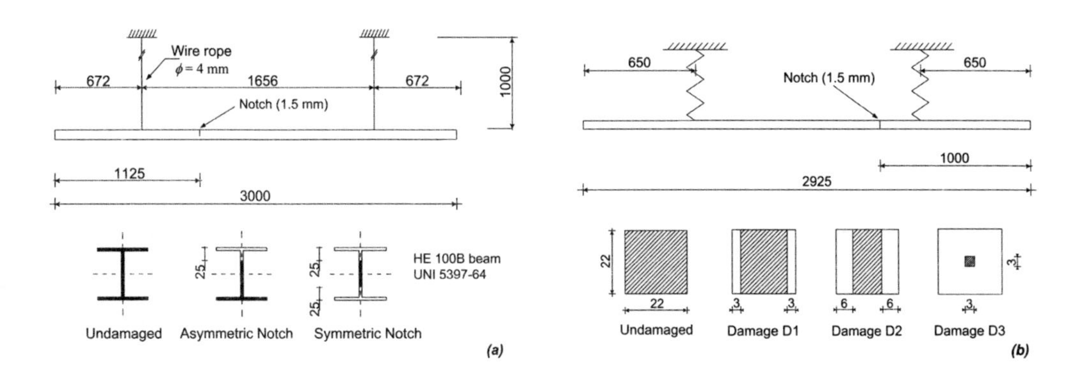

Figure 1. (a)-(b). Experimental models and damage configurations: (a) rod 1; (b) rod 2. Lengths in mm.

conditions. The excitation was introduced at one end by means of an impulse force hammer, while the axial response was measured by a piezoelectric accelerometer fixed at the centre of an end cross-section of the rod. Vibration signals were acquired by a dynamic analyzer HP35650 and then determined in the frequency domain to measure the relevant frequency response term (inertance). The well-separated vibration modes and the very small damping allowed identification of the natural frequencies by means of *single mode technique*. The damage configurations were obtained by introducing a notch of increasing depth at $s = 1.125$ m from one end. Table 1 compares the first nine experimental natural frequencies for the undamaged and damaged rod. The analytical model of the undamaged configuration generally fits well with the real case and the percentage errors are lower than 1% within the measured modes. The eigenfrequency shifts induced by the damage are relatively large with respect to the modelling errors and rod 1 provides an example for which the damage is rather severe from the beginning.

The rod was discretized in 200 equally spaced finite elements and the identification procedure was applied by considering an increasing number of natural frequencies M, $M = 1, ..., 9$. The chosen finite element mesh guarantees for the presence of negligible discretization errors during the identification process. Figures 2 and 3 show the identified stiffness coefficient when $M = 3, 5, 7, 9$ natural frequencies are considered in identification, for damage $D1$ and $D2$, respectively. Convergence of the iterative process seems to be rather fast and, typically, less than 10 iterations are sufficient to reach the optimal solution.

As it was expected from the representation formula (2.25), the reconstruction coefficient shows a wavy behavior around the reference (constant) value a_0, see Figures 2 and 3. The maximum values of the positive increments are, in some cases, comparable with the maximum reduction in stiffness, which occurs near the actual damage location $s = 1.125$ m. However, the extent of the regions with positive change in stiffness becomes less important as the number of frequencies M increases and when more severe levels of

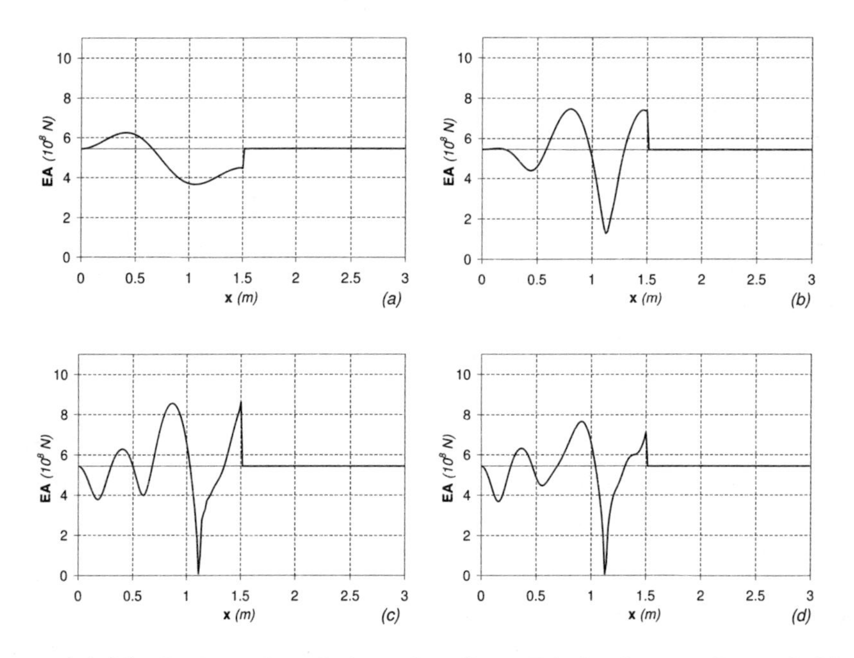

Figure 2. (a)-(d). Rod 1: identified axial stiffness EA for damage D1 with M=3 (a), M=5 (b), M=7 (c) and M=9 (d) frequencies. Actual damage location s=1.125 m.

Table 1. Experimental frequencies of rod 1 and analytical values for the undamaged configuration (the rigid body motion is omitted). Undamaged configuration: $EA = 5.4454 \times 10^8$ N, $\rho = 20.4$ kg/m, $\ell = 3.0$ m; $\Delta_n\% = 100 \cdot (f_n^{model} - f_n^{exp})/f_n^{exp}$. Damage scenarios $D1$ and $D2$; abscissa of the cracked cross-section $s = 1.125$ m; $\Delta_n\% = 100 \cdot (f_n^{dam} - f_n^{undam})/f_n^{undam}$. Frequency values in Hz.

Mode	Undamaged			Damage D1		Damage D2	
n	Exper.	Model	$\Delta_n\%$	Exper.	$\Delta_n\%$	Exper.	$\Delta_n\%$
1	861.4	861.1	0.00	805.7	-6.17	737.6	-14.37
2	1722.2	1722.2	0.00	1664.5	-3.35	1600.0	-7.10
3	2582.9	2583.3	0.02	2541.9	-1.59	2505.3	-3.00
4	3434.2	3444.4	0.30	3162.2	-7.92	3016.0	-12.18
5	4353.6	4305.5	-1.10	4332.2	-0.49	4310.2	-1.00
6	5174.4	5166.6	-0.15	4961.1	-4.12	4812.6	-6.99
7	6020.0	6027.7	0.13	5750.2	-4.48	5616.0	-6.71
8	6870.5	6888.8	0.27	6860.2	-0.15	6851.3	-0.27
9	7726.4	7749.9	0.30	7302.3	-5.49	7095.8	-8.16

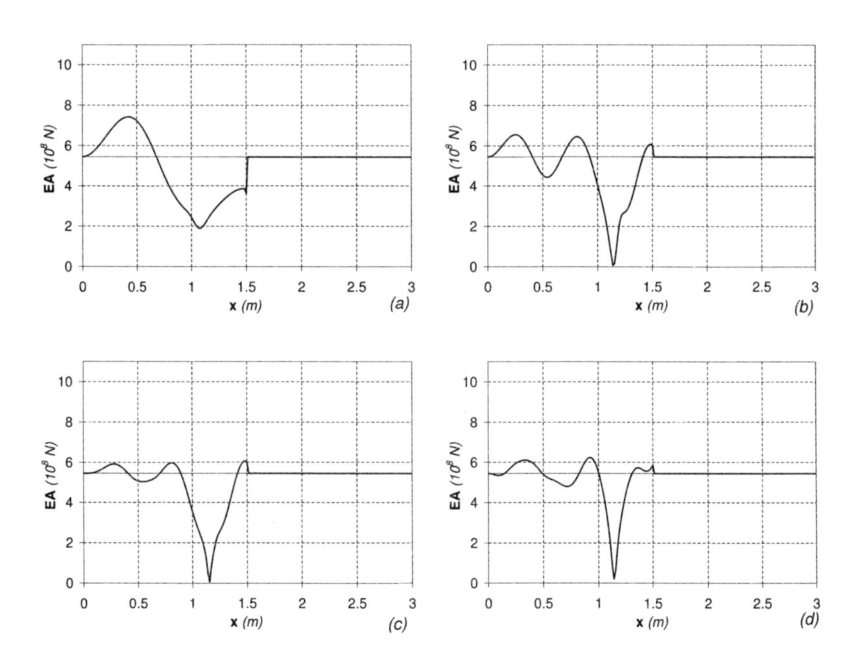

Figure 3. (a)-(d). Rod 1: identified axial stiffness EA for damage D2 with M=3 (a), M=5 (b), M=7 (c) and M=9 (d) frequencies. Actual damage location s=1.125 m.

damage are considered in the analysis.

From Figures 2 and 3 it can been seen that the reconstructed coefficient can give an indication where the damage is located. The results of identification can be slightly improved by recalling that, from the physical point of view, the coefficient $a_{dam}(x)$ clearly cannot be greater than the reference value $a_0(x)$. This suggests to *a posteriori* set the identified coefficient to be equal to $a^{(0)}(x)$ wherever $a_{dam}(x) > a^{(0)}(x)$, see also Wu (1994).

The results of most diagnostic techniques based on dynamic data strictly depend on the accuracy of the analytical model considered for the interpretation of the measurements and the severity of the damage to be identified. Rod 1 provides an example for which the analytical model (of the reference system) is very accurate and for which the damage is rather severe from the beginning. Therefore, in order to study the sensitivity of the proposed reconstruction procedure to small levels of damage, in the second experiment a steel rod of square solid cross-section with a small crack was considered (rod 2). By adopting an experimental technique similar to that used for rod 1, the undamaged bar and three damaged configurations obtained by introducing a notch of increasing depth at $s = 1.000$ m from one end, see Figure 1(b).

The analytical model turns out to be extremely accurate with percentage errors less than those of the first experiment and lower then 0.2% within the first twenty vibrating modes, cf. Table 2.

The percentage of frequency shifts caused by the damage are of order 0.1% and 0.3 –

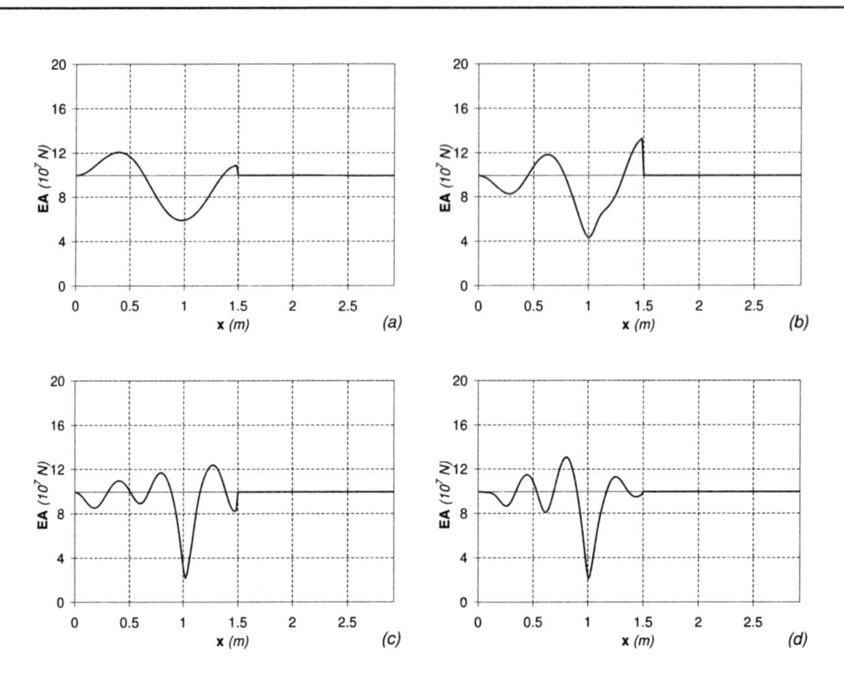

Figure 4. (a)-(d). Rod 2: identified axial stiffness EA for damage D3 with M=3 (a), M=5 (b), M=7 (c) and M=9 (d) frequencies. Actual damage location s=1.000 m.

0.4% for damage $D1$ and $D2$, respectively. Therefore, for these two configurations it is expected that modelling errors could mask the changes induced by damage. This behavior is confirmed by the identification results, see Morassi (2001). Conversely, Figures 4, 5 show that damage $D3$ is clearly identified when the first $3-5$ frequencies are measured. In this case, the results show a good stability of the identification when an increasing number of frequencies is considered in the analysis.

Finally, the diagnostic technique was also tested on free-free longitudinally vibrating beams with multiple localized damages. We refer to Morassi (2007) for more details.

2.5 The Bending Vibration Case

In the previous sections, the problem of identifying the stiffness change induced by a damage in an axially vibrating beam from frequency measurements has been discussed. Here the corresponding problem for a beam in bending vibration is considered.

The reconstruction procedure in the bending case The physical model, which will be investigate, is a simply supported Euler-Bernoulli beam. The undamped free vibration of the undamaged beam are governed by the boundary value problem

$$\begin{cases} (jv'')'' - \lambda \rho v = 0 & \text{in } [0, \ell], \\ v(0) = 0 = v(\ell), \\ j(0)v''(0) = 0 = j(\ell)v''(\ell), \end{cases} \tag{2.45}$$

Table 2. Experimental frequencies of rod 2 and analytical values for the undamaged configuration (the rigid body motion is omitted). Undamaged configuration: $EA = 9.9491 \times 10^7$ N, $\rho = 3.735$ kg/m, $\ell = 2.925$ m; $\Delta_n\% = 100 \cdot (f_n^{model} - f_n^{exp})/f_n^{exp}$. Damage scenarios $D1$, $D2$ and $D3$; abscissa of the cracked cross-section $s = 1.000$ m; $\Delta_n\% = 100 \cdot (f_n^{dam} - f_n^{undam})/f_n^{undam}$. Frequency values in Hz.

Mode	Undamaged			Damage D1		Damage D2		Damage D3	
n	Exper.	Model	$\Delta_n\%$	Exper.	$\Delta_n\%$	Exper.	$\Delta_n\%$	Exper.	$\Delta_n\%$
1	882.25	882.25	0.00	881.5	-0.09	879.3	-0.33	831.0	-5.81
2	1764.6	1764.5	-0.01	1763.3	-0.07	1759.0	-0.32	1679.5	-4.82
3	2645.8	2646.8	0.04	2644.0	-0.07	2647.0	0.05	2646.5	0.03
4	3530.3	3529.0	-0.04	3526.8	-0.10	3516.5	-0.39	3306.0	-6.35
5	4411.9	4411.3	-0.01	4408.8	-0.07	4400.0	-0.27	4250.0	-3.67
6	5293.9	5293.5	-0.01	5294.3	0.01	5295.3	0.03	5287.8	-0.12
7	6175.4	6175.8	0.01	6168.8	-0.11	6150.3	-0.41	5808.5	-5.94
8	7056.7	7058.0	0.02	7052.0	-0.07	7039.5	-0.24	6864.3	-2.73
9	7937.9	7940.3	0.03	7937.5	-0.01	7938.0	0.00	7909.5	-0.36
10	8819.9	8822.5	0.03	8809.8	-0.11	8782.0	-0.43	8340.0	-5.44
11	9702.7	9704.8	0.02	9697.3	-0.06	9682.8	-0.21	9503.3	-2.06
12	10583.8	10587.0	0.03	10582.8	-0.02	10581.3	-0.02	10514.8	-0.65
13	11464.3	11469.3	0.04	11449.0	-0.13	11410.5	-0.47	10933.5	-4.63
14	12345.2	12351.5	0.05	12339.5	-0.05	12331.5	-0.11	12158.0	-1.52
15	13224.4	13233.8	0.07	13222.8	-0.01	13322.0	+0.74	13098.0	-0.96
16	14104.0	14116.0	0.09	14087.0	-0.12	14039.0	-0.46	13543.0	-3.98
17	14985.0	14998.0	0.09	14979.0	-0.04	14964.0	-0.14	14811.0	-1.16

where $v = v(x)$ is the transversal displacement of the beam, $\sqrt{\lambda}$ is the associated natural frequency and $\rho = \rho(x)$ denotes the linear mass density. The quantity $j = j(x) \equiv EJ(x)$ is the bending stiffness of the beam. E is the Young's modulus of the material and $J = J(x)$ the moment of inertia of the cross-section. The function ρ is assumed to satisfy conditions (2.3). The bending stiffness j is such that

$$j \in C^2([0,\ell]), \quad j(x) \geq j_0 > 0 \quad \text{in } [0,\ell], \tag{2.46}$$

where j_0 is a given constant.

Under the above assumptions on the coefficients, problem (2.45) has an infinite sequence of eigenpairs $\{(v_m, \lambda_m)\}_{m=1}^{\infty}$, with $0 < \lambda_1 < \lambda_2 < ...$, $\lim_{m\to\infty} \lambda_m = \infty$ and where the mth vibration mode is assumed to satisfy the normalization condition $\int_0^\ell \rho v_m^2 = 1$ for every m, $m \geq 1$.

In analogy with the axial case, it is assumed that a structural damage can be described within the classical one-dimensional theory of beams and that it reflects on a reduction of the effective bending stiffness, without introducing changes on the mass distribution. Following the analysis presented in Section 2.2, the bending stiffness of the damaged

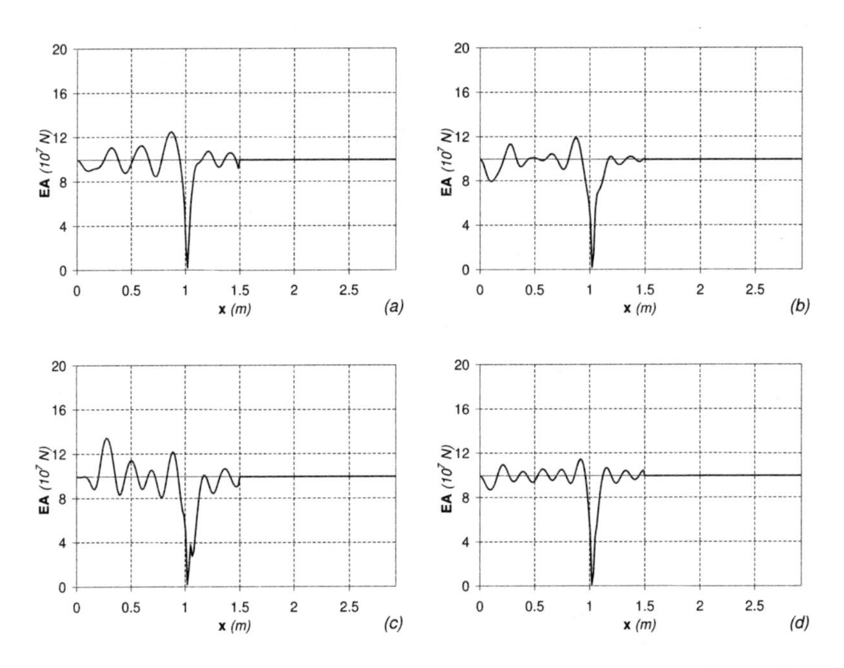

Figure 5. (a)-(d). Rod 2: identified axial stiffness EA for damage D3 with M=11 (a), M=13 (b), M=15 (c) and M=17 (d) frequencies. Actual damage location s=1.000 m.

beam is taken as

$$j_\epsilon(x) = j(x) + b_\epsilon(x), \tag{2.47}$$

where the variation b_ϵ is assumed to satisfy the following conditions:
i) (regularity)

$$b_\epsilon \in C^2([0, \ell]); \tag{2.48}$$

ii) (uniform lower and upper bound of j_ϵ) there exist a constant J_0 such that

$$j_0 \leq j_\epsilon(x) \leq J_0 \quad \text{in } [0, \ell]; \tag{2.49}$$

iii) (smallness)

$$\|b_\epsilon\|_{L^2} = \epsilon O(\|j\|_{L^2}), \tag{2.50}$$

for a real positive number ϵ.

The free bending vibrations of the damaged beam are governed by the eigenvalue problem

$$\begin{cases} (j_\epsilon v_\epsilon'')'' - \lambda_\epsilon \rho v_\epsilon = 0 & \text{in } (0, \ell), \\ v_\epsilon(0) = 0 = v_\epsilon(\ell), \\ j_\epsilon(0) v_\epsilon''(0) = 0 = j_\epsilon(\ell) v_\epsilon''(\ell). \end{cases} \tag{2.51}$$

Under the above assumptions (2.48)-(2.50), the perturbed problem has a sequence of eigenpairs $\{(v_{m\epsilon}, \lambda_{m\epsilon})\}_{m=1}^{\infty}$, with $0 < \lambda_{1\epsilon} < \lambda_{2\epsilon} < ...$ and $\lim_{m\to\infty} \lambda_{m\epsilon} = \infty$. The mth vibration mode is assumed to satisfy the normalization condition $\int_0^\ell \rho v_{m\epsilon}^2 = 1$ for every m, $m \geq 1$, and for every $\epsilon > 0$.

The present analysis will concern with perturbations of the reference beam. This condition is expressed by requiring that

$$\epsilon << 1. \tag{2.52}$$

By applying a technique similar to that shown in Section 2.2 for the longitudinal vibration case, the following asymptotic development for the mth eigenvalue holds:

$$\lambda_{m\epsilon} = \lambda_m + \int_0^\ell b_\epsilon (v_m'')^2 dx + r(\epsilon, m), \qquad m = 1, 2, ..., \tag{2.53}$$

where

$$\lim_{\epsilon\to 0} \frac{|r(\epsilon, m)|}{\|b_\epsilon\|_{L^2}} = 0. \tag{2.54}$$

As in the second order case, the main point of the proof concerns with the asymptotic behavior of the eigensolutions $\{(v_{m\epsilon}, \lambda_{m\epsilon})\}_{m=1}^{\infty}$ as $\epsilon \to 0$, see Morassi (2007) for more details.

The integral term in the right hand side of (2.53) shows that the sensitivity of the mth eigenvalue to variations of the bending stiffness depends on the square of the curvature of the mth vibration mode of the reference beam. The limit case of (2.53) for localized damages, as cracks or notches modelled by an elastic rotational spring inserted at the damaged cross-sections, was considered in Morassi (1993).

The reconstruction procedure based on equations (2.53)-(2.54) will be developed under the additional *a priori* assumption that the stiffness variation occurs on one half of the beam:

$$\text{supp } b_\epsilon(x) \subset (0, \frac{\ell}{2}). \tag{2.55}$$

In the case of an initially uniform beam, for example, the eigenpairs of the undamaged configuration are given by

$$v_m(x) = \sqrt{\frac{2}{\rho\ell}} \sin\frac{m\pi x}{\ell}, \qquad \lambda_m = \frac{j}{\rho}\left(\frac{m\pi}{\ell}\right)^4, \quad m = 1, 2, ... \tag{2.56}$$

Inserting the expressions of v_m and λ_m into equation (2.53) gives

$$\lambda_{m\epsilon} - \lambda_m = \left(\frac{m\pi}{\ell}\right)^4 \left(\frac{2}{\rho\ell}\right) \int_0^\ell b_\epsilon(x) \sin^2\frac{m\pi x}{\ell} dx + r(\epsilon, m), \quad m = 1, 2, ..., \tag{2.57}$$

where $r(\epsilon, m)$ is an higher order term on ϵ. Expressing $b_\epsilon(x)$ in terms of the functions $\{\frac{(v_m'')^2}{\lambda_m}\}_{m=1}^{\infty}$, that is

$$b_\epsilon(x) = \sum_{k=1}^{\infty} \beta_{\epsilon k} \frac{(v_k'')^2}{\lambda_k}, \tag{2.58}$$

and using the linearized form of (2.57), one has

$$\delta\lambda_{m\epsilon} = \sum_{k=1}^{\infty} A_{mk}\beta_{\epsilon k}, \qquad m = 1, 2, ..., \tag{2.59}$$

where

$$\delta\lambda_{m\epsilon} \equiv \frac{\lambda_{m\epsilon} - \lambda_m}{\lambda_m}, \qquad m = 1, 2, ... \tag{2.60}$$

If, as it was made for the axial vibration case, only the first M eigenfrequencies are considered as data in identification, then the M-approximation $\beta_{\epsilon k}^M$ of $\beta_{\epsilon k}$, see Section 2.3, has the explicit expression (2.33) and, finally, the first order approximation of the bending stiffness variation is given by

$$b_\epsilon(x) = 8j \sum_{k=1}^{M} \left(\frac{2M-1}{2M+1}\delta\lambda_{k\epsilon} - \frac{2}{2M+1} \sum_{j=1, j\neq k}^{M} \delta\lambda_{j\epsilon} \right) \sin^2 \frac{k\pi x}{\ell} . \tag{2.61}$$

This completes the study of the linearized inverse problem. The analysis of the general case is based on iterative application of the above linearized approach. The main steps of the numerical algorithm are essentially those already explained in Section 2.3 for the longitudinal vibration case. The numerical code is based on a finite element model of the beam with uniform mesh and cubic displacement shape functions. The stiffness and mass coefficients are approximated with constant value functions within the generic eth finite element. The second derivative of the eigenfunctions was estimated by using a finite difference approximation on the rotational degrees of freedom of the discrete finite element model.

Applications The above reconstruction technique was tested to detect damage on several real beams in bending vibration. The results obtained on a free-free beam with solid square cross-section, beam 1 of Figure 6, are briefly summarized in the sequel. The beam is studied under free-free boundary conditions and the finite element model includes 100 equally spaced finite elements. With this fine mesh, the first lower frequencies of the discrete model are practically indistinguishable from those of the Euler-Bernoulli model.

The specimen was suspended from above by means of two soft springs, so to simulate free-free boundary conditions. It should be recalled that the free-free beam has a double multiplicity zero eigenvalue, corresponding to two independent rigid body motions. These vibrating modes are insensitive to damage and will be omitted in the sequel. The damage consisted of two symmetric notches placed at the cross-section at 0.255 m from the left end, see Figure 6. Their depth was progressively increased by 1 mm at a time from the undamaged configuration to a final level of damage $D6$ corresponding to a depth of 6 mm on both sides of the cross-section. For each level, the lowest seven natural frequencies were measured according to an *impulse technique*, see Davini et al. (1995) for more details on the experiments. Beam 1 provides an example for which the analytical Euler-Bernoulli model is fairly good in the full range of measured frequencies, with percentage deviations which are less than 3% on the range of frequency of interest, cf. Table 3. However,

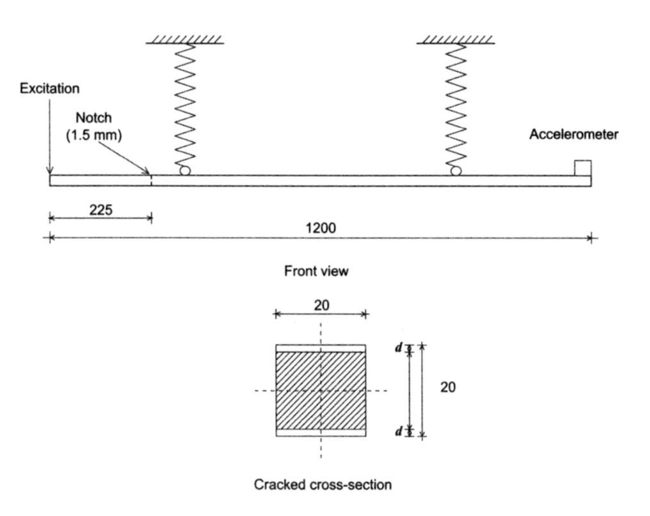

Figure 6. Experimental model of bending vibrating beam (beam 1) and damage configurations. Lengths in mm.

Table 3. Experimental frequencies of beam 1 and analytical values for the undamaged configuration (rigid body motions are omitted). Undamaged configuration: $EJ = 2627.32\ Nm^2$, $\rho = 3.083$ kg/m, $\ell = 1.200$ m; $\Delta_n\% = 100 \cdot (f_n^{model} - f_n^{exp})/f_n^{exp}$. Damage scenarios $D1$-$D6$; abscissa of the cracked cross-section: $s = 0.225$ m; $\Delta_n\% = 100 \cdot (f_n^{dam} - f_n^{undam})/f_n^{undam}$. Frequency values in Hz.

Mode	Undamaged			D1	D2	D3	D4	D5	D6
n	Exper.	Model	$\Delta_n\%$	$\Delta_n\%$	$\Delta_n\%$	$\Delta_n\%$	$\Delta_n\%$	$\Delta_n\%$	$\Delta_n\%$
1	72.19	72.19	0.00	0.00	-0.05	-0.05	-0.18	-0.35	-0.91
2	198.40	198.99	0.30	-0.04	-0.17	-0.36	-0.99	-1.87	-4.61
3	387.73	390.11	0.61	-0.06	-0.23	-0.62	-1.59	-3.29	-6.90
4	639.72	644.87	0.81	-0.05	-0.21	-0.54	-1.39	-2.49	-5.11
5	951.47	963.33	1.25	-0.02	-0.08	-0.15	-0.47	-0.84	-1.71
6	1320.56	1345.47	1.89	0.00	-0.02	-0.02	-0.04	-0.04	-0.11
7	1747.03	1791.30	2.53	-0.01	-0.02	-0.05	-0.24	-0.44	-1.07

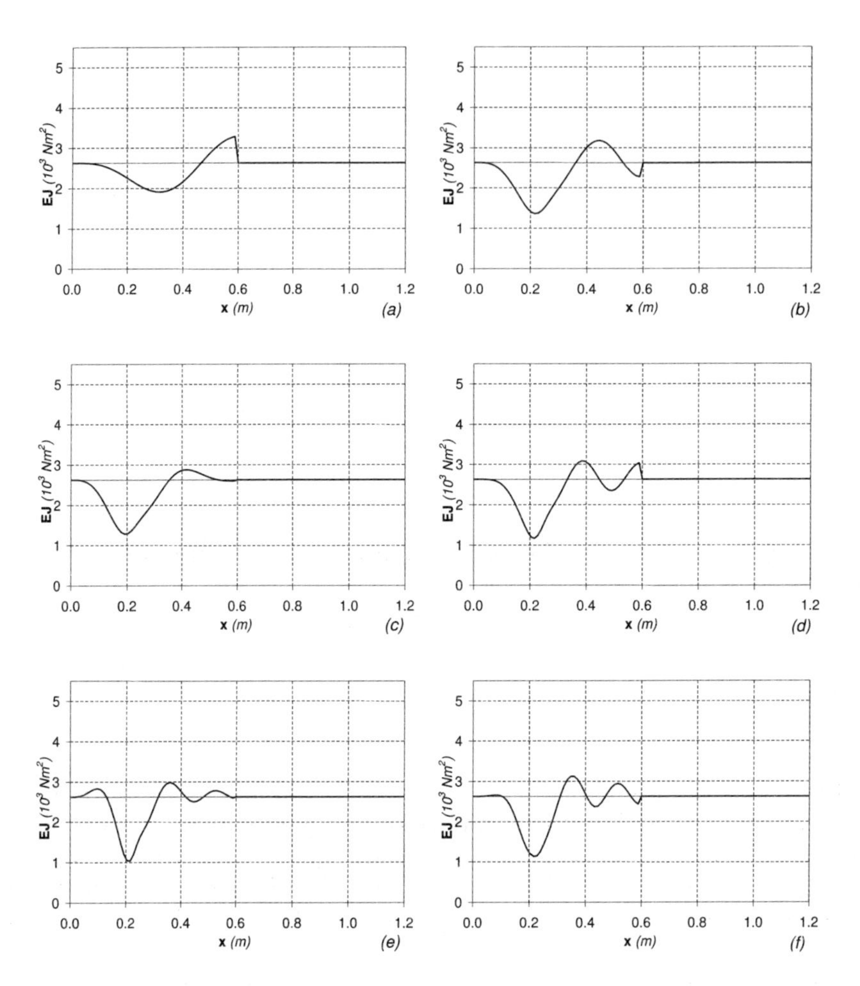

Figure 7. (a)-(f). Beam 1: identified bending stiffness EJ for damage D6 with M=2 (a), M=3 (b), M=4 (c), M=5 (d), M=6 (e) and M=7 (f) frequencies. Actual damage location s=0.225 m.

frequency variations between the undamaged and damaged configurations are very small so that, at least up to the fourth level of damage $D4$, they become mixed up with the modelling errors. In fact, the identification gives poor results up to damage level $D4$, see Morassi (2007). It is worth noting that the use of frequencies $f_5 - f_7$ that are affected by relatively large model errors, as compared with the frequency changes induced by the damage, leads to unsatisfactory stiffness distribution. Starting from level $D5$, a clear tendency emerges to a reduction of stiffness localized around the real position of the damage, cf. Figure 7. Again, it can be shown that use of higher frequencies such as f_6 and f_7 obscures this trend until the damage becomes particularly severe.

2.6 Extensions to Other Boundary Conditions and Variable Coefficients

The analysis presented in Section 2.3 is referred to an initially uniform rod under free-free boundary conditions. Aim of this part is to show how the above results can be extended to include rods under different sets of boundary conditions and rods with initial varying profile.

The longitudinal free vibration of an initially uniform rod under supported (S) boundary conditions is firstly considered. Within the notation of the previous sections, the eigenpairs of the unperturbed rod are given by

$$u_m^S(x) = \sqrt{\frac{2}{\rho \ell}} \sin \frac{m\pi x}{\ell}, \quad \lambda_m^S = \frac{a}{\rho}\left(\frac{m\pi}{\ell}\right)^2, \quad m = 1, 2, \dots . \tag{2.62}$$

Let $\{(u_{m\epsilon}^S, \lambda_{m\epsilon}^S)\}_{m=1}^{\infty}$ be the (normalized) eigenpairs of the perturbed problem

$$(a_\epsilon u_{m\epsilon}^{S\,\prime})' + \lambda_{m\epsilon}^S \rho u_{m\epsilon}^S = 0 \quad \text{in } (0, \ell), \tag{2.63}$$

$$u_{m\epsilon}^S(0) = 0 = u_{m\epsilon}^S(\ell), \tag{2.64}$$

where a_ϵ is defined by (2.6) and $b_\epsilon \equiv a_\epsilon - a$ satisfies conditions (2.7)-(2.9), for a real positive number ϵ.

Putting the expressions of $\lambda_{m\epsilon}^S$ and $u_{m\epsilon}^S$, for $m \geq 1$, into equation (2.15) (see the remarks at the end of section 2.2) gives

$$\lambda_{m\epsilon}^S - \lambda_m^S = \left(\frac{m\pi}{\ell}\right)^2 \left(\frac{2}{\rho\ell}\right) \int_0^\ell b_\epsilon(x) \cos^2 \frac{m\pi x}{\ell} dx + r(\epsilon, m), \quad m = 1, 2, ..., \tag{2.65}$$

where $\lim_{\epsilon \to 0} \frac{|r(\epsilon,m)|}{\epsilon} = 0$. Since the family $\{\Phi_m^S(x)\}_{m=1}^{\infty}$, where $\Phi_m^S(x) = \frac{(u_m^{S\,\prime}(x))^2}{\lambda_m^S} = \frac{2}{a\ell} \cos^2 \frac{m\pi x}{\ell}$, is complete in $L^2(0, \ell/2)$, one can try to find a first approximation of b_ϵ by expressing it as a linear combination of the first M functions $\{\Phi_m^S(x)\}_{m=1}^{M}$, as it was made before. Then, an iterative procedure similar to that shown in Section 2.3 can be used to estimate b_ϵ in terms of the first M eigenfrequency changes induced by the damage.

Passing to another set of boundary conditions, the mth eigenpair of an initially uniform rod with left supported end and free right end (*cantilever* C) is given by

$$u_m^C(x) = \sqrt{\frac{2}{\rho\ell}} \sin \frac{\pi(1 + 2m)x}{2\ell}, \quad \lambda_m^C = \frac{a}{\rho}\left(\frac{\pi}{2\ell}(1 + 2m)\right)^2, \quad m = 0, 1, \dots . \tag{2.66}$$

The family $\{\Phi_m^C(x)\}_{m=0}^{\infty}$, with $\Phi_m^C(x) = \frac{(u_m^{C\,\prime}(x))^2}{\lambda_m^C} = \frac{2}{a\ell}\cos^2\frac{\pi(1+2m)x}{2\ell}$, is complete in $L^2(0,\ell/2)$ and, again, the procedure can be adapted to estimate b_ϵ.

As it should be clear from the above analysis, a crucial point of the proposed procedure concerns the completeness of the family of (suitably scaled) first derivatives-squares of the longitudinal vibration modes in $L^2(0,\ell/2)$. This property easily follows from the explicit expression of the eigenpairs available in the case of a uniform rod. In the remaining of the present section, the general case of varying profile is briefly discussed. To simplify the analysis it is decided to consider the case of a rod with free ends and smooth, uniformly positive coefficients a and ρ. The method to be accounted for can be easily extended in such a way as to take general boundary conditions.

It is worth pointing out that if $(u_m(x), \lambda_m)$ is an eigenpair of the eigenvalue problem (2.1), (2.4), then $(N_m(x) = a(x)u_m'(x), \lambda_m)$ is an eigenpair of the Dirichlet eigenvalue problem

$$(a^* N_m')' + \lambda_m \rho^* N_m = 0 \qquad \text{in } (0,\ell), \tag{2.67}$$

$$N_m(0) = 0 = N_m(\ell), \tag{2.68}$$

wherein $a^* = \rho^{-1}$, $\rho^* = a^{-1}$. Now, by the general result by Borg (1949), the set $\{N_m^2(x)\}_{m=1}^{\infty}$ is complete in $L^2(0,\ell/2)$ and, recalling that $N_m(x) = a(x)u_m'(x)$, this is enough to prove the desired completeness property in the case of varying coefficient.

2.7 A comparison with a variational-type method

In this section, an identification technique based on a variational-type method will be presented and applied to damage detection in beams. The results will be compared with those obtained by the Fourier coefficient procedure illustrated in the previous sections.

The literature of variational-type methods based on eigenfrequency data is extensive, see, for example, the book by Friswell and Mottershead (1995) for a general overview and Adams et al. (1978), Hajela and Soeiro (1990), Shen and Taylor (1991), Liang et al. (1992), Hassiotis and Jeong (1993), Law et al. (1998), Kosmatka and Ricles (1999), Capecchi and Vestroni (2000), Cerri and Vestroni (2000), Vestroni and Capecchi (2000), Ren and De Roeck (2002), Sinha et al. (2002) for specific applications. Here, reference is made to the procedure adopted by Davini et al. (1993) and Davini et al. (1995) in the study of cracked beams by means of discrete models based on a special lumping of the stiffness and inertial properties of the continuous systems, and then extended by Morassi and Rovere (1997) to standard finite element models of beam structures.

Following is an outline of the identification technique in the case of a continuous Euler-Bernoulli beam in free bending vibration. The continuous model of the beam is substituted by a N degree of freedom finite element model, whose free undamped vibrations are governed by the discrete eigenvalue problem

$$\mathbf{K}^N \mathbf{v}_n^N = \lambda_n^N \mathbf{M}^N \mathbf{v}_n^N, \tag{2.69}$$

where $\lambda_n^N \equiv (2\pi f_n^N)^2$ and $\mathbf{v}_n^N$, $\mathbf{v}_n^N \neq \mathbf{0}$, $n = 1, ..., N$, are the eigenvalues and eigenvectors of the discrete system, respectively. As usual, the global stiffness matrix $\mathbf{K}^N$ and the global mass matrix $\mathbf{M}^N$ are obtained by assembling the contribution of all the N elements of the discrete model. In particular

$$\mathbf{K}^N = \sum_{e=1}^{N} \alpha_e \mathbf{K}_e, \tag{2.70}$$

where $\mathbf{K}_e$ is the stiffness matrix of the e-th finite element and $\{\alpha_e\}_{e=1}^N$ is the collection of the "stiffness multipliers".

It is assumed that the presence of a concentrate damage can be described within the framework of Euler-Bernoulli theory of beams and that it reflects into a localized reduction of the effective bending stiffness or, equivalently, a reduction of the multiplier α_e in the whole finite element. Then, one can consider the collection of α_es as descriptive of the stiffness distribution for the damaged system.

The approach to identification is of variational type and the problem becomes the following:

$$\text{to find } \{\alpha_e^{opt}\}_{e=1}^P \in \mathbb{R}^P \text{ such that } F(\alpha_1^{opt}, ..., \alpha_P^{opt}) = \min F(\alpha_1, ..., \alpha_P), \tag{2.71}$$

for $\alpha_e > 0$, $e = 1, ..., P$, where the distance between the first M experimental λ_n^{exp} and analytical λ_n^{model} eigenvalues is given by

$$F(\alpha_1, ..., \alpha_P) = \sum_{n=1}^{M} \left(1 - \frac{\lambda_n^{model}(\alpha_1, ..., \alpha_P)}{\lambda_n^{exp}}\right)^2. \tag{2.72}$$

An iterative algorithm based on an optimal gradient descent method has been used in solving the minimization problem (2.71)-(2.72), see Morassi and Rovere (1997) for more details. Since the variational problem is not convex, the success of the technique crucially depends on the choice of a good initial estimate of the stiffness multipliers to be identified. Here, the stiffness distribution of the undamaged beam has been chosen as initial point in minimization. Finally, the iterations go on until the relative variation of $F(\alpha_e) \equiv F(\alpha_1, ..., \alpha_P)$ and $\{\alpha_1, ..., \alpha_P\}$ at the kth step satisfy a chosen criterion of smallness, namely

$$\left| \frac{F(\alpha_e^{(k+1)}) - F(\alpha_e^{(k)})}{F(\alpha_e^{(k)})} \right| + \sum_{e=1}^{P} \left| \frac{\alpha_e^{(k+1)} - \alpha_e^{(k)}}{\alpha_e^{(k)}} \right| \leq 10^{-6}. \tag{2.73}$$

The variational method has been applied for damage identification in the same experimental models presented in Section 2.4 (axial vibrating rods) and in Section 2.5 (bending vibration beams). In particular, the results obtained on the free-free bending vibrating beam with solid square section of Figure 6 (beam 1 of Section 2.5) are discussed in detail in the sequel. The finite element model includes 100 equally spaced finite elements and, as before, the damage is supposed to occur on the left half of the beam.

The results of identification are fully discussed in Morassi (2007). Generally speaking, it turns out that the optimal stiffness distributions obtained by solving the variational problem (2.71)-(2.72) are very close to those deduced by the *Fourier coefficient method*, see, for example, Figure 8. Similar results have been obtained in studying all the other experimental models (rods 1 and 2 of Section 2.4).

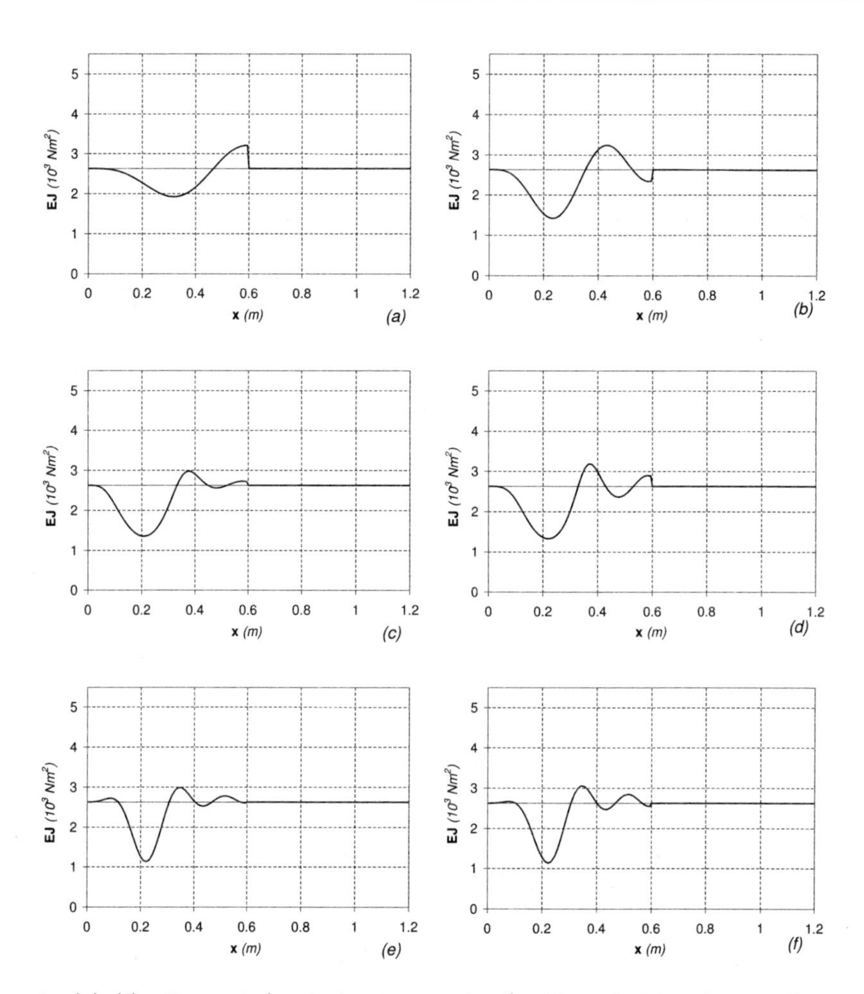

Figure 8. (a)-(f). Beam 1 (optimization method): identified bending stiffness EJ for damage D6 with M=2 (a), M=3 (b), M=4 (c), M=5 (d), M=6 (e) and M=7 (f) frequencies. Actual damage location s=0.225 m.

3 Crack Identification from Changes in Natural Frequencies

3.1 Contents

This section focuses on detecting a single crack in a vibrating rod from the knowledge of the damage-induced shifts in a pair of natural frequencies.

Differently from Section 2, where the identification of a rather general class of damages was discussed, here the damage is assumed to be an open crack. This kind of *a priori* information will play a crucial role in setting and solving the diagnostic problem. In fact, in identifying a crack two things are to be attained: the location of the crack, and its magnitude or severity. It is known that in beam-like structures the change in a natural frequency produced by a small single crack may be represented as the product of two terms, of which the first is proportional to the severity and the second depends solely on the location of the damage. This brings forth an important consequence: the ratios of the change in different natural frequencies depend only on the damage location, not on its severity, see Adams et al. (1978), Gudmundson (1982) and Morassi (1993). Hearn and Testa (1991) and Liang et al. (1992) have used this property for a damage localization analysis in beam-like structures.

Two inverse problems related to crack detection may be posed: (a) Determine the location of a crack from the ratios of the changes in the natural frequencies; (b) Determine the location and the severity of a crack from the changes in natural frequencies. In particular, if the undamaged system is completely known and if the damage is simulated by a linear spring, only two parameters need to be determined, namely the stiffness K_A of the spring and the abscissa s of the cracked cross-section. Such peculiarity in damage detection has been noted more or less explicitly elsewhere in the literature and has been emphasized by Vestroni and Capecchi (1996). Therefore, it is reasonable to investigate to what extent the measurement of the crack-induced changes in a pair of natural frequencies can be useful for identifying the damage. By using this set of data both problems (a) and (b) are generally ill-posed: if the system is symmetrical, then a crack located at any one of a set of symmetrically placed points will produce identical changes in natural frequencies. Even if the system is not symmetrical, cracks in different locations can still produce identical changes to a pair of natural frequencies.

In spite of this ill-posedeness of the diagnostic problem, there are certain situations in which the effects of the non-uniqueness of the solution may be considerably reduced thanks to a careful choice of the data.

Narkis (1994) has shown that if the damaged system is a perturbation of the virgin system, namely whenever the crack is very small, the only information required for accurate crack localization is the variation of the first two natural frequencies caused by the crack. The results were shown for uniform free-free vibrating rods and for uniform simply supported beams in bending. Quite remarkably, Narkis has found out a closed form solution for the crack location. Adams et al. (1978) have come to the same conclusions for uniform free-free rods but have not obtained an explicit relation for the crack location.

The results presented in this section develop those obtained by Narkis (1994) in different directions. It is found that for uniform free-free beams under axial vibration the knowledge of the ratio between the variations of the mth and $2m$th frequencies uniquely determines the position variable $S = \cos \frac{2m\pi s}{\ell}$, wherein s stands for the abscissa of the cracked cross-section and ℓ is the length of the rod. This way, the ensuing peculiar case equal to $m = 1$ agrees with the result achieved by Narkis (1994). Furthermore, the variations of the $2m$th and mth frequencies allow to uniquely determine the stiffness K_A of the damage-simulating elastic spring, too. In both cases simple closed form expressions are deduced for S and for K_A. As for cantilevers or beams with fixed ends it is borne out that by simultaneously employing axial frequencies related to different boundary conditions it is still possible to uniquely determine the damage parameters S and K_A. The explicit expression for the damage sensitivity of natural frequencies made out in Morassi (1993) plays a crucial role in our analysis. As a matter of fact, the methodology that has been resorted to does not call for the explicit solution of the eigenvalue problem for the damaged system, as it focuses on the sole knowledge of the eigensolutions corresponding to the integral configuration. Such method is substantially different from Narkis (1994) and, as a result, the procedure that has been put forth can be also extended to the analysis of initially non-uniform cracked beams in axial vibration. The dynamic tests performed on cracked steel rods supported the proposed method for the solution of the diagnostic problem. Analytical results agree well with experimental tests. Finally, a part of the results above are also valid for cracked beams in bending under simply supported or sliding-sliding boundary conditions.

3.2 Axially Vibrating Beams

Let us recall that the spatial variation of the infinitesimal free vibrations of an undamaged straight rod of length ℓ is governed by the differential equation

$$(au')' + \lambda \rho u = 0 \quad \text{in } (0, \ell), \tag{3.1}$$

where $u = u(x)$ describes the mode and $\sqrt{\lambda}$ is the associated natural frequency. In this section, the rod is assumed to have no material damping. As before, the quantities $a = a(x) \equiv EA(x)$ and $\rho = \rho(x)$ denote the axial stiffness and the linear-mass density of the rod. E is Young's modulus of the material and $A(x)$ the cross-section area of the rod. This analysis is concerned with rods for which the axial stiffness is strictly positive and continuously differentiable function; ρ will be assumed to be continuous and strictly positive function. The end conditions are taken as

$$a(0)u'(0) - hu(0) = 0, \tag{3.2}$$

$$a(\ell)u'(\ell) + Hu(\ell) = 0, \tag{3.3}$$

where $h \geq 0$ and $H \geq 0$ are the stiffnesses of the elastic end supports. Three important cases can be distinguished:

$$\text{Supported (S)}: \quad h = \infty = H, \quad u(0) = 0 = u(\ell), \tag{3.4}$$

$$\text{Free (F)}: \quad h = 0 = H, \quad u'(0) = 0 = u'(\ell), \tag{3.5}$$

$$\text{Cantilever (C)}: \quad h = \infty, H = 0, \quad u(0) = 0 = u'(\ell). \tag{3.6}$$

Modes and frequencies are the eigensolutions of the boundary value problem (3.1)–(3.3), and the mth eigenpair of the undamaged rod, $m \geq 0$, is denoted by $\{u_m, \lambda_m\}$. It is well known that for such a and ρ, and end conditions (3.2), (3.3), there is an infinite sequence $\{\lambda_m\}_{m=0}^{\infty}$ such that $0 \leq \lambda_0 < \lambda_1 < \ldots$ with $\lim_{m\to\infty}\lambda_m = \infty$.

Suppose that a crack appears at the cross-section of abscissa $s \in (0, \ell)$. Assuming that the crack remains always open during the longitudinal vibration, by modelling it as a massless translational spring, at $x = s$, see Freund and Herrmann (1976) and Cabib et al. (2001), the eigenvalue problem for the damaged rod is the following

$$(au_d')' + \lambda_d \rho u_d = 0 \quad \text{in } (0, s) \cup (s, \ell), \tag{3.7}$$

where, in addition to the boundary conditions (3.2), (3.3) it is necessary to consider the jump conditions

$$[u_d'(s)] = 0, \qquad K_A[u_d(s)] = a(s)u_d'(s), \tag{3.8}$$

that are to hold at the cross-section where the crack occurs. In equations (3.8), $[\phi(s)] \equiv (\phi(s^+) - \phi(s^-))$ denotes the jump of the function $\phi(x)$ at $x = s$. The expression K_A is the stiffness of the spring simulating the damage, and it can be related to the crack geometry as suggested, for example, by Freund and Herrmann (1976) and Dimarogonas and Paipetis (1983). The undamaged system corresponds to $K_A \to \infty$ or $\epsilon \equiv 1/K_A \to 0$.

The variational formulation of (3.1)–(3.3) shows that eigenvalues of the system are increasing functions of h, H and K_A, and thus decreasing functions of ϵ, so that

$$\lambda_{dm} \leq \lambda_m, \qquad \text{m=0,1,2,...} \tag{3.9}$$

Moreover, as for any constrained system, the following interlacing result holds

$$\lambda_{m-1} \leq \lambda_{dm} \leq \lambda_m, \qquad \text{m=1,2,...} \tag{3.10}$$

Following Morassi (1993), the first order variation of the natural frequencies with respect to ϵ may be found by taking

$$u_d = u + \epsilon\widetilde{u}, \qquad \lambda_d = \lambda + \epsilon\widetilde{\lambda} \tag{3.11}$$

in equations (3.7), (3.8) and (3.2), (3.3). It follows that

$$(a\widetilde{u}')' + \lambda\rho\widetilde{u} + \widetilde{\lambda}\rho u = 0 \quad \text{in } (0, s) \cup (s, \ell), \tag{3.12}$$

$$[\widetilde{u}'(s)] = 0, \qquad [\widetilde{u}(s)] = a(s)u'(s). \tag{3.13}$$

Multiplying (3.1) by $\widetilde{u}$, (3.12) by u and subtracting, and then integrating by parts in the interval $(0, \ell)$, it follows that

$$(a(s)u'(s))^2 + \widetilde{\lambda}\int_0^{\ell} \rho u^2 = 0, \tag{3.14}$$

which, with the normalizing condition $\int_0^{\ell} \rho u^2 = 1$, gives

$$\delta\lambda_m \equiv \epsilon\widetilde{\lambda_m} = -\frac{(a(s)u_m'(s))^2}{K_A}, \qquad \text{m=0,1,2,...} \ . \tag{3.15}$$

Inequalities (3.9), (3.11) and (3.15) show that the eigenvalues change with the damage. In particular, the change in a natural frequency produced by a single crack may be expressed as the product of two terms, the first of which is proportional to the severity and the second depends only on the location of the damage. This second term is the square of the axial force

$$N_m(s) \equiv a(s)u'_m(s) \tag{3.16}$$

in the mth mode shape of the undamaged rod, evaluated at the cracked cross-section.

Equation (3.15) has an important consequence: the ratios of the change in two different natural frequencies depend only on the damage location, not on its severity. This fact was probably first observed by Adams et al. (1978). That is

$$\frac{\delta\lambda_n}{\delta\lambda_m} = \left(\frac{N_n(s)}{N_m(s)}\right)^2 \equiv f(s), \tag{3.17}$$

where $s \in (0, \ell)$ and $\delta\lambda_m < 0$ (if $\delta\lambda_m = 0$ the possible crack locations coincide in the node points of the axial force N_m resulting from the mth vibrating mode of the undamaged rod). A relation similar to (3.17) has been achieved by Narkis (1994) by considering a linearization of the characteristic polynomial explicitly referred to the cracked rod, in case the damaged system should be a perturbation of the undamaged one. Adapting such a method to the analysis of non-uniform beams is likely to be difficult, while the use of the expression (3.17) makes it possible to overcome such hindrance.

The problem related to the crack location from frequency data lies in determining the solutions of the equation (3.17) for a measured value of the ratio $\delta\lambda_n/\delta\lambda_m$. It follows from equation (3.17) that all, and the only possible, locations of the crack are the abscissas of the points of the $f(x) = (N_n(x)/N_m(x))^2$ diagram intersecting with the horizontal straight line drawn parallel to the abscissa axis at a distance equal to the ratio $\delta\lambda_n/\delta\lambda_m$. On a practical level, once the free vibration problem related to the integral rod is solved, the behavior of $f(x)$ is known and therefore it is possible, via numerical method for example, to determine the solutions of equation (3.17).

Some qualitative properties of the crack location problem for initially non-uniform rods were investigated in detail in Morassi (2001). In brief, it was shown that the problem of locating a crack in a vibrating rod from knowledge of the damage-induced shifts in a pair of natural frequencies is generally ill-posed: if the system is symmetrical, then a crack at any one of a set of symmetrically placed points will produce identical changes in natural frequencies. Even if the system is not symmetrical, cracks in different locations can still produce identical changes in a pair of natural frequencies. In spite of this ill-posedeness of the diagnostic problem, it will be shown in the remainder of this section that there are certain situations in which the effects of the non-uniqueness of the solution may be considerably reduced by means of a careful choice of the data.

The simple but very common case of uniform rods, e.g. rods for which $a = a(x)$ and $\rho = \rho(x)$ are constants on the interval $(0, \ell)$ will be considered here. The most exhaustive results concern free vibrating rods (F). C_m^F is denoted by

$$C_m^F \equiv -\frac{\delta\lambda_m^F}{Bm^2}, \tag{3.18}$$

where m is a positive integer and B is the constant

$$B = \left(a\sqrt{\frac{2}{\rho\ell}\frac{\pi}{\ell}} \right)^2 . \tag{3.19}$$

It can be proved that, *if* $C_m^F > 0$, *the measurement of the pair* $\{C_m^F, C_{2m}^F\}$, $m \geq 1$, *uniquely determines the severity of the damage, e.g. the spring stiffness* K_A, *and the variable* $S = cos(2m\pi s/\ell)$ *of the damage location* s.

The eigenpairs of a free uniform rod (F) are given by

$$\lambda_m^F = \frac{a}{\rho}\left(\frac{m\pi}{\ell}\right)^2, \qquad u_m^F(x) = \sqrt{\frac{2}{\rho\ell}}\cos(\frac{m\pi x}{\ell}), \tag{3.20}$$

$m = 0, 1, 2, ...$ The rigid mode $u_0^F(x)$ obviously is always insensitive to damage. Putting the expressions of λ_m^F and $u_m^F(x)$ for $m \geq 1$ into equation (3.15) gives

$$C_m^F = \frac{1}{K_A}\sin^2(\frac{m\pi s}{\ell}) \tag{3.21}$$

and, by using standard trigonometric identities, gives

$$K_A(4C_m^F - C_{2m}^F) = 4K_A^2(C_m^F)^2. \tag{3.22}$$

Since $C_1^F > 0$, from the identity above it follows that

$$4C_1^F - C_2^F > 0, \tag{3.23}$$

and

$$\text{if} \quad C_m^F > 0 \quad \text{then} \quad 4C_m^F - C_{2m}^F > 0, \quad \text{for m} \geq 2. \tag{3.24}$$

Let it be assumed that $C_m^F > 0$. Equation (3.22) can be solved for the damage severity

$$K_A = \frac{1 - \frac{C_{2m}^F}{4C_m^F}}{C_m^F}. \tag{3.25}$$

Note that conditions (3.23), (3.24) guarantee that K_A takes positive values. By inserting the expression (3.25) of K_A into equation (3.21) the damage can be localized

$$S = \frac{C_{2m}^F}{2C_m^F} - 1, \tag{3.26}$$

where $S \in [-1, 1)$ because of inequalities (3.23), (3.24). Note that the ratio of changes in the first two natural frequency is sufficient to localize the damage (except for symmetrical positions). Finally, if $C_m^F = 0$ for a certain $m \geq 2$, then from equation (3.21) it follows that $S = 1$, that is the crack is located in one of the points of zero-sensitivity of the mth mode, and K_A remains undetermined. This establishes the assertion.

The preceding result improves known results about crack localization in different directions. One of the results obtained by Narkis (1994) is the particular case concerning the unique localization of the damage based on the knowledge of the first two frequencies of a free-free vibrating rod.

The expressions (3.25) and (3.26) for the damage parameters indicate that the pair of natural frequencies mth and $2m$th plays a crucial role when localizing the damage. In fact, provided that the mth frequency proves to be sensitive to damage, that is $C_m^F > 0$, the pair $\{C_m^F, C_{2m}^F\}$ uniquely determines the damage severity, namely the stiffness K_A. Quite surprisingly the expression for K_A turns out to be the same for all pairs of values $\{C_m^F, C_{2m}^F\}$. Finally, it is shown that the number of the possible crack locations, corresponding to the same ratio C_{2m}^F/C_m^F, increases as the order m of the modes assessed increases, which accounts for the recourse to "low" frequencies for the problem of damage localization.

The preceding result does not consider the problem that occurs when a pair of values such as $\{C_k^F, C_l^F\}$, with $l \neq 2k$, is chosen as data. It can be shown that in these cases the solution of the inverse problem is generally non-unique (even by leaving symmetrical positions aside), see Morassi (2001) for more details.

The analysis has hitherto been related to uniform beams under axial vibration with free ends and it has been borne out that the first two frequencies allow the crack (except for symmetrical positions) to be uniquely identified. Such a result does not prove true if different boundary conditions, for example (C) or (S), are being considered. For instance, in case (C), when the crack is located in the half of the rod adjacent to the fixed end, there are two distinct locations corresponding to the same ratio between the variations in the first two frequencies, as already asserted by Narkis (1994). In these situations, an alternative method of proceeding lies in resorting at the same time to frequency measurements on the cracked rod which derive from different boundary conditions.

The attention is concentrated again on initially uniform rods and the data resulting from boundary conditions of the types (S) and (F) are first considered. It can be proved that *the measurement of the* $(m+1)$*th frequency in the cracked rod under boundary conditions of the type* (F) *and of the* m*th frequency under boundary conditions of type* (S)*, for* $m \geq 0$*, determines uniquely the severity of the crack and the location variable* $S = cos(2(m+1)\pi s/\ell)$*, where* s *stands for the abscissa of the cracked cross-section.* Defining $C_m^S \equiv -\delta\lambda_m^S/B(m+1)^2$, $m = 0, 1, 2, ...$, it follows that

$$K_A = \frac{1}{C_{m+1}^F + C_m^S}, \qquad S = -1 + \frac{2}{1 + \frac{C_{m+1}^F}{C_m^S}}. \tag{3.27}$$

(If $C_m^S = 0$ then $S' = -1$.) In particular, it turns out that the damage is uniquely determined (except for symmetrical positions) by the measurement of the pair $\{C_0^S, C_1^F\}$.

A similar result holds true also when frequency measurements are derived from (F) and (C) boundary conditions. In this case it can be shown that *from the knowledge of the* m*th frequency in the cracked rod under boundary conditions* (C) *and of the* $(1+2m)$*th frequency under boundary conditions* (F) *it is possible to uniquely determine the stiffness* K_A *and the position variable* $S = cos((1+2m)\pi s/\ell)$, $m = 0, 1, 2, ...$ By proceeding as exemplified above, if $C_m^S > 0$ then it follows that

$$K_A = \frac{1 - \frac{C_{1+2m}^F}{4C_m^C}}{C_m^C}, \qquad S = 1 - \frac{C_{1+2m}^F}{2C_m^C}, \tag{3.28}$$

for $m = 0, 1, 2, ...$

3.3 Applications

In the preceding section it has been elucidated how to employ the measurement of a pair of axial frequencies of a cracked rod so as to asses the location as well as the severity of the damage. Aiming to account for the prospective practical use of the results above within the analysis of real cases, the present section is devoted to outlining some applications of experimental character.

Before bringing forward the results it is seen fit to make some remarks. The conclusions drawn in the Section 3.2, and the respective identification technique, have been inferred from qualitative and quantitative properties underlying the analytical model ruled by the equations (3.7)-(3.8) for the cracked rod in the case of not severe damage. At this point, it is known that as a rule the mono-dimensional analytical model, which is based on the classical theory regarding beams under axial vibration and on the macroscopic description of the notch, provides an efficient assessing of the frequencies in the lower section of the spectrum, whereas it gradually lacks accuracy on increasing of the order of vibrating modes. Such feature suggests that the employment of lower frequencies in the application, for example, of formulas (3.25), (3.26) should guarantee a more accurate assessing of the damage parameters. From this point of view, opting the first pair of frequencies turns out to be optimal, even though under certain circumstances (see, for example, the second experiment in Morassi (2001), one of the frequencies in question may be subject to serious modelling errors. If such is the case, it is safer to resort to frequency pairs of higher order also.

The second aspect concerns the perturbative character of the present analysis. The work hypothesis, as stated beforehand in Section 3.2, is that the crack should be small, namely the cracked configuration should be a perturbation of the undamaged one. If such hypothesis is reasonable from the practical point of view, in that it is crucial to be in a position to identify the damage right as it arises, on the other hand the crack-induced variations in the frequencies prove to be small only in case of low vibrating modes. As a matter of fact, it can be noted in expression (3.15) that the frequency variation caused by the crack increases as the order of the mode does. Such a mutual relation is in agreement with a general property proved in Morassi (1997) according to which the high frequency spectrum of the cracked rod splits in two branches corresponding to the asymptotic course of the spectrum of the two rod segments adjacent to the cracked cross-section (under suitable boundary conditions). The asymptotic separation of the spectrum is more heavily marked in case of severe damage. Also the latter aspect suggests the employment of lower frequencies for the damage identification.

Summing up, the technique applied to real cases is expected to yield the more reliable results when the less severe the damage and, in the meantime, the lower the order of the frequencies considered are. Furthermore, it is worth noting that a relevant factor is

Table 4. Determination of the crack location in rod 1 by using the pair $\{C^F_m, C^F_{2m}\}$, $m = 1 - 3$, as data in formula (3.25). Experimental (s_{exper}) estimates of the crack location in the interval $(0, \ell/2)$. Actual crack location $s = 1.125$ m.

m	Damage D1	Damage D2
	s_{exper}	s_{exper}
1	1.146	1.150
2	0.340	0.416
	1.160	1.084
3	0.204	0.227
	0.796	0.773
	1.204	1.227

Table 5. Determination of the spring stiffness K_A in rod 1 by using the pair $\{C^F_m, C^F_{2m}\}$, $m = 1 - 4$, as data in formula (3.26). Experimental (K_{exper}) estimates of the spring stiffness. Stiffness values in N/m.

m	Damage D1 $K_{exper} \times 10^9$	Damage D2 $K_{exper} \times 10^9$
1	2.51753	11.88132
2	2.34800	15.50771
3	4.11534	26.26323
4	2.38855	15.86864
Actual value	2.28723	9.43470

represented by possible measurement and modelling errors the technique, as shall be seen, seems to be sensitive to. Here, an application to a cracked axially vibrating bar will be presented, see Morassi (2001) for more details. The experimental model is the steel bar under free-free (F) boundary conditions shown in Figure 1(a) (rod 1 of Section 2.4). The specimen was damaged by saw-cutting the transversal cross-section and the width of each notch was equal to 1.5 millimeters. Because of the small level of the excitation, during the dynamic tests the notch remains always open. Table 1 compares the experimental natural frequencies for the undamaged and damaged rod. The severity and the location of the damage have been achieved by applying the formulas (3.25), (3.26). The results of the identification are summed up in Tables 4 and 5. With reference to the localization of the notched cross-section the accuracy of the method proves to be satisfactory. Moreover, our expectations find further confirmation, as the assessing of K_A proves not reliable when the crack is very severe, and such is the case in the configuration D2, see Table 5.

3.4 Bending Vibrating Beams

In the previous section the inverse problem of identifying a crack in an axially vibrating beam from frequency measurements has been discussed. Here a cracked beam in bending will be considered. To start, a simply supported uniform Euler-Bernoulli beam with an open crack at the cross-section of abscissa s is considered. According to Freund and Herrmann (1976), the crack is represented by the insertion of a massless rotational elastic spring at the damaged cross-section. The stiffness K_B of the spring may be related in a precise way to the geometry of the damage. Denoting as usual the Young's modulus of the material by E and the linear mass-density by ρ, the mth eigenpair $\{v_{dm}(x), \lambda^S_{dm}\}$ of the bending vibrations of the cracked beam satisfies the following boundary value problem

$$(EIv_d'')'' = \lambda^S_d \rho v_d \quad \text{in } (0,s) \cup (s,\ell), \tag{3.29}$$

$$v_d = 0, \quad v_d'' = 0 \quad \text{at } x = 0 \text{ and } x = \ell, \tag{3.30}$$

where the jump conditions

$$[v_d(s)] = [v_d''(s)] = [v_d'''(s)] = 0, \qquad EIv_d''(s) = K_B[v_d'(s)], \tag{3.31}$$

hold at the cross-section where the crack occurs. In the equations above I and A represent the moment of inertia and the area of the cross-section, respectively.

If the crack is small, namely K_B is large enough, on proceeding as in Section 3.2 and with the above notation, the first order variation of the mth eigenvalue with respect to $1/K_B$ is given by

$$\delta\lambda^S_m = -\frac{(M^S_m(s))^2}{K_B}, \qquad \text{m=0,1,2,...}, \tag{3.32}$$

where $M^S_m(s) \equiv -EIv_m''(s)$ is the bending moment at the cross-section of abscissa s in the mth (normalized) bending mode of the undamaged beam.

At this stage the problem of identifying the position and severity of the crack from the knowledge of the changes in a pair of natural frequencies can be posed. In the above mentioned paper by Narkis (1994), it was shown that knowledge of the first two frequency changes induced by the damage suffices to identify uniquely the crack location (except for symmetrical positions). Here an improvement of such result is presented. Denote by C^S_m the quantity

$$C^S_m \equiv -\frac{\delta\lambda^S_m}{B(1+m)^2}, \tag{3.33}$$

where $m \geq 0$ and B is the constant

$$B = \left(EI\sqrt{\frac{2}{\rho\ell}}\left(\frac{\pi}{\ell}\right)^2\right)^2. \tag{3.34}$$

It can be proved that, *if* $C^S_m > 0$*, the measurement of the pair* $\{C^S_m, C^S_{2m}\}$*,* $m \geq 0$*, determines uniquely the severity of the damage, e.g. the spring stiffness* K_B*, and the variable* $S = \cos(2(m+1)\pi s/\ell)$ *of the damage location* s*.*

A result similar to what attained at the end of Section 3.2 is valid for cracked beams in bending as well, namely *the measurement of the* m*th frequency of the cracked beam under* *simply supported boundary conditions (S) and of the $(m+1)$th frequency for sliding-sliding boundary conditions (Sl) (e.g., $v' = v''' = 0$ at $x = 0$ and $x = \ell$) determines uniquely the stiffness K_B of the rotating spring and the position variable $S = cos(2(m+1)\pi s/\ell)$.*

Defining $C_m^{Sl} \equiv -\delta\lambda_m^{Sl}/Bm^4$ and keeping the conventional meaning of the symbols, $m = 0, 1, 2$, it is possible to attain the following expressions for the damage parameters

$$K_B = \frac{1}{C_{m+1}^{Sl} + C_m^S}, \qquad S = -1 + \frac{2}{1 + \frac{C_m^S}{C_{m+1}^{Sl}}}. \tag{3.35}$$

The latter expression is valid if $C_{m+1}^{Sl} > 0$. If $C_{m+1}^{Sl} = 0$ then $S = 1$.

4 Crack Identification from Changes in Node Positions

4.1 Contents

In this section we are concerned with using another kind of spectral data for a damage localization problem for vibrating rods: we seek to detect a single notch from a knowledge of the damage induced shifts in the nodes of the mode shapes of the rod.

In most studies of dynamic methods for damage localization, researchers have used changes in natural frequency as the diagnostic tool. In last years, attention in the field has been turned to the use of changes in the mode shapes to detect damage. Yuen (1985) showed that for a cantilever beam there is a systematic change in the first mode shape with change in the location of a notch. For estimating the location and depth of cracks, Rizos et al. (1990) developed a method which uses the amplitudes of a principal mode at two points on a beam, in conjunction with an analytical expression for the dynamic response. Pandey et al. (1991) demonstrated that changes in the curvature of mode shapes may be useful for damage detection in beams. Application of these diagnostic techniques is somewhat limited because modes shapes are relatively insensitive to damage; in practical situations there are further complications caused by the sensitivity of the identification procedure to measurement errors and noise, see, for example, Hearn and Testa (1991). However, mode shapes still remain valuable for correlating analytical and experimental frequencies. Mode shapes are particularly useful for detecting damage in symmetrical (undamaged) structures. Notches at any one of a set of symmetrically placed points of a symmetrical structure produce identical changes to natural frequencies; but they will produce different changes to mode shapes.

Very little work has been done on using changes in the nodes of the mode shapes to identify localized damage. Natke and Cempel (1991) observed that mode shapes are more sensitive near nodes (or nodal lines) that at maximum amplitudes, and showed how the presence of a hole in a plate may dramatically change its nodal lines. Chen and Garba (1998) investigated, both analytically and experimentally, the modal pattern change of a large space structure due mainly to damage of one of its members. They observed that large changes in mode shapes and shift of the nodes occur close to the location of the damage. Indichandy et al. (1987) observed similar location properties of mode shapes in an experimental investigation of an offshore platform. From a mathematical point of view, inverse nodal problems are relatively new; research started with the general papers by McLaughlin (1988) and Hald and McLaughlin (1989) concerning the reconstruction of a vibrating rod or membrane from nodal positions. Their point of view was different from ours: they were identifying a complete system rather than detecting localized damage. However, our inspiration for considering nodal points as data for damage detection came from a remark they made in Hald and McLaughlin (1990): *One possible application of our theory is to determine regions of fatigue in metals. An example, that is important in practice, is fatigue in turbine blades. One effect of fatigue is a dramatic shift of the nodal lines.*

In the following section we consider a specific problem: the detection of a single notch in an axially vibrating thin rod. The notch is represented by using a spring, as in Freund and Herrmann (1976). We show that nodes move toward the notch. By this we mean that every node located to the left of the notch in the undamaged configuration moves to the right, and every node on the right moves to the left. This means that for every principal mode, which has at least two nodes, there is exactly one neighbouring pair which move toward each other, and the notch is located between them.

Experimentally, nodal positions are easier to measure than are mode shapes; in fact, their require, roughly speaking, just a detection of modal component sign, rather than a set of measurements of magnitudes, see Gladwell and Morassi (1999) for some experimental applications.

4.2 Axially Vibrating Beams

In this section an investigation of the effect of a crack on the nodes of free axial vibration modes of a beam is presented. The spatial variation of the infinitesimal undamped free vibration of the undamaged and cracked rod are governed by the eigenvalue problems (3.1)–(3.3) and (3.7), (3.2), (3.3), (3.8), respectively.

The eigenfunctions of the undamaged system exist only for certain values of λ, the eigenvalues. Denote a generic eigenvalue by λ_0. However, for any (real) values of λ it may be found the solution, $\vartheta(x) = \vartheta(x, \lambda)$, of (3.1) subject only to (3.2). The solution will be determined apart from an arbitrary multiplicative constant which, if necessary, may be chosen to satisfy some norming condition. Similarly, the solution $\varphi(x) = \varphi(x, \lambda)$ of (3.1), (3.3) may be found. The eigenvalues λ_0 of the undamaged system are the values of λ for which $\vartheta(x, \lambda)$, $\varphi(x, \lambda)$ are multiples of each other, i.e. $\vartheta(x, \lambda_0) = c\varphi(x, \lambda_0)$, where $c \neq 0$ is a real constant.

Again, the eigenfunctions of the damaged rod will exist only for certain values of λ, the eigenvalues of the damaged system. To the left of the damaged section, the eigenfunction $u_d(x)$ will be proportional to $\vartheta(x, \lambda)$, and to the right it will be proportional to $\varphi(x, \lambda)$. Thus to see how the nodes of the eigenfunctions to the left of s change, the zeros of $\vartheta_0(x, \lambda_0) = \vartheta_0(x)$ and $\vartheta(x, \lambda) = \vartheta(x)$ can be compared; and to see how the nodes to the right of s change, the zeros of $\varphi(x, \lambda_0)$ and $\varphi(x, \lambda)$ can be compared. *It will be proved that nodes move toward the damaged location.* To show this, adaptations of two Sturm's theorems are needed.

Lemma 4.1. *Let $\vartheta_0(x)$, $\vartheta(x)$ be solutions of* (3.1) *with parameter λ_0 and λ, respectively, where $\lambda_0 > \lambda$. Between two consecutive zeros of $\vartheta(x)$, there is at least one zero of $\vartheta_0(x)$.*

Lemma 4.2. *Let $\vartheta_0(x)$, $\vartheta(x)$ be the solutions of* (3.1), (3.2) *with parameter λ_0 and λ, respectively, where $\lambda_0 > \lambda$. If $\vartheta(x)$ has m zeros in an interval $(0, b]$, $\vartheta_0(x)$ has at least n zeros in the same interval, and the mth zero of $\vartheta_0(x)$ is less than the nth zero of $\vartheta(x)$.*

The analysis of Section 3.2 shows that the damage cannot increase the natural frequencies of the system. *If the damage actually does decrease a natural frequency, and this will certainly happen if $u'(s) \neq 0$ for that mode, then, from Lemma 4.1 and Lemma 4.2, the nodes of the modes to the left of the damage, at s, like the zeros of $\vartheta(x)$, will move to the right. A similar argument shows that the nodes to the right of s will move to the left.* This means that for every principal mode, which has at least two nodes, there is exactly one neighboring pair which move toward each other, and the crack is located between them.

If the damage is small, i.e. $\epsilon = 1/K_A$ is small enough, the first order change in the position of the nodes with respect to ϵ may be determined. To do this, the first order change in the zeros of $\vartheta(x)$ to the left of s will be estimated. If x_0 is a zero of ϑ_0, then, to first order in ϵ, the corresponding zero of ϑ will be $x = x_0 + \epsilon\widetilde{x}$. By using a perturbation argument it can be shown that the change in position of the nodes is given by

$$\epsilon\widetilde{x} = \frac{\epsilon(a(s)u_0'(s))^2 \int_0^{x_0} \rho u_0^2}{a(x_0)(u_0'(x_0))^2}, \tag{4.1}$$

where u_0 is the eigenfunction of the undamaged rod. The corresponding result for the change in position of nodes to the right of s is

$$\epsilon\widetilde{x} = \frac{-\epsilon(a(s)u_0'(s))^2 \int_{x_0}^{\ell} \rho u_0^2}{a(x_0)(u_0'(x_0))^2}. \tag{4.2}$$

4.3 Bending Vibrating Beams

The result obtained in previous section left an important question unsolved, and namely the capability to extend the identification procedure to cracked beams in bending vibration. Although desirable for the application of diagnostic methods, the existence of a similar property is all but obvious. In fact, the monotonicity property shown in Section 4.2 is a consequence of the oscillation and separation theorems for solutions of the Sturm-Liouville operator governing the axial vibration of a thin rod. General properties of this kind become more involved for the fourth-order differential operator of the bending vibrations of a beam, see, for instance, the classical paper by Leighton and Nehari (1958). In particular, on adapting the Comparison Theorem 5.1 of Leighton and Nehari (1958), one can show that the bending analogue of the monotonicity property found for the second-order case involves conjugate points (associated to one end of the beam) and crack location, that is *conjugate points move toward the crack.* Conjugate points are points of the beam axis given by the intersection of two special solutions of the differential operator governing the bending vibrations. Unfortunately, unlike nodal points, conjugate points lack a clear physical interpretation; therefore, since their variations cannot be measured, damage location cannot be deduced. Considering the effect of a crack on the nodal points of flexural modes, the insertion of a crack in some areas of the beam axis was found to cause some nodes to shift away from the damaged area. This effect does not apply to axial modes. Despite no monotonicity property is found to exist between nodes and crack location, a parametrical analysis of the bending problem proved useful to characterize the areas of the beam axis where a crack causes nodes always shift along the same direction, for varying damage severity, in originally uniform beams with general boundary conditions. In the examined cases, the signs of nodal shifts in the first three/four vibration modes proved essential to assess crack location with good accuracy, see Dilena and Morassi (2002) for more details.

5 Extensions

5.1 Crack Identification from Antiresonance Measurements

Cracks at any one of a set of symmetrically placed points of a symmetrical structure produce identical changes to natural frequencies; so, as it was shown in Section 3.2, the measurement of a (suitable) pair of natural frequencies cannot eliminate symmetrical solutions in the damage location problem. Here it will be shown how an appropriate use of natural frequencies and antiresonance frequencies may overcome the non-uniqueness of the damage location problem due to structural symmetry.

To show this, attention will be focused on axially vibrating beams. To fix the ideas, a free uniform rod with a small open crack at the cross-section of abscissa s is considered. First order variations of natural frequencies with respect to the damage were derived in Section 3.2, equation (3.15). Antiresonances correspond to zeros of frequency response functions (frf) $H(\sqrt{\lambda}, x_i, x_o)$, where x_i, x_o are the abscissas of the excitation point and measurement point, respectively. When $x_i = x_o$, the zeros of the frf $H(\sqrt{\lambda}, x_i, x_i)$ are the frequencies of a beam in which the longitudinal displacement at the cross-section of abscissa x_i is hindered. In particular, if $x_i = x_o = 0$, then the antiresonances of the frf $H(\sqrt{\lambda}, 0, 0)$ are the square root of the eigenvalues λ_m^C of the cantilever rod (3.6). Therefore, under the assumption of small crack, namely K_A large enough, on proceeding as in Section 3.2 and with the same notation, the first order variation of the (square of the) mth antiresonance of the frf $H(\sqrt{\lambda}, 0, 0)$ with respect to $1/K_A$ may be evaluated by equation (3.15).

Denoting $C_m^C \equiv -\delta\lambda_m^C / B(1 + 2m)^2$, with $B = (a\pi\sqrt{2/\rho\ell}/2\ell)^2$, $m = 0, 1, 2, ...$, it turns out that $C_m^C = cos^2((1 + 2m)\pi s/2\ell)/K_A$ and therefore equation (3.28) gives a closed form expression of the stiffness K_A and of the position variable $S = cos((1 + 2m)\pi s/\ell)$ of the damage location s. Considering $m = 0$, *it turns out that the damage is uniquely determined by the measurement of the first resonance of the rod and the first antiresonance of the frf* $H(\sqrt{\lambda}, 0, 0)$. It is worth noticing that the required data can be extracted from the frf measurement of $H(\sqrt{\lambda}, 0, 0)$ only.

Extension of the above result to other boundary conditions and to initially non-uniform rods, together with the study of the flexural case and some experimental applications, are fully discussed in Dilena and Morassi (2004).

5.2 Detection of Multiple Cracks

The physical model, which will be mainly investigated in this section, is a simply supported uniform Euler-Bernoulli beam with two cracks of equal severity located at cross sections of abscissa s_1 and s_2, with $0 < s_1 < s_2 < \ell$, where ℓ is the beam length.

Assuming that cracks remain always open during the flexural vibration, the infinitesimal bending vibrations of the cracked beam are governed by the eigenvalue value problem (3.29), (3.30), where the jump conditions (3.31) hold at the cross sections of abscissa $x = s_1$ and $x = s_2$ where cracks occur. As in Section 3.2, every crack is represented by inserting a massless rotational spring of stiffness K_B.

If cracks are small, namely $\epsilon \equiv 1/K_B$ is small enough, then the first order variation of the natural frequencies with respect to ϵ may be evaluated. On proceeding as in Section 2.5 and with the above notation one has

$$\delta\lambda_m^S = -\frac{(M_m(s_1))^2}{K_B} - \frac{(M_m(s_2))^2}{K_B}, \qquad \text{m=1,2,...}, \tag{5.1}$$

where $M_m(x) \equiv -EIw_m''(x)$ is the bending moment at the cross-section of abscissa x in the mth (normalized) bending mode of the undamaged beam.

Since only three parameters need to be determined, namely s_1, s_2 and K_B, it will be investigated to what extent the measurement of the first three natural frequencies may be useful for identifying the damage.

The system is symmetrical with respect to $x = \ell/2$ and therefore a crack located at any one of a set of symmetrically placed points will produce identical changes in natural frequencies. Without affecting the character of generality of the analysis, it can be assumed that $0 < s_1 < s_2 \le \ell/2$. By inserting the expressions of the eigenmodes $w_m(x)$ of the undamaged beam, for m=1,2,3, into equation (5.1) and by using standard trigonometric identities, the following system of three non-linear equations is obtained

$$\alpha + \beta = 2 - 2KC_1^S, \quad \alpha^2 + \beta^2 = 2 - KC_2^S, \quad \alpha^3 + \beta^3 = 2 - \frac{K}{2}(3C_1^S + C_3^S), \tag{5.2}$$

where

$$\alpha \equiv \cos\left(\frac{2\pi s_1}{\ell}\right) \in [-1, 1), \quad \beta \equiv \cos\left(\frac{2\pi s_2}{\ell}\right) \in [-1, 1) \tag{5.3}$$

and

$$C_m^S \equiv -\frac{\delta\lambda_m^S}{\frac{2}{\rho\ell}\lambda_m^S}, \qquad \text{m=1,2,3.} \tag{5.4}$$

Since $0 < s_1 < s_2 \le \ell/2$, it turns out that $C_m > 0$ (to simplify notation, the superscript S will be omitted) and functions $\alpha = \alpha(s_1)$, $\beta = \beta(s_2)$ are one-to-one correspondences, with $\alpha \ne \beta$. By using algebraic identities in equations (5.2), the following polynomial equation in the unknown K_B is obtained

$$K_B\left(4C_1^3K_B^2 + 3C_1(C_2 - 4C_1)K_B + \left(\frac{15}{2}C_1 - 3C_2 + \frac{C_3}{2}\right)\right) = 0. \tag{5.5}$$

Apart from the trivial solution $K_B = 0$, which clearly does not correspond to the assumption of small damage, it can be proved that the polynomial of second degree enclosed in square brackets in equation (5.5) always has two real positive (possibly equal) roots for every set of data $\{C_1, C_2, C_3\}$, see Morassi and Rollo (2001), that is there exist two values of the stiffness K_B of the spring simulating the damage

$$K_{B1,2} = \frac{1}{8C_1^2}\left[-3(C_2 - 4C_1) \pm \left(9(C_2 - 4C_1)^2 - 16C_1\left(\frac{15}{2}C_1 - 3C_2 + \frac{C_3}{2}\right)\right)^{\frac{1}{2}}\right], \tag{5.6}$$

where indexes 1 and 2 correspond to + sign and - sign respectively.

By inserting the expression (5.6) of K_B into first two equations of (5.2), the location of the damage may be determined. Note that the role of variables α and β in system (5.2) is completely interchangeable, namely, if the ordinate triple $\{K, \alpha, \beta\}$ is a solution of the diagnostic problem, also $\{K, \beta, \alpha\}$ is a solution. Then, it is enough to determine the position variable α. By using equations (5.2) the following polynomial equation of second degree in the variable α is obtained

$$g(\alpha) \equiv 2\alpha^2 - 2\alpha(2 - 2K_BC_1) + (2 - 2K_BC_1)^2 - (2 - K_BC_2) = 0. \tag{5.7}$$

It was shown by Morassi and Rollo (2001) that the polynomial $g(\alpha)$ in equation (5.7) always has two distinct real roots α_1, α_2 belonging to the interval $[-1, 1)$, namely

$$\alpha_{1,2} = \frac{1}{2}\left[(2 - 2K_BC_1) \pm \left(-4K_B^2C_1^2 + 8K_BC_1 - 2K_BC_2\right)^{\frac{1}{2}}\right] \tag{5.8}$$

where indexes 1 and 2 represent + sign and - sign before the square root respectively.

Finally, the complete set of solutions of the system (5.2) with reference to cracks located on the left half of the beam is given by $\{K_{B1}, s_1(K_{B1}), s_2(K_{B1})\}$, $\{K_{B2}, s_1(K_{B2}), s_2(K_{B2})\}$. That is, two cracks of same severity K_{B1} (evaluated via expression (5.6) with + sign) located at the cross-sections of abscissa $s_1(K_{B1})$, $s_2(K_{B1})$ (evaluated via expressions (5.8) and (5.3)) produce changes in the first three natural frequencies identical to those induced by two cracks of same severity K_{B2} (expression (5.6) with - sign) located at the cross-sections of abscissa $s_1(K_{B2})$, $s_2(K_{B2})$ (expressions (5.8) and (5.3)).

The present method can be adapted to identify two small cracks of the same severity in an axially vibrating beam with free ends.

5.3 Crack Identification in Dissipative Systems

Consider a thin straight elastic rod of length ℓ with a crack at the cross section of abscissa $s \in (0, \ell)$. The beam has constant cross-section of area A, Young's modulus E and uniform linear mass density ρ. For definiteness, the left end of the beam is assumed to be free, whereas its right end is constrained by a lumped linear damper of viscous damping coefficient $c > 0$. Assuming that the crack remains always open during motion and modelling it as a massless translational spring at $x = s$ of stiffness K_A, the infinitesimal free longitudinal vibrations of the damaged rod (F-D) are governed by the following dimensionless eigenvalue problem

$$u_d'' - \lambda_d u_d = 0 \qquad \text{in } (0, \sigma) \cup (\sigma, 1), \tag{5.9}$$

$$u'_d(0) = 0, \qquad u'_d(1) + \gamma\sqrt{\lambda_d}u_d(1) = 0, \tag{5.10}$$

$$[u_d(\sigma)] = \epsilon u'_d(\sigma), \qquad [u'_d(\sigma)] = 0, \tag{5.11}$$

where $u_d = u_d(z)$ and where the dimensionless quantities z, σ, γ and ϵ are defined as follows

$$z = \frac{x}{\ell}, \quad \sigma = \frac{s}{\ell}, \quad \gamma = c(EA\rho)^{-\frac{1}{2}}, \quad \epsilon = \frac{EA}{K_A \ell}. \tag{5.12}$$

The undamaged system corresponds to $K_A \to \infty$ or $\epsilon \to 0$. Under the assumption that $\gamma < 1$, the mth eigenpair $\{u_m, \lambda_m^2\}$ of the undamaged rod is given by $\sqrt{\lambda_m} = \ln\left((1-\gamma)/(1+\gamma)\right)/2 + im\pi \equiv \xi + im\pi$, $u_m(x) = A_m \cosh(\sqrt{\lambda_m} z)$, m $= 0, \pm1, \pm2, ...$, where i is the imaginary unit and A_m is an arbitrary complex number. For $\gamma > 1$ the mth eigenpair is given by $\sqrt{\lambda_m} = \ln((\gamma-1)/(1+\gamma))/2 + \mathrm{i}(1+2m)\pi/2$, $u_m(z) = A_m \cosh(\sqrt{\lambda_m} z)$, $m = 0, \pm1, \pm2, ...$ Note that if $\gamma = 1$ there is no spectrum at all. To fix ideas, and considering that this is a common situation in practice, the condition $\gamma < 1$ will be assumed hereafter. It can be shown that also for the eigenvalue problem (5.9), (5.10), (5.11) there is an infinite sequence of simple eigenvalues.

Similarly to the ideal undamped case, see Section 3.2, the first order change in an eigenvalue produced by a single small crack may be expressed as the product of two terms, the first of which is proportional to the severity of damage and the second only depends on the location of damage

$$\delta\lambda_m = \epsilon(u'_m(\sigma))^2, \quad \mathrm{m} = 0, \pm1, \pm2, ..., \tag{5.13}$$

where the normalizing condition $\int_0^1 u_m^2 + \gamma u_m^2(1)/2\sqrt{\lambda_m} = 1$ was taken into account. However, concerning the problem of damage identification, there is a substantial difference between the undamped case and the dissipative case considered here. In fact, the mode shape of the undamaged rod is now a complex value function and this fact has a strong influence on the damage detection procedure, as it will be shown in the following.

Again, under the assumption that the undamaged system is completely defined and with damage simulated as above, only the stiffness K_A and the abscissa s of the cracked cross-section need to be determined. Therefore, changes in a suitable pair of eigenvalues will be considered in identifying damage. By introducing the expression of the mth eigenpair of the undamaged rod into equation (5.13) gives

$$C_m = \epsilon\left[\sinh^2(\xi\sigma)\cos^2(m\pi\sigma) - \cosh^2(\xi\sigma)\sin^2(m\pi\sigma) + \mathrm{i}\cdot\sinh(\xi\sigma)\cosh(\xi\sigma)\sin(2m\pi\sigma)\right] \tag{5.14}$$

where $C_m \equiv \delta\lambda_m/\lambda_m$ and $\xi \equiv \mathrm{Re}\sqrt{\lambda_m}$.

In order to identify damage, it is convenient to deal with the imaginary part of C_m's, $m \neq 0$. Let it be assumed that $\mathrm{Im}C_m \neq 0$ for a certain $m \neq 0$. Then, by using standard trigonometric identities, from expression (5.14) it follows that

$$S \equiv \cos(2m\pi\sigma) = \frac{\mathrm{Im}C_{2m}}{2\mathrm{Im}C_m}, \tag{5.15}$$

namely, *if* $ImC_m \neq 0$, *the measurement of the pair* $\{ImC_m, ImC_{2m}\}$, $m \neq 0$, *uniquely determines the variable* $S \equiv cos(2m\pi\sigma)$ *of the normalized damage location* σ. Note that the sign of m does not affect the localization result. Moreover, for those values of σ such that $|S| = 1$, it turns out that $\text{Im}C_m = 0$, and then $S \in (-1, 1)$. Therefore there are exactly $2|m|$ possible crack locations σ_k, $k = 1, ..., 2|m|$, corresponding to the same ratio $\text{Im}C_{2m}/\text{Im}C_m$. These damage locations are symmetrically placed with respect to the mid-point of the rod, namely to every generic crack location σ_k corresponds its symmetric $\sigma_k^{sym} = \sigma_{2|m|-k+1}$, $k = 1, ..., 2|m|$. Note that, for $|m| = 1$, crack location can be uniquely determined from equation (5.15), except for a symmetrical position.

By inserting the expression of a possible damage location, let say σ_k, $k = 1, ..., 2|m|$, into the expression of $\text{Im}C_m$, the corresponding damage severity ϵ_k can be determined

$$\text{Im}C_m = \epsilon_k \sinh(\xi\sigma_k)\cosh(\xi\sigma_k)\sin(2m\pi\sigma_k). \tag{5.16}$$

Since $\sin(2m\pi\sigma_{2|m|-k+1}) = -\sin(2m\pi\sigma_k)$, $k = 1, ..., 2|m| - 1$, and $\xi < 0$, equation (5.16) gives $2|m|$ values of σ_k, half of them are negative, the remaining half being positive. More precisely, $\epsilon_k\epsilon_{k+1} < 0$, $k = 1, ..., 2|m| - 1$. Then, since the only physically plausible values of ϵ_k are the positive ones, half of the possible $2|m|$ damage locations identified from equation (5.15) can be discarded. This fact has an important consequence in the simplest situation where $|m| = 1$, because it implies the unique determination of crack location from the knowledge of ratio $\text{Im}C_2/\text{Im}C_1$. *It is worth noticing that there is no direct analogue of this property in the undamped case.*

Suppose now that $\text{Im}C_m = 0$ for a certain $m \neq 0$. From expression (5.14) it easily follows that $S = 1$, i.e. the possible damage locations are $\sigma_k = k/2|m|$, $k = 1, ..., 2|m| - 1$. Damage severity can be identified by substituting $\sigma = \sigma_k$, $k = 1, ..., 2|m| - 1$, into the expression of the real part of C_m. A direct calculation shows that

$$\text{Re}(C_m) = \frac{\epsilon_k}{2}((-1)^k \cosh(2\xi\sigma_k - 1)). \tag{5.17}$$

The expression within brackets is positive for even k and negative for odd k. Then, $\epsilon_k\epsilon_{k+1} < 0$, $k = 1, ..., 2|m| - 2$, and reasoning as before $(|m| - 1)$ or $|m|$, depending on the case, of the $(2|m| - 1)$ possible damage locations $\sigma_k = k/2|m|$ can be discarded.

The case when both ends of the rod are constrained by a lumped linear damper of equal viscous coefficient c can be investigated by a similar procedure. For clamped-damped boundary conditions, a quite common situation in the applications, the simultaneous use of eigenvalues related to different boundary conditions enables for the identification of the crack. Finally, it can be shown that similar results hold true for the case $\gamma > 1$, see Dilena and Morassi (2003).

6 Concluding Remarks

This paper was concerned with a class of diagnostic methods for damage detection in beams, either under longitudinal or bending vibration, based on measurement of damage-induced changes in the vibratory behavior of the system. In particular, techniques that make use of changes in natural frequencies and antiresonances, and in nodes of mode shapes have been investigated in detail. The prediction of the theory and reliability of the proposed techniques were checked on the basis of results of several dynamic tests performed on real beams, and some of them are also presented and discussed in the paper.

As a conclusion, some remarks about open problems and possible directions of future research are listed in the following.

1) It has been mentioned (see Section 2.1) that the identification procedure presented in Section 2 is only applicable under the a priori assumption that the damage belongs to one half of the rod or beam segment (see condition (2.21) or (2.55)). This hypothesis is rather restrictive and it would be worth removing it for practical applications, see Dilena and Morassi (2007) for some work on this direction which uses a suitable combination of natural frequency and antiresonant frequency measurements as data.

2) There are several open problems associated with the diagnostic procedure proposed in Section 2, such as the convergence of the iterative procedure (2.35)–(2.40) and the connection of the method with general results of the inverse eigenvalue theory. Moreover, the relationship between the Fourier coefficient method and variational procedures based on frequency measurements of the type described in Section 2.7 requires further investigation. Concerning this last issue, an interesting least-squares approach for solving second order inverse eigenvalue problems with finite data has been recently proposed by Rohrl (2005).

3) The diagnostic techniques based on frequency measurements presented in Section 3 hold under the assumption that the severity of the damage is "small". In fact, this assumption allows for a linearization of the general inverse problem in a neighborhood of the undamaged configuration, with great simplification of the analytical treatment. It would be worth determining the minimal sets of data (analogous, for example, to those presented in Section 3.2) that allow for the unique identification of deep cracks.

4) In the case of localized damages, the results shown in this paper hold on the assumption that the notch, or the crack, is open during vibration (see, for example, the Freund-Herrmann crack model of equation (3.8)). It would be worth considering also the identification of cracks that are partially closed, even in simple structures, see Testa (2008) for a contribution and a list of references on this subject.

5) Finally, all the problems considered in this research involve one-dimensional systems and, in fact, it was already mentioned that a rather extensive literature on damage detection in vibrating rods or beams is available. Conversely, little is known about two-dimensional systems, such as, for example, vibrating membranes and plates. This issue is of great importance for applications and the author believes that it will be one of the main challenges of damage detection methods in the near future.

Bibliography

R.D. Adams, P. Cawley, C.J. Pye and B.J. Stone. A vibration technique for non-destructively assessing the tntegrity of structures. *Journal of Mechanical Engineering Science*, 20(2):93–100, 1978.

G. Borg. Eine Umkehrung der Sturm-Liouvilleschen Eigenwertaufgabe. Bestimmung der Differentialgleichung durch die Eigenwerte. *Acta Mathematica*, 78:1–96, 1946.

G. Borg. On the completeness of some sets of functions. *Acta Mathematica*, 81:265–283, 1949.

H. Brezis. *Analisi funzionale*, Napoli: Liguore Editore, 1986.

E. Cabib, L. Freddi, A. Morassi, and D. Percivale. Thin notched beams. *Journal of Elasticity*, 64:157–178, 2001.

D. Capecchi and F. Vestroni. Monitoring of structural systems by using frequency data. *Earthquake Engineering and Structural Dynamics*, 28:447–461, 2000.

M.N. Cerri and F. Vestroni. Detection of damage in beams subjected to diffused cracking. *Journal of Sound and Vibration*, 234(2):259–276, 2000.

J.-C. Chen and J.A. Garba. Structural damage assessment using a system identification technique. *Structural Safety Evaluation Based on System Identification Approaches*, H.G. Natke and C. Cempel, Editors, Wiesbaden: Vieweg Braunschweig, 474–492, 1988.

T.G. Chondros and A.D. Dimarogonas. Identification of cracks in welded joints of complex structures. *Journal of Sound and Vibration*, 69(4):531–538, 1980.

S. Christidies and A.D.S. Barr. One-dimensional theory of cracked Bernoulli-Euler beams. *International Journal of the Mechanical Science*, 26(11/12):639–648, 1984.

C. Davini, F. Gatti and A. Morassi. A damage analysis of steel beams. *Meccanica*, 28:27–37, 1993.

C. Davini, A. Morassi and N. Rovere. Modal analysis of notched bars: tests and comments on the sensitivity of an identification technique. *Journal of Sound and Vibration*, 179(3):513–527, 1995.

M. Dilena and A. Morassi. Identification of crack location in vibrating beams from changes in node positions. *Journal of Sound and Vibration*, 255(5):915–930, 2002.

M. Dilena and A. Morassi. Detecting cracks in a longitudinally vibrating beam with dissipative boundary conditions. *Journal of Sound and Vibration*, 267:87–103, 2003.

M. Dilena and A. Morassi. The use of antiresonances for crack dectection in beams. *Journal of Sound and Vibration*, 276(1-2):195–214, 2004.

M. Dilena and A. Morassi. Damage detection in discrete vibrating systems. *Journal of Sound and Vibration*, 289(4-5):830–850, 2006.

M. Dilena and A. Morassi. Structural health monitoring of rods based on natural frequency and antiresonant frequency measurements. Submitted, 2007.

A.D. Dimarogonas and S.A. Paipetis. *Analytical Methods in Rotor Dynamics*, London: Applied Science, 1983.

L.B. Freund and G. Herrmann. Dynamic fracture of a beam or plate in plane bending. *Journal of Applied Mechanics*, 76:112–116, 1976.

M.I. Friswell and J.E. Mottershead. *Finite Element Model Updating in Structural Dynamics*, Dordrecht: Kluwer Academic Publishers, 1995.

M.I. Friswell. *Damage identification using inverse methods.* In A. Morassi and F. Vestroni, editors, *Dynamic Methods for Damage Detection in Structures*, CISM Courses and Lectures. Springer Verlag, Wien-New York, 2008.

G.M.L. Gladwell. *Inverse Problems in Vibration*, Dordrecht: Martinus Nijhoff Publishers, 1986.

G.M.L. Gladwell. *Inverse Problems in Vibration*, Second Edition, Kluwer Academic Publishers, 2004.

G.M.L. Gladwell and A. Morassi. Estimating damage in a rod from changes in node positions. *Inverse Problems in Engineering*, **7**(3):215–233, 1999.

P. Gudmundson. Eigenfrequency changes of structures due to cracks, notches or other geometrical changes. *Journal of Mechanics and Physics of Solids*, **30**(5):339–353, 1982.

P. Hajela and F.J. Soeiro. Structural damage detection based on static and modal analysis. *AIAA Journal*, 28(6):1110–1115, 1990.

O. Hald. The inverse Sturm-Liouville problem and the Rayleigh-Ritz method. *Mathematics of Computation*, 32(143):687–705, 1978.

O. Hald and J.R. McLaughlin. Solutions of inverse nodal problems. *Inverse Problems*, 5:307–347, 1989.

O. Hald and J.R. McLaughlin. Examples of inverse nodal problems. *Inverse Methods in Action*, P. Sabatier, Editor, Springer-Verlag, New-York, 1990.

S. Hassiotis and G.D. Jeong. Assessment of structural damage from natural frequency measurements. *Computer & Structures*, 49(4):679–691, 1993.

G. Hearn and R.B. Testa. Modal analysis for damage detection in structures. *ASCE Journal of Structural Engineering*, 117:3042–3063, 1991.

V.G. Indichandy, G. Ganapathy and R.P. Srinivasa. Structural integrity monitoring of fixed off-shore platforms. *IABSE Colloquium*, Bergamo (Italy), 237–261, 1987.

R. Knobel and B.D. Lowe. An inverse Sturm-Liouville problem for an impedance. *Z. Angew Math. Phys.*, 44:433–450, 1993.

J.B. Kosmatka and J.M. Ricles. Damage detection in structures by modal vibration characterization. *ASCE Journal of Structural Engineering*, 125(12):1384–1392, 1999.

S.S. Law, Z.Y. Shi and L.M. Zhang. Structural damage detection from incomplete and noisy modal test data. *ASCE Journal of Engineering Mechanics*, 124(11):1280–1288, 1998.

W. Leighton and Z. Nehari. On the oscillation of solutions of self-adjoint linear differential equations of the fourth order. *Transactions of the American Mathematical Society*, 89:325–377, 1958.

S.P. Lele and S.K. Maiti. Modelling of transverse vibration of short beams for crack detection and measurement of crack extension. *Journal of Sound and Vibration*, 257(3):559–583, 2002.

R.Y. Liang, J. Hu and F. Choy. Quantitative NDE technique for assessing damages in beam structures. *ASCE Journal of Engineering Mechanics*, 118(7):1468–1487, 1992.

J.R. McLaughlin. Inverse spectral theory using nodal points as data - A uniqueness result. *Journal of Differential Equations* 73(2):354–362, 1988.

A. Morassi. Crack-induced changes in eigenparameters of beam structures. *ASCE Journal of Engineering Mechanics*, 119(9):1798–1803, 1993.

A. Morassi. A uniqueness result on crack location in vibrating rods. *Inverse Problems in Engineering*, 4:231–254, 1997.

A. Morassi and N. Rovere. Localizing a notch in a steel frame from frequency measurements. *ASCE Journal of Engineering Mechanics*, 123(5):422–432, 1997.

A. Morassi. Identification of a crack in a rod based on changes in a pair of natural frequencies. *Journal of Sound and Vibration*, 242:577–596, 2001.

A. Morassi and M. Rollo. Identification of two cracks in a simply supported beam from minimal frequency measurements. *Journal of Vibration and Control*, 7:729–739, 2001.

A. Morassi. Damage detection and generalized Fourier coefficients. *Journal of Sound and Vibration*, 302:229–259, 2007.

Y. Narkis. Identification of crack location in vibrating simply supported beams. *Journal of Sound and Vibration*, 172:549–558, 1994.

H.G. Natke and C. Cempel. Fault detection and localisation in structures: a discussion. *Journal of Mechanical Systems and Signal Processing*, 5(5):345–356, 1991.

A.K. Pandey, M. Biswas and M.M. Samman. Damage detection from changes in curvature mode shapes. *Journal of Sound and Vibration*, 145(2):321–332, 1991.

H.J. Petroski. Comments on "Free vibration of beams with abrupt changes in cross-section". *Journal of Sound and Vibration*, 92(1):157–159, 1984.

W.X. Ren and G. De Roeck. Structural damage identification using modal data. I: Simulation verification. *ASCE Journal of Structural Engineering*, 128(1):87–95, 2002.

P.F. Rizos, N. Aspragathos and A.D. Dimarogonas. Identification of crack location and magnitude in a cantilever beam from the vibration modes. *Journal of Sound and Vibration*, 138(3):381–388, 1990.

N. Rohrl. A least-squares functional for solving inverse Sturm-Liouville problems. *Inverse Problems*, 21:2009–2017, 2005.

M.-H.H. Shen and J.E. Taylor. An identification problem for vibrating cracked beams. *Journal of Sound and Vibration*, 150:457–484, 1991.

J.K. Sinha, M.I. Friswell and S. Edwards. Simplified models for the location of cracks in beam structures using measured vibration data. *Journal of Sound and Vibration*, 251(1):13–38, 2002.

R. Testa. *Characteristics and detection of damage and fatigue cracks*. In A. Morassi and F. Vestroni, editors, *Dynamic Methods for Damage Detection in Structures*, CISM Courses and Lectures. Springer Verlag, Wien-New York, 2008.

W.T. Thomson. Vibration of slender bars with discontinuities in stiffness. *Journal of Applied Mechanics*, 16:203–207, 1949.

F. Vestroni and D. Capecchi. Damage evaluation in cracked vibrating beams using experimental frequencies and finite element models. *Journal of Vibration and Control*, 2:69–86, 1996.

F. Vestroni and D. Capecchi. Damage detection in beam structures based on frequency measurements. *ASCE Journal of Engineering Mechanics*, 126:761–768, 2000.

F. Vestroni. *Structural identification and damage detection*. In A. Morassi and F. Vestroni, editors, *Dynamic Methods for Damage Detection in Structures*, CISM Courses and Lectures. Springer Verlag, Wien-New York, 2008.

H.F. Weinberger. *A first course in partial differential equations*, New York: Dover Publications Inc., 1965.

Q. Wu. Reconstruction of integrated crack function of beams from eigenvalue shifts. *Journal of Sound and Vibration*, 173(2):279–282, 1994.

Q. Wu and F. Fricke. Determination of blocking locations and cross-sectional area in a duct by eigenfrequency shifts. *Acoustical Society of America*, 87(1):67–75, 1990.

M.M.F. Yuen. A numerical study of the eigenparameters of a damaged cantilever. *Journal of Sound and Vibration*, 103(3):301–310, 1985.

Characteristics and Detection of Damage and Fatigue Cracks

Rene B. Testa*

Department of Civil Engineering and Engineering Mechanics,
Columbia University, New York, NY, USA

Abstract. A simple, pragmatic approach to the detection of damage is outlined with a closer look at the detailed effects of individual cracks. The approach focuses on the frequency changes caused by damage and on determining the location, but not the severity, of the damage whether locally or globally. Laboratory application illustrates the detection of fatigue cracks in a simple structure. A second application looks at crack closure that may develop in fatigue and its potential influence on detection.

1 Introduction

Most types of damage in materials and structures are accompanied by stiffness deterioration which shows up in the dynamic response whether in the speed of wave propagation or in the vibration response. Such changes in the dynamic response caused by changes in structural parameters were subjects of early interest in the aerospace community (Fox and Kapoor 1968, and Farshad 1974) and applications to structures (Cawley and Adams 1979).

Various ASTM and other standards have made use of this to test for the quality and deterioration of building materials (Hearn and Testa 1991 and 1993) . These tests are generally aimed at finding deterioration throughout the material such as would occur in concrete from freeze-thaw damage. Distributed deterioration is the substance of continuum damage mechanics (Cauvin and Testa 1999) and these damage detection techniques might be employed to evaluate the continuum damage tensor components that describe the decrease in elastic moduli. The dynamic methods, while suitable to detect changes in moduli and thus to measure damage, are not recommended in ASTM for evaluation of the moduli themselves which are sensitive to many other conditions of testing.

In a structure, the problem of detection is more often associated with damage of the isolated type, usually a crack or loss of section. Such damage also affects stiffness, which is something that is often not as well predicted as strength because there exist various elements, such as partitions and secondary bracing, or other non-structural components, that are not considered among the primary resisting elements but which do contribute to stiffness. As a consequence, the dynamic response of a structure may not be modeled as precisely as its static response. Today, with potent

* The author is indebted to former colleagues and students who have contributed substantially to the original work in this paper. In particular, Professor George Hearn, University of Colorado, Dr. Wuzhen Zhang, Amman and Whitney, NY, and Dr. Chyun-Huey Kao, NY.

analytical tools available to investigate the dynamic response of bridges and other important structures for seismic adequacy, computer models of a structure need to be calibrated and refined. This is accomplished by monitoring the motion of the structure and comparing measured frequencies and modes to those predicted by the model. At the same time, estimates of damping can be obtained to assist in the prediction of dynamic response. Such motion monitoring has been used not only to improve analytical models but also to aid in assessing effects of changes in design. This latter application, it may be noted, is also the problem of damage detection.

In broad terms, this is the area of system identification which uses the dynamic response to evaluate defining parameters of a system. Much has been written in recent years on this subject for application to structural health monitoring, far more than can be listed here, and especially its application in cases where incomplete data is available to fully characterize a system. Equally significant is the work on methods of analysis such as wavelet theory (Hou et al 2000), empirical mode decomposition (Huang et al 1998, and Yang et al 2004), neural network theory (Masri et al 2000), and various sensitivity measures and optimization techniques using genetic algorithms (Mares and Surace 1996, Friswell et al 1998, Rao et al 2004). The work that follows is not intended so much to address these broad and important questions in the general problem of damage detection, nor the larger questions of sensitivity and uniqueness, but it outlines a simple, pragmatic approach to the detection of damage and presents a closer look at the detailed effects of individual cracks. In the first part, the basic premise and method are outlined with application to a simple structure in a laboratory test (Hearn and Testa 1991); the second part gives details of the characterization of a fatigue crack regarding its effects on frequency and damping, and its detection and location in the presence of crack closure (Testa et al 2000, and Testa and Zhang 1999 and 2002).

2 Damage Locator Model

Whether in a material or a structure, identification of damage involves securing information on four basic items: the very existence of damage, the type of damage, its location, and its extent or severity. The first and third of these are paramount because once existence and location are ascertained, the type and extent of damage can be pursued effectively. Therefore, this work focuses only on detecting the occurrence of damage and locating it. Moreover, the focus in this discussion is on elastic damage as opposed to excessive or permanent deformation that results from yielding. In such cases, the damage is reflected in stiffness changes (Figure 1) as opposed to permanent deformation and the same is true whether the damage is distributed as in continuum damage mechanics or isolated in the form of cracks or other disintegration. If one looks at the dynamic response, changes in damping characteristics due to damage may also be reflected.

In a linear structure, the dynamic characteristics are seen in the parameters of the natural modes - mode shape, frequency, and damping – and the magnitude of change in these as a consequence of damage or deterioration depends on the four characteristics of the damage listed above. A single deterioration event will affect each vibration mode differently, having a strong effect on certain modes and a weak effect on others. This dissimilarity of effect on various modes, since it can be predicted, is the basis for the identification of damaged members. Quantitative relations between damage and the resulting changes in the modal parameters of frequency and mode shape can be developed from a perturbation of the equation of motion for the system when damage is expressed as a reduction in stiffness properties. By contrast, while sources of damping in structures are understood in a general way (e.g., slip in bolted connections), quantitative relations between deterioration and the level of damping in structures are not available and, therefore, while change in modal damping may be indicative of damage occurrence, it may not be so useful in locating the damage. Therefore, the undamped elastic structure is considered in the following.

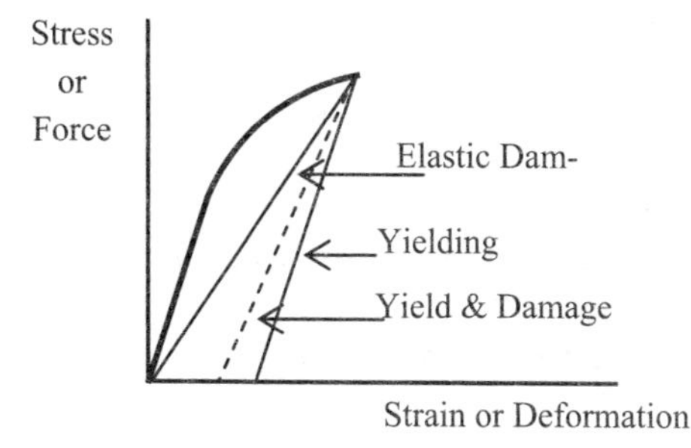

Figure 1. Elastic Damage in 1D

In a 3D continuum, elastic damage reflected in stiffness degradation is described by a damaged elastic modulus tensor of fourth rank **E'** which relates to the undamaged moduli **E** through a fourth order damage tensor **D** (Cauvin and Testa 1999) by

$$\mathbf{E}' = (\mathbf{I} - \mathbf{D}):\mathbf{E} \tag{2.1}$$

$$\delta\mathbf{E} = \mathbf{D}:\mathbf{E} \tag{2.2}$$

in which **I** is the unit fourth order tensor. Only in very special case can **D** be represented by a single scalar damage variable D. Similarly, in a structure the global stiffness matrix **K** is composed of element stiffness matrices $\mathbf{k}_i$ for each of the members **i**, and damage in such a member is reflected in a decrease of the elements of $\mathbf{k}_i$ which can be described by a damage matrix α_i for member **i** :

$$\delta\mathbf{k}_i = \alpha_i\,\mathbf{k}_i \tag{2.3}$$

The elements of the damage matrix depend on the type, severity and location of the damage within member **i**. However, often the structural role of a member relies predominantly on one of the stiffnesses (e.g. axial, bending, torsional) and damage that compromises that stiffness characterizes the member's response. Then, it may be possible to assume that a scalar damage variable α_i which describes the modification of the dominant stiffness also describes the effect of damage on the overall member stiffness:

$$\delta\mathbf{k}_i = \alpha_i\,\mathbf{k}_i \tag{2.4}$$

The eigenvalue problem for free vibration of the undamped structure is governed by the relation

$$(\mathbf{K} - \omega^2\mathbf{M})\mathbf{\Phi} = 0 \tag{2.5}$$

in which **M** is the modal mass matrix and **Φ** the mode shape. In general, one may consider perturbations in global stiffness $\delta\mathbf{K}$, frequency $\delta\omega$, mass $\delta\mathbf{M}$, and mode shape $\delta\mathbf{\Phi}$. Changes in mass will be neglected for the types of damage of interest here, and changes in mode shape are generally not great, nor are they easily measured in sufficient detail in field applications, and they too will be ignored. In that case, Equation 2.5 written for any mode φ_n gives

$$\delta\omega_n^2 = (\phi_n^T\delta\mathbf{K}\,\phi_n) / (\phi_n^T\mathbf{M}\,\phi_n) \tag{2.6}$$

where a superscript T denotes the transpose. The numerator may be written in terms of deformations in member i for mode n denoted by $\mathbf{\Delta}_i(\varphi_n)$ and the member stiffness perturbations to give

$$(\phi_n^T\delta\mathbf{K}\,\phi_n) = \Sigma_i\,[\mathbf{\Delta}_i^T(\varphi_n)\,\delta\mathbf{k}_i\,\mathbf{\Delta}_i(\varphi_n)] \tag{2.7}$$

If there is damage in member i alone, the summation in this equation is removed. If furthermore, the assumption in Equation 2.4 is invoked, then

$$\delta\omega_n^2 = \alpha_i\,[\mathbf{\Delta}_i^T(\varphi_n)\,\mathbf{k}_i\,\mathbf{\Delta}_i(\varphi_n)]/(\phi_n^T\mathbf{M}\,\phi_n) \tag{2.8}$$

This shows the dependence of the change in frequency n from damage in one member i on both location and severity of the damage. It also shows that the modal frequency ω_n will be affected more when damage occurs in a member i which has greater strain energy induced in that mode; the quantity in square brackets is the strain energy induced in the member for mode n.

The value of the simplifying assumption in Equation 2.4 is that the damage severity appears as a multiplier in the expression for its effect on each frequency. Therefore, if the ratio of frequency changes from a damage event in member i is considered, the severity of damage α_i vanishes from the result and the ratio depends only on the member damaged or, in effect, the location of damage in the structure. For modes n and m,

$$\frac{\delta\omega_n^2}{\delta\omega_m^2} = \frac{[\mathbf{\Delta}_i^T(\varphi_n)\,\mathbf{k}_i\,\mathbf{\Delta}_i(\varphi_n)]/(\phi_n^T\mathbf{M}\,\phi_n)}{[\mathbf{\Delta}_i^T(\varphi_m)\,\mathbf{k}_i\,\mathbf{\Delta}_i(\varphi_m)]/(\phi_m^T\mathbf{M}\,\phi_m)} \tag{2.9}$$

Moreover, in the ratio of frequency changes resulting from one damage event, only initial undamaged properties of the structure are needed to evaluate the changes. Therefore, in this simplified model, one may predict the expected frequency changes for a damage scenario and compare with measured changes in the actual structure until a match is found. A pragmatic approach should be used in the selection of modes to monitor for various damage locations, recognizing that damage in certain members will have more pronounced effect on modes which involve that member strongly in terms of the strain energy in the member.

If r modes are monitored in a structure, judiciously selected to detect potential damage in some group of members of the structure, changes from the undamaged or reference state can be measured and r ratios of measured frequency changes can be generated. Using Equation 2.9, r ratios of predicted frequency changes for any postulated damage scenario can also be generated.

Any one of the predicted frequency changes $\delta\omega^2$ may be used as the common denominator for all the ratios, preferably the largest one so that one of the r ratios will be unity. By comparison with the measured frequency change ratios, the most likely of the postulated damage scenarios may be identified. Quantitatively, this may be done by calculating the mean square deviation of the measured ratios relative to the predicted values; the scenario which gives the least deviation is the likeliest damage location.

This simple method was applied to a laboratory test frame (Hearn and Testa 1991) in which two stages of fatigue crack damage were observed under cyclic loading. For that small frame, six modes were monitored for both frequency and damping. In the case of first damage at the welded connection of a single member, the damage location was clearly identified whereas the case of multiple damage was more problematic and required additional reasoning beyond the direct application of the method outlined here in order to identify the damaged members. An additional observation in those tests was that the modal damping, as small as it is in such a frame, showed significant change with damage events and thus might serve as a damage indicator even if the absence of analytical predictability prevents its use to locate damage.

Such a damage locator model offers a pragmatic way of monitoring especially sensitive locations in structures. The method does not address questions of completeness or uniqueness of the solution, nor does it consider the type and number of modes that should be monitored except for the observation that one should choose modes which induce greater strain energy in members of interest. Optimization techniques might be applied to aid in the process of selecting modes to give the best results. Moreover, because the method does not consider the severity of damage but only its location, and because very often, the damage of interest results from fatigue cracking in a structure, there is some question about the effect of the crack condition on the application of the locator method. The following section explores this question.

3 Fatigue cracks with closure

Various aspects of the effects of cracks on the vibration response of beams have been addressed (e.g. Morassi 1993) and the fact that fatigue cracks may be partially closed so that they may be obscured in a crack detection model is one aspect to be considered (Zhang and Testa 1999, Testa et al 2000, Testa and Zhang 2002). In the following, experimental results are presented to show to show the role that crack closure may play in crack detection and at the same time, they give insight to the mechanism of crack closure and of the damping induced by a fatigue crack.

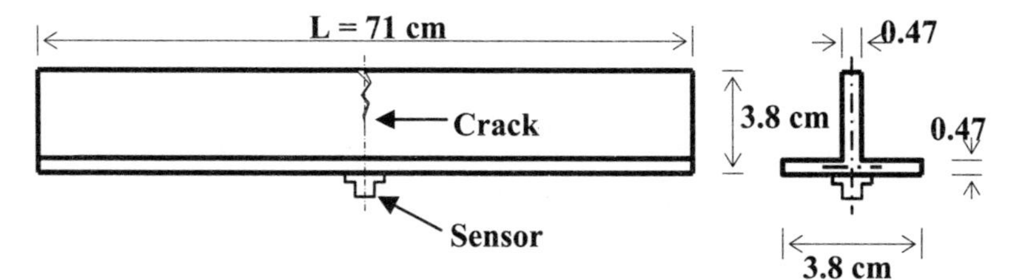

Figure 2. Beam test

Steel Tee beams, Figure 2, were fatigue loaded in flexure with a central point load P to propagate a crack from the edge of the stem of the Tee where an initial small notch was introduced. The load was controlled so that, as the crack grew, the maximum stress intensity factor at the crack tip (K_{max}) was held fixed in each beam as the load was cycled to give a stress intensity range, K_p , from zero to that fixed value; i.e. during crack propagation, the stress intensity range is a constant K_p and $K_{max} = K_p$. The stress intensity range (K_p) determines, in this test, the degree of crack closure that will be present for the given material (steel with 317 MPa yield strength) and geometry at each crack length. At any given crack length, the crack with closure may also be forced open statically by a load producing a stress intensity K_{max} exceeding the propagation stress intensity K_p. The degree of crack closure will then be affected by the new maximum stress intensity (K_{max}). Therefore, the degree of crack closure which is indicated by the value of the load P_o to open the crack elastically, depends on the crack length (a), the stress intensity at which the fatigue crack was propagated (K_p), and the maximum stress intensity (K_{max}) to which it has been subjected

Four fatigue crack propagation histories were used in the tests, namely: K_p = 590, 694, 833 and 903 MPa-mm$^{1/2}$. At each increment of 2.54 mm in crack length, a plot of static load (P) vs. crack mouth opening displacement (CMOD) was generated to establish the degree of crack closure (load P_o) but with care not to exceed the stress intensity K_p. At the same time, both frequency and damping for several fundamental modes were measured. On reaching a crack length of 2.54 cm at the specified cyclic loading K_p, fatigue loading was stopped and the crack was opened statically using successively higher values of K_{max} with modal testing after each step. In the modal tests, the beam was suspended by flexible threads at the nodal points of each fundamental mode being measured so as to approach free-free end conditions. Modal frequencies are readily identified by FFT of the free vibration response in each test, but the very low values of modal damping present in the beam require more precise measure than afforded by the FFT. A method based on the decay of modal energy was used as described in (Zhang & Testa 1999).

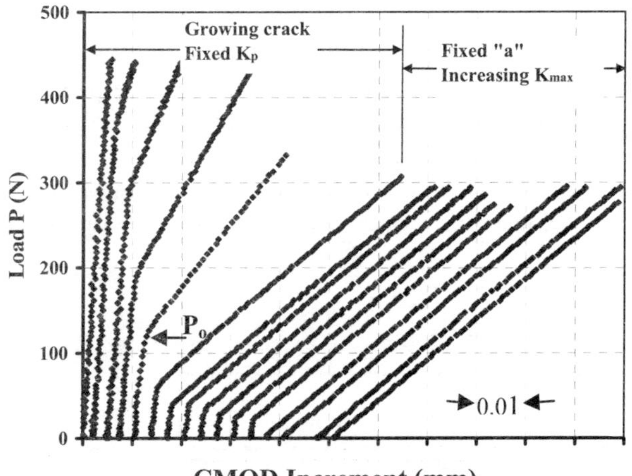

Figure 3. Closure effect on CMOD

Figure 3 shows the load P_o to overcome crack closure as indicated by the point at which the load vs CMOD bends sharply. Before the knee in the curve, the crack is not fully effective in reducing stiffness and after the knee it is fully open. The dependence of P_o on crack length a can be seen from the first few curves for increasing crack length with constant stress intensity $K_{max} = K_p$ and the dependence on K can be seen in the final curves when K_{max} is being increased with constant a.

The modal test results are given for the first vertical mode, it being the one most affected by the location of the fatigue crack in these tests. Observed frequency changes during crack propagation when there is crack closure of varying degrees in the four test cases are shown in Fig.4. The modal damping in these beams does not change appreciably during the crack propagation phase and remains at a very low level which is not plotted in Figure 4. Both frequency and damping vary significantly as the crack closure is removed by forcing the crack to open through increasingly high values of stress intensity as shown in Figure 5. While neither frequency nor damping change very much during crack growth with closure, both frequency and damping undergo substantial and rapid change as the crack goes from one with closure to one that is open. There is a fairly well defined transition range in terms of maximum stress intensity which the crack must experience to go from relatively ineffective to fully effective.

It is also interesting to observe that at the transition, there is very dramatic increase in the modal damping, in fact, two or three orders of magnitude, something that may provide additional avenues for damage detection. Moreover, the higher the degree of crack closure, the greater the peak damping at the transition. It is noted as well, that the transition value of K_{max} is not fixed but changes with the history of the fatigue crack. With the fatigue crack propagated at lower stress intensity levels, the transition point moves to lower values of K_{max}. In other words, less static load is needed to open the crack and make it fully effective dynamically.

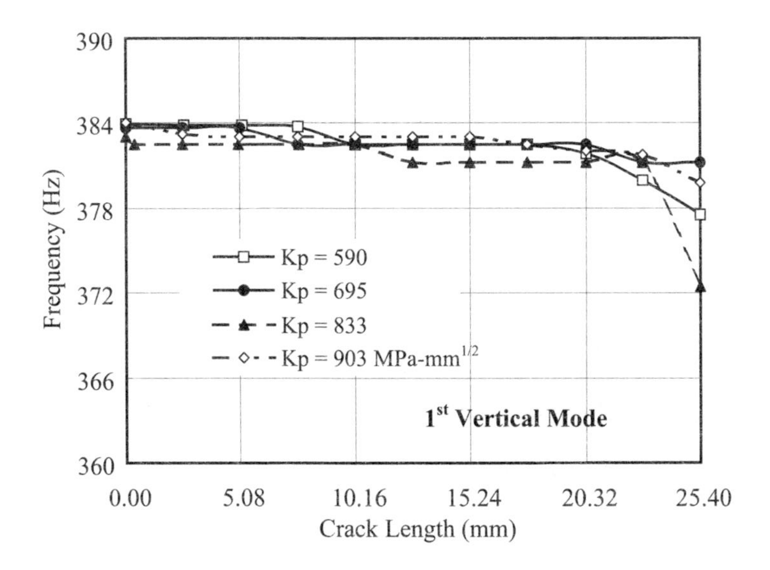

Figure 4. Frequency with crack closure

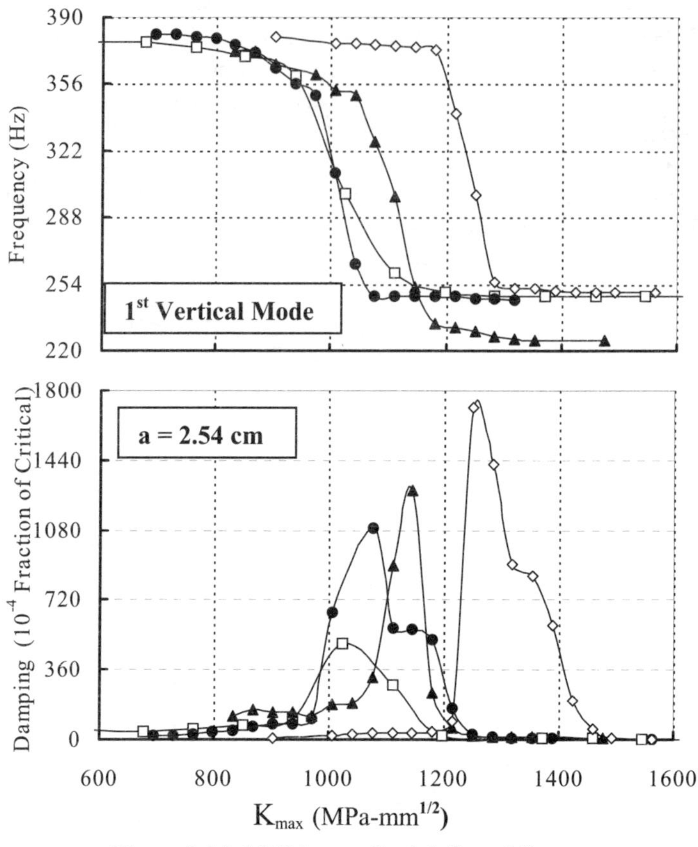

Figure 5. Modal Values as Crack is Forced Open

Because the frequency change tends to be obscured in the presence of crack closure, as Figure 4 shows, there is some question of the ability of the simple method outlined earlier to detect such fatigue cracks. Or, equivalently, if the frequency change is much smaller than the fully open crack would generate, how sensitive is the proposed method. Investigation of this question also addresses the premise that the damage might be detected and located without regard to the level of damage.

4 Locating cracks with closure

In applying the simple crack locator model of Section 2, the single beam structure can be divided into finite elements; determining the damaged element locates the damage on the beam. The frequency changes for the two extreme conditions of closed and open cracks are considered. In the modal tests presented in Section 3, only the first vertical mode results were given. For the damage locator model, the first and second modal frequencies for vertical, (V), lateral (H), and torsional (T) vibrations of the Tee beam are monitored. The measured changes in frequency relative to the undamaged state are shown in Figure 6 in the form of ratios to the change observed in the first vertical mode for both crack states (with and without crack closure) of the 25.4 mm fatigue crack at midspan.

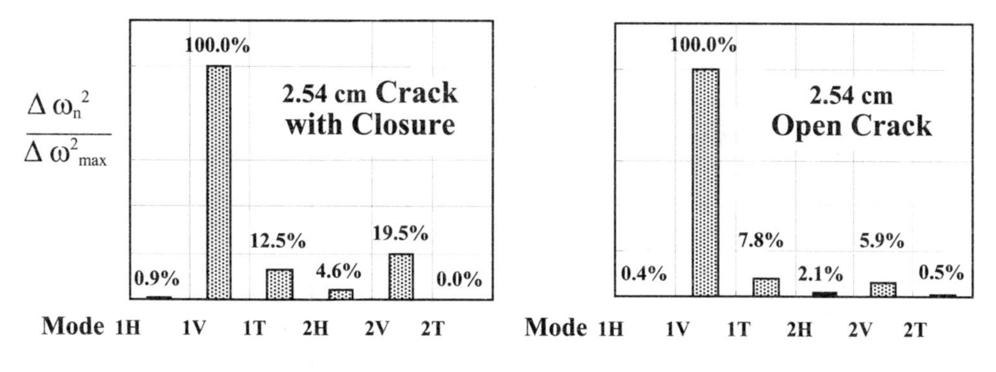

Figure 6. Measured frequency ratios in cracked beam

These relative changes are compared with the predicted values when damage is assumed at various likely locations. In the case of the present Tee beam, the predictions were made for damage in any one of seven finite elements into which the beam was divided along its span. In practice, only four of the elements need be considered because of symmetry. By the same token, it is clear that the locator does not distinguish between symmetrical locations of potential damage. Figure 7 shows the predicted frequency change ratios for two postulated damage locations. The actual damage (crack) is in element 4. Clearly, even in this simple case, the ensemble of relative frequency changes do not show the damage location obviously and irrefutably, although the predictor for element damage location (the correct location) does match the graphs of observed changes in Figure 6 better that the predictor for damage in element 2.

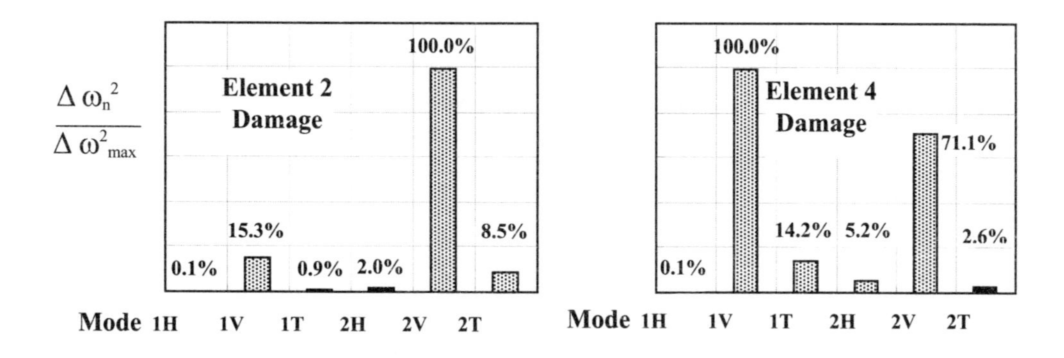

Figure 7. Predicted frequency ratios for specified damage in beam

If one applies the RMS deviation criterion for locating the damaged element, the locator model is quite clear in this simple case. Table 1 gives the RMS values of the deviations of the six frequency change ratios monitored in the beam. For both the closed and open crack conditions, RMS deviation indicates that the correct element of the beam is the damaged one. Although the frequency effects of the crack were considerably obscured by crack closure, the simple crack locator model shows remarkable sensitivity in this example.

Table 1. RMS deviations of six beam frequency changes

Crack Location Beam Element No.	RMS Deviation Closed Crack	RMS Deviation Open Crack
1	0.34	0.35
2	0.31	0.32
3	030	0.31
4	0.09	0.13

5 Conclusion

There is practical value in a simple model that can give information on damage location even if it is not the full characterization of a structure that one seeks in the general problem of system identification. The model outlined here has the requisite simplicity and it shows sensitivity in an application where the damage effects are obscured by crack closure, something that may well be encountered in a structure with fatigue cracking. The simplification introduced in the description of damage rendered the model independent of the severity of the damage at a location, a feature that was observed in the application to the cracked beam.

Many other studies have modeled the effects of a crack in a beam, usually treating the crack as a slot. While the results presented here show a great potential difference between the slot and a crack if there is closure insofar as effects on frequency and damping are concerned, there is not such a pressing distinction between the two in the results of the application of the present crack locator model.

Because of the effect of crack closure on stiffness, which appears in Figure 3 to give a bilinear load-deflection response but which is actually nonlinear near the knee of the curve, there are also nonlinear effects in the vibration response. These have been addressed further in (Testa and Zhang 2002).

Further work on the application of this simplified model requires consideration of the proper selection of modes to be monitored for a specific diagnostic, as well as questions of uniqueness.

References

Capecchi, D. and Vestroni, F. (1999). Monitoring of Structural Systems by Using Frequency Data. *Earthquake Engineering and Structural Dynamics*, vol 28 447-461.

Cauvin, A., and Testa, R.B. (1999). Damage Mechanics: Basic Variables in Continuum Theories. *International Journal of Solids and Structures*, vol 36, 747-761.

Cawley, P. and Adams, R.D. (1979). The Localization of Defects in Structures from Measurements of Natural Frequencies. *Journal of Strain Analysis*, vol 14, 49-67.

Farshad, M. (1974). Variations of Eigenvalues and Eigenfunctions in Continuum Mechanics. *AIAA Journal*, vol 12, 560-561.

Fox, R.L. and Kapoor, M.P. (1968). Rates of Change of Eigenvalues and Eigenvectors. *AIAA Journal*, vol 6, 2426-2429.

Friswell, M.I., Penny, J.E.T. and Garvey, S.D. (1998). A combined Genetic and Eigensensitivity Algorithm for Location of Damage in Structures. *Computers and Structures*, vol 69, 547-556.

Hearn, G. and Testa, R.B. (1991). Modal Analysis for Damage Detection in Structures. *Journal of Structural Engineering*, vol 117, 3042-3063.

Hearn, G., and Testa, R.B. (1993). Resonance Monitoring of Building Assemblies for Durability Tests. *Journal of Testing and Evaluation*, vol 21, 285-295.

Hou, Z., Noori, M. and St. Amand, R. (2000). Wavelet-Based Approach for Structural Damage Detection. *Journal of Engineering Mechanics*, vol 126, 677-683.

Huang, N.E., Shen, Z., Long, S.R., Wu, M.C., Shih, H.H., Zheng, Q., Yen, N.C., Tung, C.C., and Liu, H.H. (1998). The Empirical Mode Decomposition and the Hilbert Spectrum for Nonlinear and Non-Stationary Time Series Ananlysis, *Proceedings of the Royal Society of London*, vol 454, 903-995.

Mares, C. and Surace, C. (1996). An Application of Genetic Algorithms to Identify Damage in Elastic Structures. *Journal of Sound and Vibration*, vol 195, 195-215.

Masri, S. F., Smyth, A. W., Chassiakos, A. G., Caughey, T. K. and Hunter, N. F. (2000). Application of Neural Networks for Detection of Changes in Nonlinear Systems. *Journal. of Engineering Mechanics*, vol. 126, 666-676.

Morassi, A. (1993). Crack-Induced Changes in Eigenparameters of Beam Structures. *Journal of Engineering Mechanics*, vol. 119, 1798-1803.

Rao, M.A., Srinivas, J. and Murthy, B.S.N. (2004). Damage Detection in Vibrating Bodies Using Genetic Algorithms. *Computers and Structures*, vol 82, 963-968.

Testa, R.B., Zhang, W., Smyth, A.W. and Betti, R. (2000). Detection of Cracks with Closure. Proceedings of Structural Materials Technology IV - an NDT Conference. *Technomic Publishing Company*, 405 - 410.

Testa R.B. and Zhang, W. (2002). Modeling Crack Closure Effects on Frequency and Damping. *Proceedings of the 15th ASCE Engineering Mechanics Conference*, Columbia University.

Yang, J. N., Lei, Y., Lin, S. and Huang, N. (2004). Hilbert-Huang Based Approach for Structural Damage Detection. *Journal of Engineering Mechanics*, vol 130, 85-95.

Zhang, W., and Testa, R.B. (1999). Closure Effects on Fatigue Crack Detection. *Journal of Engineering Mechanics*, vol. 125, 1125-1132.

The reflection of the fundamental torsional mode from cracks and notches in pipes*

A. Demma, P. Cawley, and M. Lowe
Department of Mechanical Engineering, Imperial College, London SW7 2BX, United Kingdom

A. G. Roosenbrand
Shell Global Solutions, P.O. Box 38000, 1030 BN Amsterdam, The Netherlands

(Received 1 May 2002; accepted for publication 12 April 2003)

A quantitative study of the reflection of the $T(0,1)$ mode from defects in pipes in the frequency range 10–300 kHz has been carried out, finite element predictions being validated by experiments on selected cases. Both cracklike defects with zero axial extent and notches with varying axial extents have been considered. The results show that the reflection coefficient from axisymmetric cracks increases monotonically with depth at all frequencies and increases with frequency at any given depth. In the frequency range of interest there is no mode conversion at axisymmetric defects. With nonaxisymmetric cracks, the reflection coefficient is a roughly linear function of the circumferential extent of the defect at relatively high frequencies, the reflection coefficient at low circumferential extents falling below the linear prediction at lower frequencies. With nonaxisymmetric defects, mode conversion to the $F(1,2)$ mode is generally seen, and at lower frequencies the $F(1,3)$ mode is also produced. The depth and circumferential extent are the parameters controlling the reflection from cracks; when notches having finite axial extent, rather than cracks, are considered, interference between the reflections from the start and the end of the notch causes a periodic variation of the reflection coefficient as a function of the axial extent of the notch. The results have been explained in terms of the wave-number-defect size product, ka. Low frequency scattering behavior is seen when $ka<0.1$, high frequency scattering characteristics being seen when $ka>1$. © *2003 Acoustical Society of America.* [DOI: 10.1121/1.1582439]

PACS numbers: 43.20.Fn, 43.20.Mv [DEC]

I. INTRODUCTION

The presence of defects in pipelines is a major concern in the oil and chemical industries and NDT techniques are required to assess the integrity of pipes in service. Conventional ultrasonic testing such as local thickness gauging uses bulk waves and only tests the region of pipe below the transducer. An alternative ultrasonic technique is to use guided waves for long-range inspection. This can be done by using a pulse-echo arrangement from a single location on a pipe. Using this configuration waves propagate along the length of the pipe in both directions from the point where the wave is excited and all of the reflected signals are detected and analyzed. It is therefore important to quantify the reflection coefficient from defects of different size and shape.

The interaction of cylindrically guided waves with discontinuities in the geometry of the waveguide is a topic that has stimulated a great deal of interest.[1–10] The ability of guided waves to locate cracks and notches in pipes has been documented.[1–5] Furthermore the effect of defect size on the reflection and transmission characteristics has been investigated by many researchers.[6–11] There are some exact closed-form analytical solutions to elastodynamic scattering problems in the case of bulk waves. Closed form solutions for scattering of bulk waves can be obtained at high frequencies or at low frequencies[12] but in the midfrequency range numerical methods are necessary.

The guided wave scattering problem presents an enhanced complexity of the solution compared with bulk waves due to the presence of at least two modes at any given frequency. The large number of possible wave modes in a pipe is illustrated in the group velocity dispersion curves for a 3 inch schedule 40 steel pipe in Fig. 1. The software DISPERSE[13] has been used to trace the dispersion curves. In order to refer to different modes in cylindrical systems consistently, we use here a modified version of the system used by Silk and Bainton[3] which tracks the modes by their type, their circumferential order and their consecutive order. The modes are labeled $L(0,n)$, $T(0,n)$, and $F(m,n)$ and they, respectively, refer to longitudinal, torsional, and flexural modes. The first index m gives the harmonic number of circumferential variation and the second index n is a counter variable. Figure 1 shows that the modes travel at different speeds and are dispersive so that the original wave packet is distorted as it travels through the given structure. Much of the effort in flaw detection in pipes has been concentrated on the generation of a single mode in order to reduce the difficulty of dealing with many modes.[14] In a long range test it is also convenient to use a mode in a nondispersive region,[15] though a new technique which enables compensation for the signal spread caused by dispersion in a long-range guided wave testing is now being studied.[16]

Most of the previous investigators studied the effect of defects with small axial extent on the reflection of longitudinal waves.[7–9,17] Quantitative studies of the reflection due to notches[8] examined the correlation between the circumferential and through-thickness extent of notches and the amplitude of the reflected mode when a longitudinal $L(0,2)$ mode

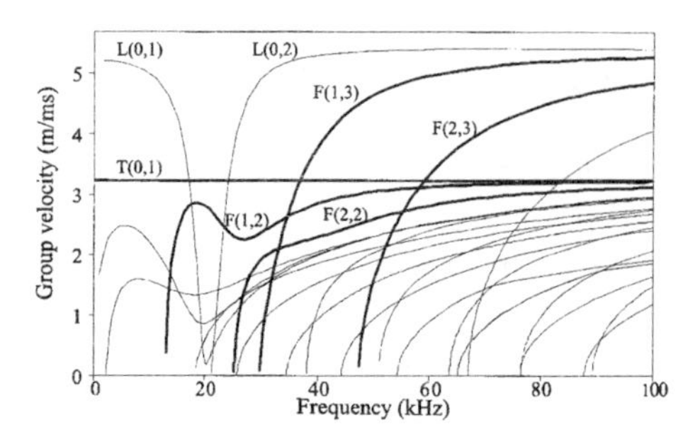

FIG. 1. Group velocity dispersion curves for 3 inch diameter pipe. Modes of interest are shown in bold.

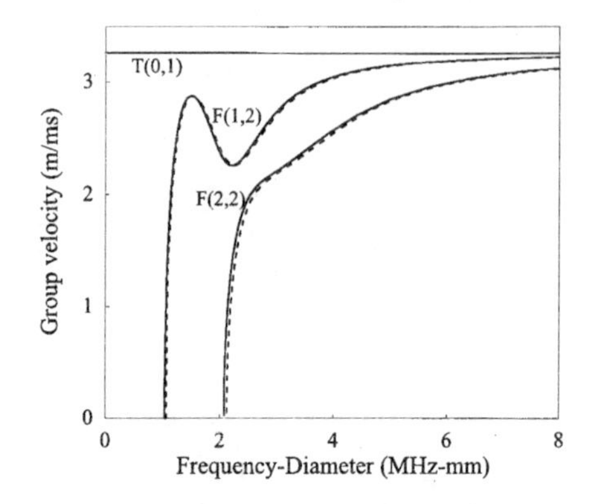

FIG. 2. Group velocity dispersion curves for $T(0,1)$, $F(1,2)$, and $F(2,2)$ modes in 24 inch (solid lines) and 3 inch pipes (dashed lines) as a function of frequency-diameter product.

is generated in the waveguide. Mode-conversion studies[7] when the $L(0,2)$ mode is launched in the test pipe enabled discrimination between axially symmetric reflectors such as circumferential welds and nonaxially symmetric defects. Guided wave tuning of nonaxisymmetric waves has also been used for defect detection in tubing.[11] Recently, work on crack characterization using guided circumferential waves has been published.[18] Guided circumferential waves have the limitation that they do not propagate along the length of the pipe so the range of the inspection is limited to a small percentage of the pipe.

In principle both axisymmetric and nonaxisymmetric modes can be used for long range inspection. Axisymmetric modes are in general preferable because they are easier to excite and have relatively simple acoustic fields. Initial practical testing was done using the longitudinal $L(0,2)$ mode.[7–9] However, more recent testing has employed the torsional mode.[10] This has the advantage that, in contrast to the $L(0,2)$ mode, the $T(0,1)$ mode propagation characteristics are not affected by the presence of liquid in the pipe and there is no other axially symmetric torsional mode in the frequency range, so axially symmetric torsional excitation will only excite the $T(0,1)$ mode, whereas when the $L(0,2)$ mode is used, the transducer system must be carefully designed to suppress the $L(0,1)$ mode.

No information is currently available on the reflection and mode conversion characteristics when a torsional mode is incident at a defect. The aim of this paper is to determine the reflection coefficients from cracks and notches of varying depth, circumferential and axial extent when the $T(0,1)$ mode is travelling in the pipe. An experimental and modelling study on the reflection caused by a series of defect geometries in both 3 inch and 24 inch pipes is presented. It is hoped later to generalize the findings to other pipe sizes.

II. GUIDED MODE PROPERTIES

Wave propagation properties in pipes are extremely complicated. Figure 1 shows the group velocity dispersion curves for a 3 inch, schedule 40 steel pipe. They were calculated using the program DISPERSE,[13] developed at Imperial College, and they show the velocity of propagation of the wave packets. The curves shown in bold are the modes which were used for the study. The dispersion curves of the $F(1,2)$ and $F(2,2)$ modes scale approximately with the frequency-diameter product in the frequency region of interest and the $T(0,1)$ velocity curve remains constant at all frequency values. This is shown in Fig. 2 which plots the group velocities of these modes as a function of frequency-diameter product for 3 inch (5.5 mm wall thickness) and 24 inch (20 mm wall thickness) pipes. The small difference between the curves is due to the fact that the diameter-to-thickness ratio is different for the two pipe sizes. This study considered an incident $T(0,1)$ mode in all cases, but mode conversion to the nonsymmetric modes could occur at nonaxisymmetric features. The frequency regions under examination were 10–50 kHz and 40–100 kHz for the 24 inch and the 3 inch pipes, respectively. The lowest frequencies of the range were chosen to be the frequencies usually used in practical testing (45–65 kHz for the 3 inch pipe and 10–20 kHz for the 24 inch pipe). It was also of interest to study the sensitivity of the $T(0,1)$ mode at higher frequencies.

The mode shape of the torsional $T(0,1)$ mode is not frequency dependent and it is completely nondispersive at all frequencies; its group velocity is the bulk shear velocity. No other torsional mode is present in the frequency range which was used for both finite element models and experiments. The mode shape of the $T(0,1)$ mode in a 3 inch pipe is shown in Fig. 3. This shows the profile of the tangential displacement through the thickness of the pipe wall. No axial or radial displacements are present in this mode. It can be seen that the tangential displacements are approximately constant through the wall thickness, indicating that defects will be detectable anywhere in the cross section of the pipe and there may also be potential for defect sizing. Since the $T(0,1)$ mode shape is also constant with frequency, pure mode excitation was obtained by simply imposing the mode shape at the center frequency.[19] The torsional mode in the 24 inch pipe has a very similar mode shape so all the comments about the $T(0,1)$ mode in the 3 inch pipe are also valid for the 24 inch pipe.

When an axisymmetric mode is incident at an axisymmetric defect, only the axisymmetric propagating modes ex-

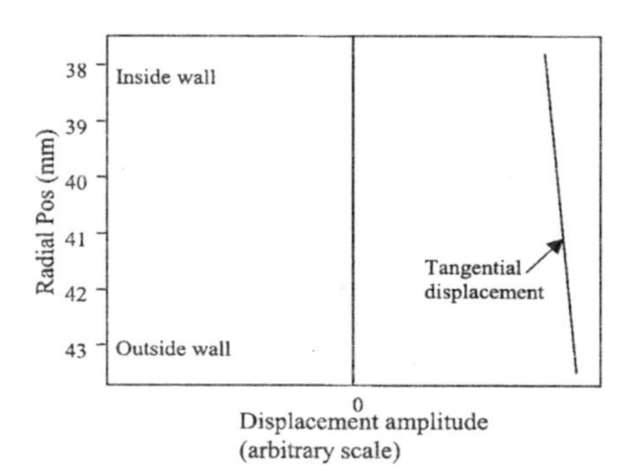

FIG. 3. $T(0,1)$ mode shape in a 3 inch pipe at 45 kHz. Radial and axial displacements are zero.

isting at the frequency of interest will contribute to the reflected field. Since we worked below the $T(0,2)$ cutoff frequency, there was no mode conversion at an axisymmetric defect and $T(0,1)$ was simply reflected or transmitted past the defect. However $T(0,1)$ can convert to the $F(1,2)$, $F(1,3)$, and $F(2,2)$ modes when an asymmetric feature exists in the pipe. The conversion is dependent on the similarity in particle motion of the incident and mode converted modes. Figures 4(a)–4(c) show the mode shapes of the $F(1,2)$, $F(1,3)$, and $F(2,2)$ modes, respectively, at 45 kHz. The $F(1,2)$ and $F(2,2)$ modes have dominant tangential displacement so the conversion to these modes from $T(0,1)$ is strong. Their torsional behavior becomes even more dominant at higher frequencies as illustrated in Figs. 5(a) and (b) which show the mode shapes at 100 kHz. Both tangential and longitudinal displacements are significant in the case of the $F(1,3)$ mode at 45 kHz so this mode has both torsional and extensional behavior at this frequency. The torsional behavior becomes less significant for this mode as the frequency increases [see the mode shape for the $F(1,3)$ mode at 100 kHz in Fig. 5(c)]. It is also important to understand the mode shape characteristics at low frequency; Fig. 5(d) shows that the $F(1,2)$ mode shape at 25 kHz is characterized by longitudinal and tangential displacements of the same order.

III. EXPERIMENTAL SETUP

Laboratory experiments were performed on a set of 3 inch schedule 40 steel pipes to determine the reflection sensitivity of the torsional $T(0,1)$ mode to a series of notches with different dimensions. The specimens which were used were 3.1 meter long pipes and the notches were machined at 2.3 meters from end A using a 3.2 mm diameter slot drill cutter (see Fig. 6). The pipe was tested horizontally; it was radially clamped at end B using 8 bolts evenly distributed around the circumference and rested on a steel support positioned approximately at the midpoint of the pipe. The support produced a negligible reflection of the ultrasonic signal; the reflection from end B was not monitored so the nature of the clamping there was unimportant. The cutter was aligned along a radius of the pipe so the circumferential extent of the cut was changed by simply rotating the pipe about its axis of symmetry. Two rings of 16 dry coupled transducers were clamped to the pipe at 0.9 meters from end A and left in place during the cutting so all of the experiments on one pipe were done without removing the transducers. The use of two rings enabled us to distinguish the direction of propagation of the reflected signal (left or right with respect to the ring position).[10] Using piezoelectric transducers equally spaced to produce tangential displacement, only torsional modes are generated.[10] In order to generate only axially symmetric

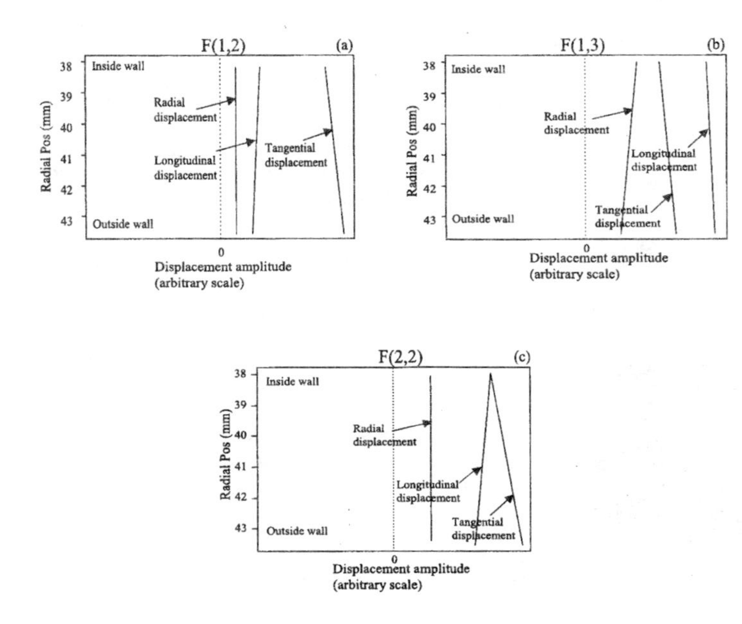

FIG. 4. Displacement mode shapes in a 3 inch pipe at 45 kHz for $F(1,2)$ (a), $F(1,3)$ (b), and $F(2,2)$ mode (c).

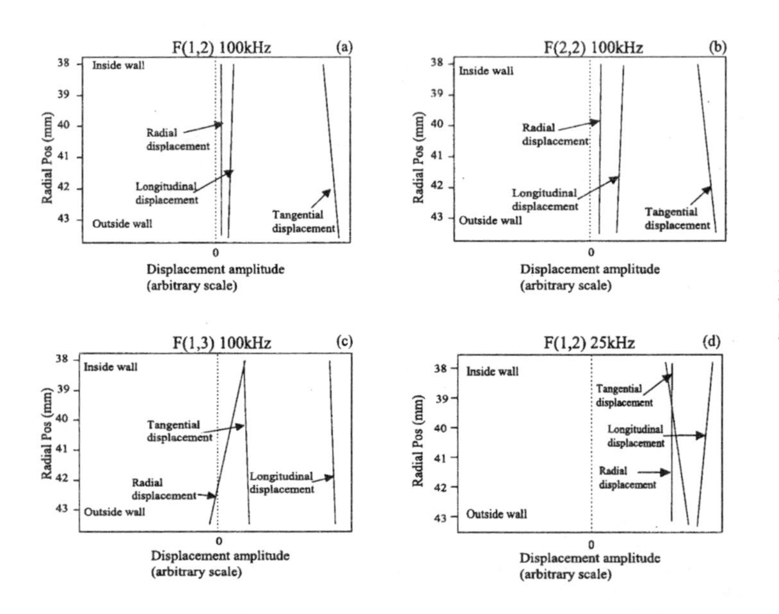

FIG. 5. Displacement mode shapes in a 3 inch pipe at 100 kHz for $F(1,2)$ (a), $F(2,2)$ (b), and $F(1,3)$ mode (c) and at 25 kHz for $F(1,2)$ mode (d).

modes the number of elements in the ring has to be greater than n where $F(n,1)$ is the highest order flexural mode whose cutoff frequency is within the bandwidth of the excitation signal.[20] A Guided Ultrasonics Ltd. Wavemaker 16 instrument was used to generate a six cycle Hanning-windowed toneburst at many frequency values from 40 to 70 kHz. The same instrument was also used for signal reception. In order to obtain a reference measurement, reflections from end A were recorded before any machining was done. By doing that we verified that the vibration induced from the cutting left the contact characteristics of the transducer rings on the pipe unchanged. Experiments were conducted on four separate pipes, using the following notches: (a) a through-thickness notch with axial extent equal to about 3.5 mm and varying circumferential extent; (b) a notch extending over the full circumference of the pipe with axial extent equal to about 3.5 mm and varying notch depth; (c) a notch extending over 25% of the circumference of the pipe with axial extent equal to about 3.5 mm and varying notch depth; (d) a notch extending over the full circumference of the pipe with notch depth equal to 20% of the thickness and varying axial extent.

FIG. 6. Experimental setup.

IV. FINITE ELEMENT MODELS

The finite element (FE) method has been extensively and successfully used to study the interaction between guided waves and defects in structures.[7–9,15,21–25] In general a three-dimensional solid model is required to perform a numerical analysis of the interaction between guided waves and discrete defects. However 3D models are computationally expensive so when possible we use simplified models.[7] In some cases it is possible to study a specific 3D model by using a combination of two 2D models. Here we considered three types of models:

(1) *Membrane*: This is a three-dimensional analysis in which the pipe wall is represented by membrane elements. These allow defects removing a fraction of the pipe circumference to be modelled, but the defects remove the full wall thickness. The axial extent of the defect can be varied.

(2) *Axisymmetric*: This is a two-dimensional analysis which can deal with defects which remove part of the wall thickness, but are axially symmetric and so cover the whole pipe circumference. The axial extent of the defect can be varied.

(3) *Three dimensional*: Previous work[7–9] had indicated that the $L(0,2)$ reflection coefficient from a defect was linearly proportional to its circumferential extent. The reflectivity of a defect of a given depth and circumferential extent could therefore be predicted from results of the axisymmetric analysis by obtaining the reflection coefficient for an axisymmetric defect having the depth of interest and scaling by the fraction of the circumference covered by the defect. The three-dimensional models were used to demonstrate the va-

lidity of making predictions using combinations of results from the different types of two-dimensional model when the torsional $T(0,1)$ mode is incident.

In each case a length of the pipe was modelled and a notch was introduced at some distance along it. The input wave signal was excited in the model by prescribing time-varying displacements at one end of the pipe. Narrow band signals were used, typically composed of 10 cycles of the desired center frequency modulated by a Gaussian window. The tangential displacements were then monitored at a location between the excitation end and the notch, thereby detecting both the incident wave on its way to the notch and the reflected wave returning. The reflection ratio was calculated by dividing the amplitude of the reflected mode by the amplitude of the incident mode in the frequency domain.

In both axisymmetric and 3D models the excitation was achieved by prescribing the displacements to match the exact mode shape at the end of the pipe. The displacement distribution through the thickness was derived using Disperse.[13] In the membrane model the torsional wave was excited by prescribing the displacement all around one circumference; displacement distribution through thickness is not possible using this model. The notches were introduced either by disconnecting adjacent elements or by removing elements.

A. Membrane models (full depth, part circumference)

For the membrane models the geometry of the pipes was discretized using membrane finite elements. This type of element describes a two-dimensional rectangular region and it models the stresses which lie in the plane: the two orthogonal in-plane direct stresses and the in-plane shear stress. It assumes plane stress conditions, therefore incorporating the capacity of the plate to change thickness when the in-plane stresses are applied. Although it is a two-dimensional element by shape and (local) behavior, it may be used in a three-dimensional model, meaning simply that it can be orientated arbitrarily in a three-dimensional space. In a three-dimensional model its nodes have displacement degrees of freedom in the three coordinate directions and it has mass which participates in any coordinate direction.

The justification for using membrane elements is the nature of the mode shapes of all of the modes which were studied. As already seen in Figs. 3–5, the pipe wall shows extensional or torsional behavior for all of the modes (especially at high frequency). Therefore the modes of interest could be modelled reasonably well by using the membrane elements, each element lying at the center of the pipe wall. It is not possible to use membrane models in order to simulate modes in which there is local bending of the pipe wall, nor to model part-through notches.

Half of the circumferential extent of a length of pipe was modelled as in previous work with the $L(0,2)$ mode[7,8] but in this case we assumed one plane of antisymmetry. A wave was excited at the end of the pipe, a defect was simulated and the signals were monitored on a line between the excitation and the defect. A mesh of identically sized linear (four noded) quadrilateral membrane elements was used. The models of the 24 inch pipe represented a 3.6 m length, using 900 elements along the length and 40 elements around the circumference. As a result each element was about 4 mm long and 24 mm along the circumference. The models of the 3 inch pipe also represented a 3.6 m length, using 900 elements along the length and 40 elements around the circumference. Hence each element was about 4 mm long and 3 mm in the circumferential direction. Convergence studies indicated that the meshes gave reliable results. The zero-length notches were modelled by disconnecting adjacent elements; thus although the elements on each side of the notch had nodes at coincident locations, they were not connected. The notches with nonzero lengths were modelled by removing elements from the model and were therefore rectangular in shape. The notches were introduced adjoining one of the planes of antisymmetry so that the geometric model described half of the pipe and half of the notch, and the center of the notch was therefore at the plane of antisymmetry.

The excitation of the axially symmetric modes (circumferential order 0 modes) was achieved by prescribing identical tangential displacement histories at all of the nodes at the end of the pipe. On reception of the multimode reflected signal, the order 0 modes were extracted simply by adding the tangential displacements at all of the monitoring nodes around the circumference of the pipe. For the order 1 modes, a phase delay of $\theta/2\pi$ was added to each signal before summing them,[7,8] where θ is the angular distance from the center of the notch. Thus a separate processing calculation was performed in order to extract the amplitude of each type of mode. In principle three order 0 modes $[L(0,1),L(0,2),T(0,1)]$ and three order 1 modes $[F(1,1),F(1,2),F(1,3)]$ can exist in the high frequency range. However $L(0,1)$ and $L(0,2)$ are both longitudinal and are not excited, and $F(1,1)$ does not propagate in the membrane model because it is characterized by through-wall bending behavior. In most cases the processing already described was therefore sufficient to identify separately the remaining $T(0,1)$, $F(1,2)$, and $F(1,3)$ modes. At some frequencies the $F(1,2)$ and $F(1,3)$ modes have very similar group velocity so it was not possible to separate them in the time domain. Since the signals were rather narrow-band, the processing could reasonably have been performed directly on the raw time records. However, for better accuracy, the calculations were performed in the frequency domain.

Typical time records from the simulations are shown in Figs. 7(a) and 7(b). Both of these records are for the 24 inch pipe with a notch which extends around 25% of the circumference, and for the $T(0,1)$ mode incident at 50 kHz; the difference between them is in the processing. Figure 7(a) shows the signal with processing for order 0. This shows clearly the incident signal on its way towards the notch and then the reflected $T(0,1)$ mode. Figure 7(b) shows the signal when the same raw results are processed to extract the order 1 modes. This now shows the reflection of the $F(1,2)$ mode. Also, the incident mode has vanished in this plot because it is order 0 and is therefore not detected by the order 1 processing.

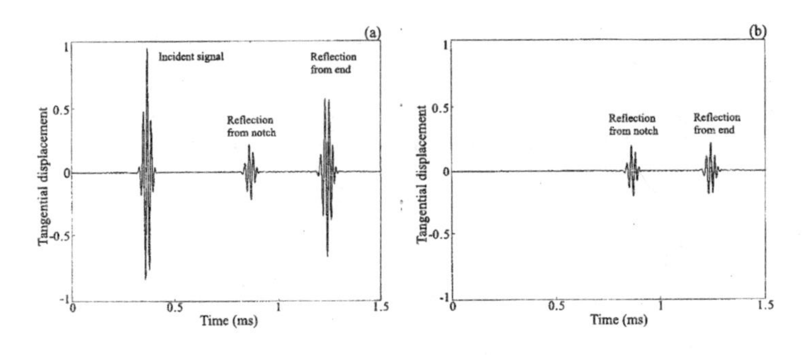

FIG. 7. Predicted time record for membrane model of 24 inch pipe with notch extending around 25% of circumference and $T(0,1)$ mode incident. Results processed to show order 0 modes (a) and order 1 modes (b).

B. Axisymmetric models (full circumference, part depth)

The axisymmetric elements describe a two-dimensional region representing a radial–axial section through an axially symmetric structure. Thus in application to the pipe models they represent a section through the pipe wall on a plane which contains the axis of the pipe. They model the three in-plane stresses (the radial and axial direct stresses and the radial-axial shear stress) and also the circumferential stress. Thus radial displacements of the elements are correctly coupled to circumferential stresses. It follows that any geometric shape which is modelled must be axially symmetric. Therefore, although it is possible to model notches which extend part-way through the pipe wall (unlike the membrane models which only represent through-wall notches), any notch is assumed to extend around the full circumference. Since the geometry is axially symmetric, there is no mode conversion between modes of different circumferential orders, so only modes of the same order as the excitation signal can propagate. In our models only the $T(0,1)$ torsional mode can be reflected from the axisymmetric notch because this is the only torsional mode existing in the working frequency range.

The excitation, monitoring and notch locations were arranged along the pipe similarly to the membrane models discussed earlier. Identically sized linear (four noded) quadrilateral axisymmetric elements were used. The wall thickness of the 24 inch and 3 inch pipes were taken to be 20 mm and 5.5 mm, respectively. The models of the 24 inch pipe represented a 2.4 m length, using 1000 elements along the length and 10 elements through the thickness. Therefore each element was 2.4 mm long and 2 mm in the thickness direction. The models of the 3 inch pipe with the $T(0,1)$ mode represented a 2.4 m length, using 2000 elements along the length and 10 elements through the thickness. Therefore each element was 1.2 mm long and 0.55 mm in the thickness direction. Again convergence studies showed that this discretization was satisfactory. Two other FE models of the 3 inch pipe with an 80% defect depth were run at 150 kHz and 240 kHz center frequency in order to obtain the reflection coefficient at high frequencies. A finer mesh was needed in the 240 kHz case according to the practical rule of having at least eight elements per wavelength of the smallest wavelength of the excitation signal. The ratio between the frequency spectrum of the reflected signal and the spectrum of the input was then derived for each FE run. Using four FE models with different center frequencies (40 kHz, 100 kHz, 150 kHz, and 240 kHz) we covered the frequency bandwidth from 15 kHz to 270 kHz. Higher frequencies were not considered because of the presence of the second torsional mode $T(0,2)$ [the cutoff frequency for the $T(0,2)$ torsional mode is 300 kHz in a 3 inch pipe].

The effect of the axial extent of the notch on the reflection coefficient of an axisymmetric notch was also studied. Several FE models with 20% defect depth and different axial extent have been studied. The notches were created in the FE simulations by simply removing elements in the mesh. When the axial extent was long enough, the reflection of the start of the notch was separated from the reflection from the end of the notch.

C. 3D models (part circumference, part depth)

The three-dimensional modelling used 8-node "brick" elements to discretize a length of pipe fully in three dimensions. The brick elements are cuboid in shape and are used to represent, piecewise, the whole volume of the structure. Potentially this offers the advantage of being able to model notches which have limited extent both in the circumferential and the through-wall directions. However, there is a heavy computational penalty because of the large numbers of elements which are necessary for three-dimensional models. We here modelled a pipe with a part-circumference and part-thickness notch. The purpose was to compare the reflection coefficient obtained with the one derived by combining the results of membrane and axisymmetric models in order to confirm the assumption which has been made that the reflectivity from such notches can be inferred from the combined results of the membrane and axisymmetric models.

Half of the circumferential extent of a 2.4 m length of the 3 inch pipe was modelled, assuming one plane of antisymmetry. The excitation, notch and monitoring locations were arranged similarly to the membrane models. The pipe was modelled with 600 elements along the length, 40 elements around the circumference, and three elements through the 5.5 mm wall thickness. The $T(0,1)$ mode at 100 kHz was excited and monitored. The excitation was achieved by prescribing the displacements to match the exact mode shape at the end of the pipe. The $T(0,1)$ mode was monitored by adding the circumferential displacements at all of the nodes

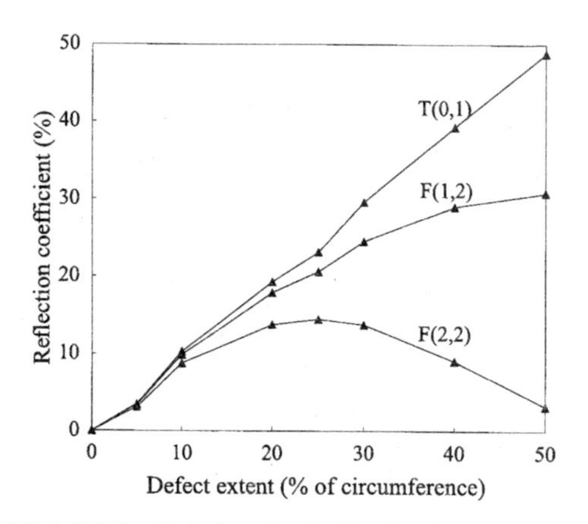

FIG. 8. Variation of reflection ratio with defect circumferential extent for zero axial length, full wall thickness defect. Results are from membrane model with $T(0,1)$ incident in 3 inch pipe at 100 kHz.

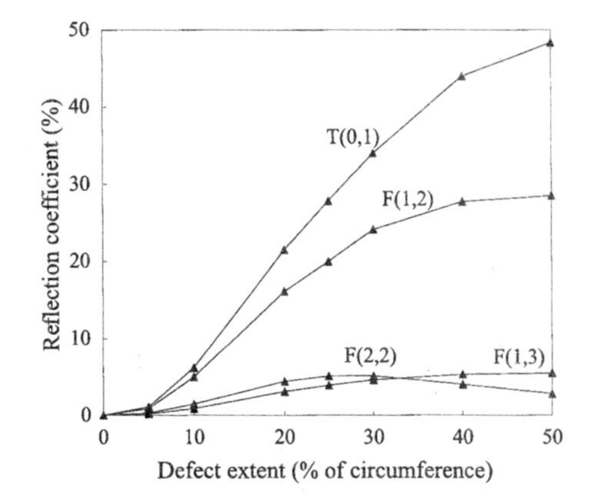

FIG. 9. Variation of reflection ratio with defect circumferential extent for zero axial length, full wall thickness defect. Results are from membrane model with $T(0,1)$ incident in 3 inch pipe at 45 kHz.

around one circumference at the monitoring location.

Before tackling the problem of the notch with limited extent in both circumferential and depth directions, some analyses of an axially symmetric part-depth notch and a through-wall part-circumference notch were studied using the 3D model and compared with the related 2D models. The results from the 3D mesh compared well with those obtained from the 2D models, so validating the discretization.

A 3D model of a part-depth (33%), part-circumference (25%) notch was analyzed using 100 kHz $T(0,1)$ excitation. A notch axial length of 180% of the wavelength ($\lambda=32.6$ mm) was chosen, so the defect was a square patch (60 mm×60 mm) 1.8 mm deep.

V. STUDY OF PARAMETERS AFFECTING REFLECTION AND MODE CONVERSION

The numerical results are presented below. We study separately the influence of parameters such as pipe dimension, frequency of the excitation, circumferential extent of the defect, depth of the defect and axial extent of the notch. The section is divided into three sections in which we describe the results for the three different FE models.

A. Through thickness defects with zero axial extent (membrane model)

Using the results of the membrane models we studied the effect of the circumferential extent of a defect on the backscattering of the torsional wave at different frequencies. We first present the results of the simulation for the 3 inch pipe case and later offer a comparison with the 24 inch case.

Practical testing of 3 inch pipes using torsional excitation is usually performed at relatively low frequency (45–65 kHz).[10] We start by presenting the results at higher frequency (100 kHz) because of the simplicity of this case and we will focus later on the frequencies used in practical testing. The reflection coefficient on the 3 inch pipe with $T(0,1)$ input at 100 kHz is shown in Fig. 8, the computed points being joined by straight lines.

The $T(0,1)$ reflection coefficient is approximately linear with circumferential extent, the $F(1,2)$ reflection coefficient has a maximum at 50% of the circumferential extent, and the $F(2,2)$ has a maximum at 25% of the circumferential extent and a minimum corresponding to 50% of the circumferential extent. The results obtained in this case are very similar to the results obtained in previous work on the $L(0,2)$ reflection coefficient in a 3 inch pipe at 70 kHz.[7] No significant presence of the $F(1,3)$ mode is noticeable at this frequency.

The variation of the $T(0,1)$ reflection coefficient with circumferential extent at 100 kHz shown in Fig. 8 indicates some divergence from the overall linear trend; this is probably due to numerical problems. In principle a finer mesh could be used to investigate this but the model which was used here is close to the limits of the available resources so this has not been pursued. In any case the extent of the scatter is small in absolute terms, and is not important from the point of view of practical application.

The interpretation of the 3 inch FE simulation at lower frequencies is more complicated than at high frequency because of changes in the mode shapes of the relevant modes. In the intermediate frequency range (from 30 to 70 kHz) $F(1,3)$ is seen in the reflected signals along with the other modes. This is because there is significant circumferential motion in its mode shape over this frequency range as shown in Fig. 4(b), whereas this displacement component is smaller at higher frequency [see Fig. 5(c)]. Figure 9 shows the results at a frequency of 45 kHz. At all circumferential extents the amplitude of the $F(1,3)$ reflection is about 20% of that of the $F(1,2)$ mode. Figure 10 shows the $T(0,1)$, $F(1,2)$, and $F(1,3)$ reflection coefficients at frequencies from 25 kHz to 100 kHz for the 25% circumferential extent case. As expected from the mode shapes, the ratio between the $F(1,3)$ and $F(1,2)$ reflection coefficients becomes smaller at higher frequencies. At low frequency (25 kHz) the $T(0,1)$ reflection coefficient is small. Figure 10 also shows that the $F(1,2)$ mode reflection coefficient at low frequency is very small compared with the other results at higher frequencies. This is

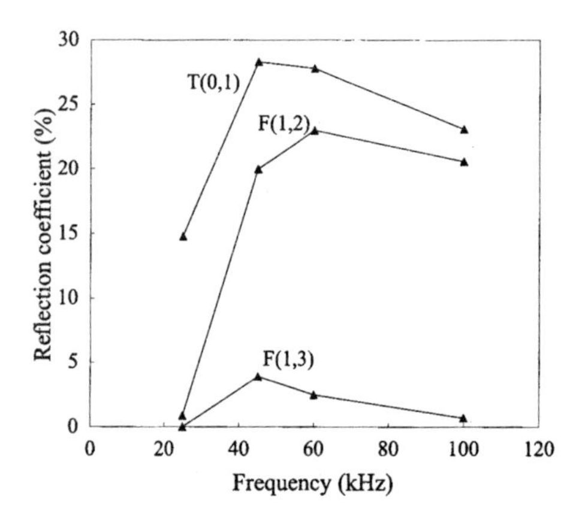

FIG. 10. Variation of reflection ratio with frequency for $T(0,1)$ mode input in 3 inch pipe using membrane model with 25% notch circumferential extent.

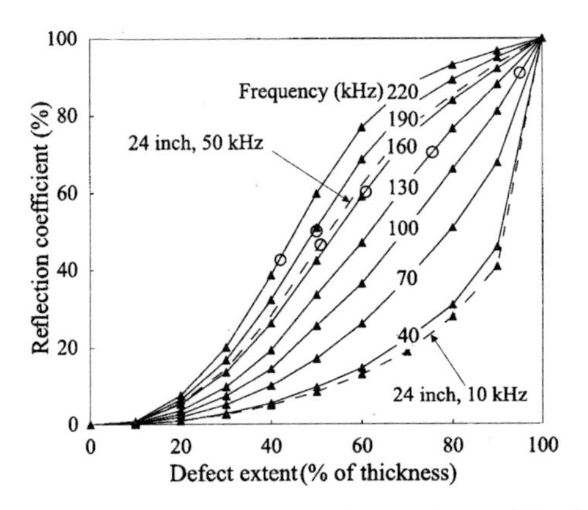

FIG. 12. Variation of reflection ratio with defect depth for zero axial length at various frequencies, axisymmetric defect. Results are from axisymmetric model with $T(0,1)$ incident in 3 inch (solid lines) and 24 inch pipes (dashed lines). The empty circles indicate the depth value for which $ka=1$ at each frequency.

due again to the $F(1,2)$ mode shape at low frequency. From Fig. 5(d) it can be seen that at 25 kHz the $F(1,2)$ mode has a longitudinal displacement bigger than the tangential displacement so this mode does not have predominantly torsional behavior at low frequency.

In the case of a 24 inch pipe the general behavior of the reflection coefficient seen in the 3 inch pipe at 100 kHz case is confirmed (see Fig. 11). No presence of the $F(1,3)$ mode was observable in the 24 inch FE simulations that were run.

B. Part thickness axisymmetric defects (axisymmetric model)

The results of the axisymmetric FE modelling are plotted in Figs. 12–14. Since the defect is axisymmetric, there is no mode conversion between the axisymmetric (T) modes and the flexural (F) modes. Figure 12 shows the trend of the $T(0,1)$ reflection coefficient for both 3 inch and 24 inch pipes as a function of defect depth at different excitation frequencies. If we consider one pipe dimension at one frequency value we observe that the reflection coefficient increases with defect depth in a nonlinear manner. The shape of the curves is similar to that predicted for a 3 inch pipe at 70 kHz using the $L(0,2)$ mode in earlier published work.[8] If we now consider one pipe dimension at different frequencies of excitation, we can see that as the frequency decreases, the curves become increasingly concave, the reflectivity at low defect depths decreasing markedly. It therefore becomes more difficult to detect shallower defects as the test frequency is decreased. It is also of interest to observe that when the frequency-thickness product is similar, the reflection is similar. Specifically, the reflection coefficients for a 3 inch pipe with 5.5 mm wall thickness at 40 kHz (frequency-

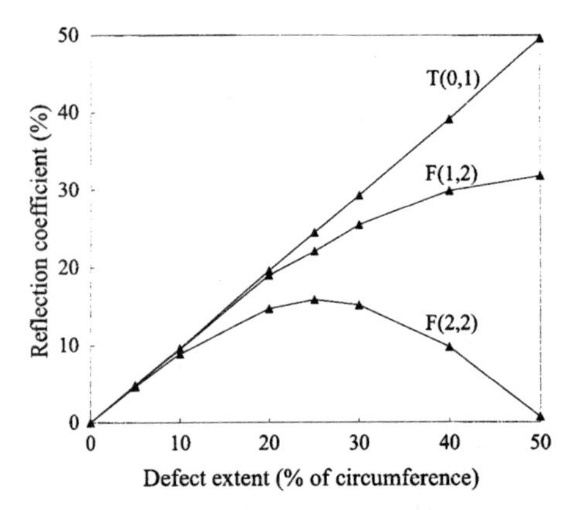

FIG. 11. Variation of reflection ratio with defect circumferential extent for zero axial length, full wall thickness defect. Results are from membrane model with $T(0,1)$ incident in 24 inch pipe at 50 kHz.

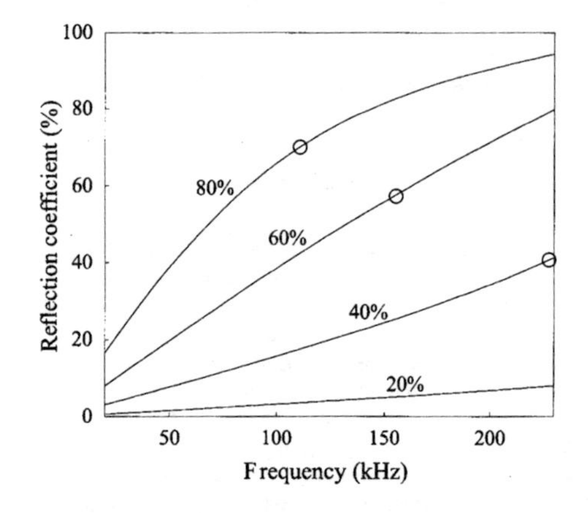

FIG. 13. Variation of reflection ratio with frequency for zero axial length at various defect depths, axisymmetric defect. Results are from axisymmetric model with $T(0,1)$ incident in 3 inch pipe. The empty circles indicate the frequency at which $ka=1$ for each defect depth.

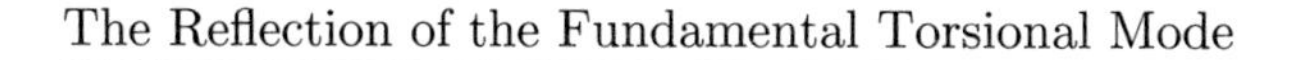

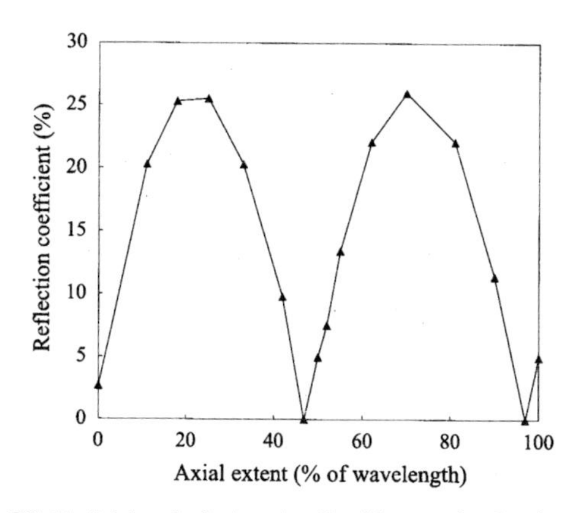

FIG. 14. Variation of reflection ratio with axial extent when there is an axisymmetric defect with 20% thickness depth. Results are from axisymmetric model with $T(0,1)$ incident in 3 inch pipe at 100 kHz.

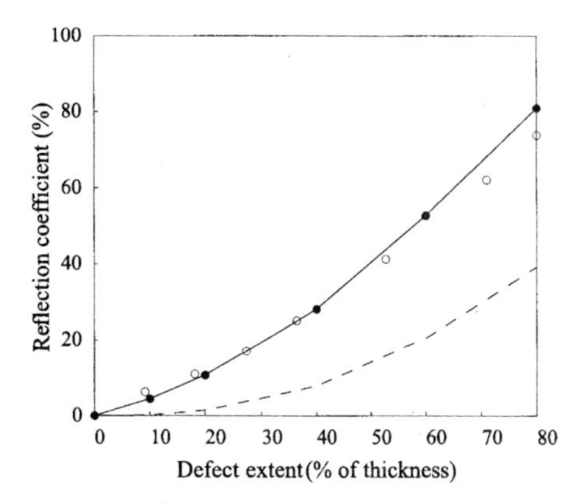

FIG. 15. Variation of reflection ratio in incident mode with defect depth for 3 mm axial length, axisymmetric defect. FE (–●–) and experimental (○) results are for $T(0,1)$ incident in 3 inch pipe at 55 kHz. The crack case (zero axial extent) is also displayed for comparison (dashed line).

thickness product 220 kHz mm) are similar to the reflection coefficients for a 24 inch pipe with 20 mm wall thickness at 10 kHz (frequency-thickness product 200 kHz mm). The same observation is valid at high frequency-thickness values; the 3 inch pipe at 160 kHz (880 kHz mm) has a reflection coefficient similar to that of the 24 inch pipe at 50 kHz (1000 kHz mm). Figure 13 shows the reflection coefficient in the frequency domain when a 3 inch pipe with various depth notches is excited with the torsional $T(0,1)$ mode. The reflection coefficient increases with frequency at all depth values. In the 80% depth case the rate of change decreases as the frequency increases.

The effect of the axial extent of the defect on the reflectivity of the $T(0,1)$ mode in a 3 inch pipe is plotted in Fig. 14. The reflection coefficient has a maximum at an axial extent of about a quarter wavelength and a minimum at about half-wavelength. Similar behavior has been reported for the s_0 mode in a plate.[25]

C. 3D models

The results of the 3D analysis are given in Table I. The results demonstrate excellent agreement between the predictions of the 2D axisymmetric analysis with those of the 3D analysis; for the case of an axisymmetric, 33% depth defect, the predicted reflection coefficients are 23.5% and 23.7% with the 2D axisymmetric and 3D models, respectively. Similarly good agreement is shown between the membrane and 3D models for the case of a through wall defect with 25% circumferential extent, the predicted reflection being 21.1% with both models.

Finally, Table I shows the results for the 3D analysis of a notch with 25% circumferential extent, 33% depth and axial extent of 180% of the wavelength. The 3D analysis predicts a reflection ratio of 5.2% for this case. This is closely matched by the multiplication of the axisymmetric and membrane results which give 5.0%. This confirms the validity of combining the axisymmetric and membrane results to predict the 3D case.

VI. EXPERIMENTAL VALIDATION OF NUMERICAL MODELLING

A series of experiments was carried out in order to validate the results obtained from the FE modelling. From previous modelling it was clear that the axial extent can be an important parameter affecting the reflection coefficient from notches, so when simulating the laboratory test cases using FE models it was necessary to introduce a notch with the precise axial extent used in the experiments. Part-thickness, part-circumference defects were modelled using a combination of 2D axisymmetric and 2D membrane models.

Figure 15 shows the measured and predicted reflection coefficients at 55 kHz for a series of axisymmetric notches with 3 mm axial extent in which the notch depth was varied, together with the prediction for a crack (zero axial extent).

TABLE I. Comparison of $T(0,1)$ reflection ratio in a 3 inch pipe at 100 kHz from 3D model with combined results from axisymmetric and membrane models.

Axial extent [% of λ(32.6 mm)]	180	180	180
Circumferential extent (%)	100	25	25
Through-wall extent (%)	33	100	33
Reflection ratio (%): Axisymmetric model (2D)	23.5		
Reflection ratio (%): Membrane model (2D)		21.1	
Reflection ratio (%): 3D model	23.7	21.1	5.2
Reflection ratio predicted by combining 2D analyses			5.0

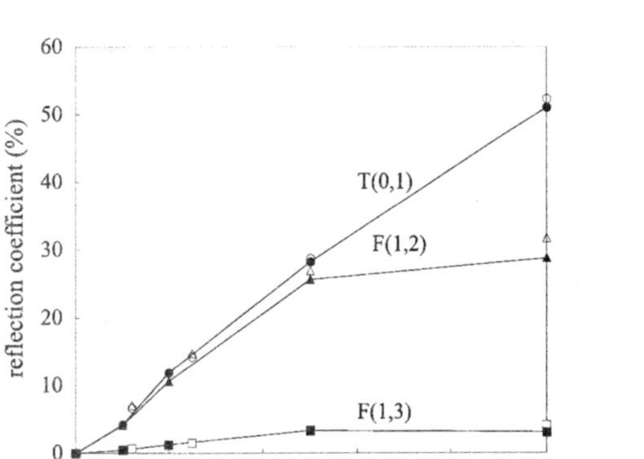

FIG. 16. Variation of reflection ratio with circumferential extent for 3 mm axial length, through thickness defect. FE (lines with solid symbols) and experimental (empty symbols) results are for $T(0,1)$ incident in a 3 inch pipe at 55 kHz.

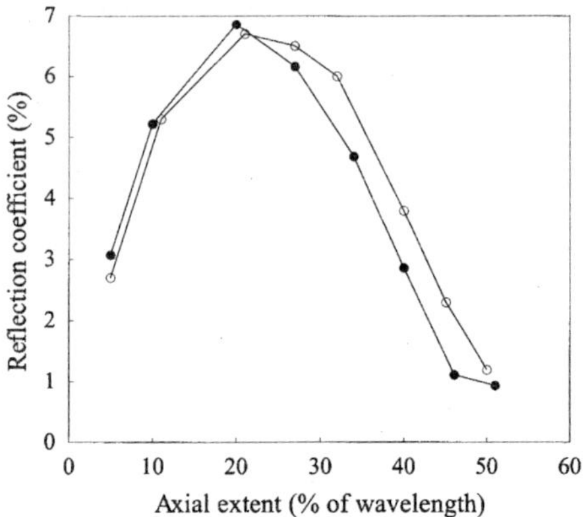

FIG. 18. Variation of reflection ratio in incident mode with axial extent for a 20% depth, 25% circumferential extent defect. FE (●) and experimental results (○) are for $T(0,1)$ incident in 3 inch pipe at 55 kHz.

The experimental measurements and finite element predictions can be seen to agree well. The reflection coefficients for the notch case are higher than those for the crack case. Moreover the concavity of the curve is reduced with respect to the crack case.

Figure 16 shows a comparison between experimental and FE results in the case of through-thickness notches with 3 mm axial extent and varying circumferential extent when the $T(0,1)$ mode is excited at 55 kHz. Again there is good agreement between the experimental and modeling results.

Figure 17 shows the dependence of the reflection coefficient on the frequency in the case of a through-thickness defect with 25% circumferential extent and axial extent equal to 3 mm. The reflection coefficient for $T(0,1)$ decreases when the frequency is increased from 50 kHz to 70 kHz in both experiments and predictions. Good agreement

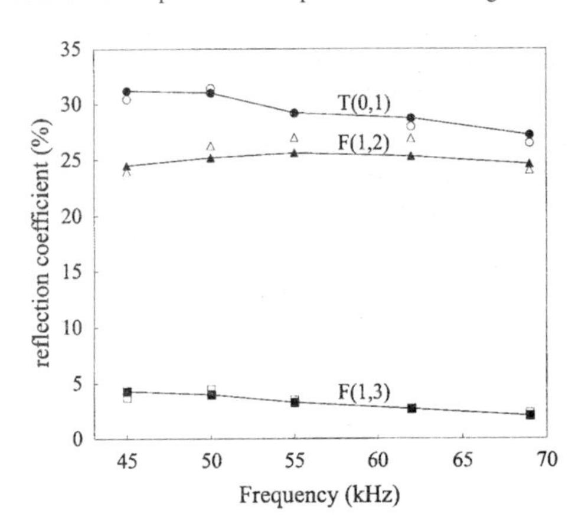

FIG. 17. Variation of reflection ratio with frequency for 3 mm axial length, through thickness defect extending over the 25% of the circumference of a 3 inch pipe. Both FE (lines with solid symbols) and experimental results (empty symbols) are displayed.

between experiments and finite elements is also found in this case.

Figure 18 shows the behavior of the reflection coefficient when the axial extent is changed. The results are derived for a 20% depth defect extending over 25% of the circumference at 55 kHz center frequency excitation. The form of the experimental results agrees with the predictions, the peak of the reflection coefficient occurring when the axial extent is about a quarter wavelength and the minimum at about half wavelength.

VII. DISCUSSION

A defect in a solid body represents a scatterer for elastic waves and so in principle it can be detected and characterized by its effect on an incident pulse of ultrasonic wave motion. Elastodynamic scattering problems can be solved in closed form in the high frequency regime using either the Kirchoff approximation or the geometrical diffraction theory and at low frequency applying the Rayleigh approximation.[12] In the midfrequency range these methods are not valid because the wavelength of the incident pulse is of the same order as the characteristic dimension of the defect and preferable alternatives are numerical methods such as finite element, boundary element or finite difference. Scattering of bulk waves from a crack in an infinite medium has been studied by many investigators and much of this work has been reviewed by Kraut[26] and Datta.[27] Elastodynamic ray theory has been thoroughly studied by Achenbach *et al.*[28] in order to construct scattering fields generated by cracks.

The interaction of Lamb waves with cracks and notches in plate structures is a topic that has received a great deal of interest in recent years.[9,24,25,29–31] Wave scattering of guided elastic waves from cracks and notches in isotropic hollow cylinders has also been investigated.[6–9,11,17,18,32] The scattering of guided waves is more difficult than the bulk wave case as a consequence of the higher number of modes to be considered and the complexity of the mode shapes. When a

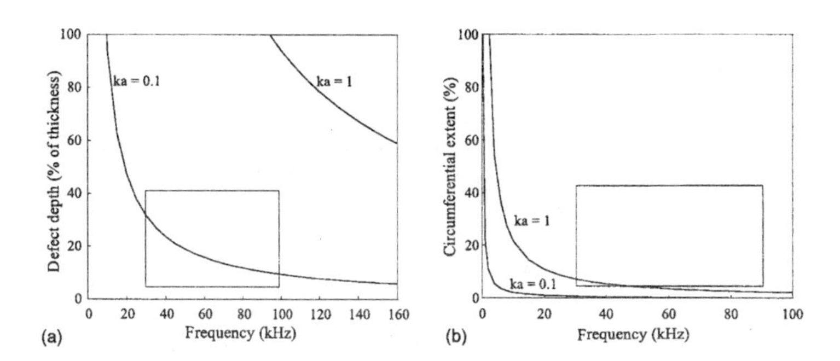

FIG. 19. Scattering regime regions in the case of the axisymmetric defect in 3 inch pipe with 5.5 mm wall thickness (a) and through thickness defect (b). The boxes indicate the practical testing regions.

single mode in a waveguide encounters the defect there is coupling between the incident mode and all of the propagating and nonpropagating modes of the waveguide.[33] If the location of the detector of the wave is placed far enough from the defect, only the propagating modes will have significant amplitude. In fact only a few of the propagating modes contain a significant amount of scattered energy so the problem can be further simplified by only considering these modes.[7]

A. Analysis of the sensitivity of torsional mode to cracks (zero axial extent)

It is well known that three scattering regimes can exist depending on the dimension of the scatterer in relation to the wavelength of the ultrasonic wave.[12] A parameter frequently used to define and distinguish the scattering regimes is ka where a is the characteristic dimension of the defect and k is the wave number defined by

$$k=\frac{2\pi f}{V}, \qquad (1)$$

where f is the frequency and V is the phase velocity. In the specific case of a torsional wave, the velocity is constant with frequency so the wave number simply increases linearly with frequency. When $ka<0.1$ the problem can be approximated using a low frequency or quasistatic approach. At values for which $ka>1$ it is possible to use a high frequency approximation (however $ka=1$ cannot be considered as a strict boundary between the different regimes). For part-thickness, axisymmetric defects the characteristic length is the defect depth, while for through-thickness, part-circumference defects the characteristic length is the circumferential extent. Figures 19(a) and (b) show the scattering regimes for axisymmetric and through thickness cracks in the case of a 3 inch pipe with 5.5 mm wall thickness. In both figures a rectangular region which is important from the point of view of practical testing is also highlighted. It is evident that in the case of axisymmetric cracks we are in the low–intermediate frequency regime, whereas in the case of through thickness defects we are in the intermediate–high frequency regime.

1. Discussion on through thickness cracks

The motivation of separately studying the reflection characteristics of the through-thickness, part-circumference, and the part-depth axisymmetric defects is the difference in geometry and scattering regime of the two classes of crack. The reflection coefficients of the 24 inch pipe with $T(0,1)$ input at 10 kHz and 50 kHz as a function of circumferential extent are shown in Fig. 20. The circumferential extent is expressed as a percentage of the pipe circumference (1980 mm). The $T(0,1)$ reflection coefficients are very similar at the two frequencies and they increase roughly linearly with circumferential extent, especially at the higher circumferential extents (corresponding to higher ka). A somewhat smaller reflection is obtained for a small defect (5% of circumference) at low frequency (10 kHz). The same behavior is noticeable in the 3 inch case.

Let us first give an explanation of how the frequency of the test influences the reflection coefficient from through thickness cracks and we will then explain why the dimension of the defect is an important parameter. Lowe *et al.*[7,8] showed that the $L(0,2)$ reflection coefficient varies roughly linearly with respect to the circumferential extent of the defect. They presented a calculation in order to discover whether the reflection coefficients could be estimated from a simple assumption of the opening profile of the crack. The profile of the opening displacement of the FE nodes along

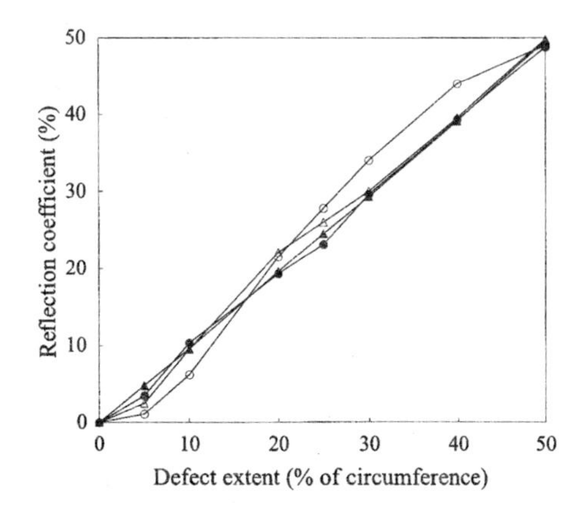

FIG. 20. Variation of reflection ratio in incident mode with defect circumferential extent for zero axial length, full wall thickness defect. Results are from membrane model with $T(0,1)$ incident in 24 inch pipe at 10 kHz (empty triangle) and 50 kHz (solid triangle) and in 3 inch pipe at 45 kHz (empty circle) and 100 kHz (solid circle).

the crack face was almost constant in the case of the $L(0,2)$ mode at 70 kHz incident on a crack. Clearly when the crack opening displacement (COD) is constant along the defect, the reflection coefficient of the axisymmetric mode is equal to the extent of the crack as a fraction of the circumference, leading to a linear behavior of the reflection coefficient as a function of circumferential extent. As they explained, the COD is approximately constant over the crack face when the wavelength of the incoming wave is sensibly smaller than the circumferential extent of the crack (i.e., $ka>1$). Under such conditions the crack face simply behaves as a free surface which reflects the incoming wave. When ka diminishes, the crack opening displacement can no longer be considered as constant. The variation of the COD with frequency was observed by Lowe in the case of a s_0 wave incident on a crack in a plate.[9] He showed that the COD changes with frequency and there is a transition between the high frequency COD and the quasistatic COD. From that analysis it is clear that only at relatively high frequency ($ka>1$) can the COD reasonably be approximated by a constant displacement along the face of the crack.

It is possible to interpret the results at 5% of the circumferential extent for both the 3 inch pipe at 45 kHz and the 24 inch pipe at 10 kHz (see Fig. 20) by saying that the ka value in these cases is smaller than 1, therefore the reflection coefficient is smaller than its linear approximation. Another explanation of the results at low frequency and small circumferential extent can be proposed. The COD is zero at the boundary points where the crack starts and ends; consequently in the regions of the crack near these boundary points the COD goes from zero to a finite value. When the defect is big enough in terms of circumferential extent, the COD might be considered to be constant and the linear approximation for the reflection coefficient would be valid. However, when the defect is relatively small, the boundary effects described would dominate the COD. In this case the COD cannot be considered as constant over the length of the crack and the reflection coefficient will not be satisfactorily approximated by a linear fit.

Another issue to be tackled is the understanding of mode conversion. Following the analysis of Ditri[6] and Lowe *et al.*,[7] the strength of conversion to each mode by a circumferential crack may be estimated from the degree to which the crack opening profile matches the stresses in the mode. Let us first consider the simple case in which only one of the modes of a given circumferential order existing at the frequency of interest has a stress profile similar to the incident mode. In this case it is possible to evaluate the reflection coefficient of this mode converted mode by using a spatial Fourier decomposition of the displacement around the circumference at the location of the crack. As shown by Lowe *et al.*[7] this spatial Fourier transform gives the excitation strengths of the mode converted waves. This is applicable for example to the $F(1,2)$ mode in a 3 inch pipe when the $T(0,1)$ mode is incident at 100 kHz. When more than one mode (or none of the modes) with a given circumferential order has a mode shape similar to the stress profile of the incident mode, the simple spatial Fourier decomposition is no longer valid. Using the spatial Fourier decomposition and supposing that only one of the possible order two modes is involved in the mode conversion, the value of reflection coefficient at a circumferential extent of 50% would be zero, as explained by Lowe *et al.*[7] In fact in all of the modelled cases in this paper this value is not zero [see $F(2,2)$ at 50% circumferential extent in Figs. 8, 9, and 11]. This is due to the fact that more than one order two mode is involved in the mode conversion phenomenon.

2. Discussion on axisymmetric cracks

In the case of axisymmetric cracks the scattering regime in which we are interested is the low-intermediate frequency shown in Fig. 19(b). In Fig. 13 the amplitude reflection coefficient for a series of crack depths is plotted as a function of frequency, the $ka=1$ points being indicated with empty circles. We can notice that the curve of reflection coefficient versus frequency is roughly linear until it approaches the value $ka=1$. It is also interesting to notice that at $ka=1$ the reflection coefficient is approximately equal to the percentage depth of the defect. We then see a small change in behavior going from the intermediate to the high frequency scattering regime. The reflection coefficient curve is less steep at high frequency which can be explained by saying that the reflection coefficient tends to a value which is the asymptotic value at high frequency. This behavior has been encountered in previous work on scattering from cracks in plates.[9,25] Unfortunately in our case it was not possible to increase the test frequency in order to derive the value of the reflection coefficient at very high frequency because another torsional mode appears at 300 kHz [$T(0,2)$].

In Fig. 12 we showed the amplitude reflection coefficient of the $T(0,1)$ mode for axisymmetric cracks as a function of defect depth at different frequency values. The value at which the high frequency scattering regime is reached is indicated in the figure by an empty circle. Around $ka=1$ the curve changes its shape from convex to concave. This behavior can again be attributed to a change in scattering regime.

However, the reflection coefficient is not only ka dependent. Figure 12 shows that if we consider one pipe size with two defect depths and the same ka value, a deep defect at low frequency produces larger reflection coefficient than a shallow defect at high frequency. For example, the reflection coefficient for an 80% depth defect at 40 kHz is 31% whereas that from a 20% defect depth at 160 kHz is 5.3%. This is probably because the geometry of the system, and hence the stress distribution, is very different for shallow and deep defects; the reflection coefficient would be dependent on ka only at small defect depths (e.g., below 10%). However, it has not been possible to check this because of limitations on the finite element mesh.

B. Effect of axial extent

Figure 14 shows the amplitude reflection coefficient for a series of axisymmetric notches with 20% defect depth and different axial extent at 100 kHz. The notch case is shown schematically in Fig. 21. The simplest case to be considered is that in which the notch has an axial extent which is long enough to separate in time the first reflection coming from

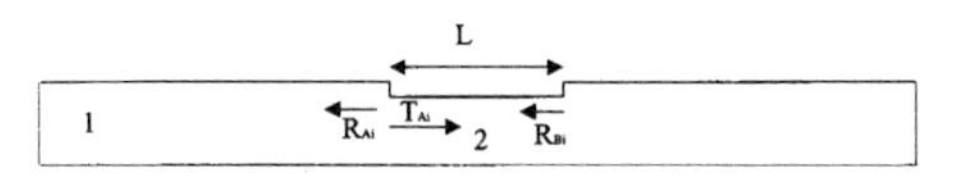

FIG. 21. Example notch case to explain reflection and transmission characteristics at the start and at the end of the notch.

the start of the notch and the second reflection coming from the end of the notch. The reflection coefficient from a step down in a pipe in which we removed 20% of the thickness is plotted in Fig. 22 as a function of frequency. It was simply obtained by dividing the FFT of the reflected signal by the FFT of the input signal. It is clear that this reflection is almost independent of frequency. No mode conversion is present in this case since the defect is axisymmetric. Furthermore the $T(0,1)$ mode is not dispersive so there is no change in velocity due to the change in the thickness of the pipe wall. If we consider that the tangential displacement is almost constant through the thickness and that the radius of the pipe is very much bigger than the thickness, we can approximate the value of reflection coefficient obtained at a step in a pipe by using the formula

$$R=\frac{1-\alpha}{1+\alpha} \tag{2}$$

in which $\alpha=A_2/A_1$, where A_1 is the cross-section area before the notch and A_2 is the cross section at the notch location. The signal arriving at the start of the notch has a component which is reflected and a component which is transmitted. The same phenomenon is repeated at the end of the notch. Let us now consider all of the reflections inside the notch (back and forth):

$$R_{A1}=\frac{1-\alpha}{1+\alpha}, \tag{3}$$

$$T_{A1}=\frac{2}{1+\alpha}, \tag{4}$$

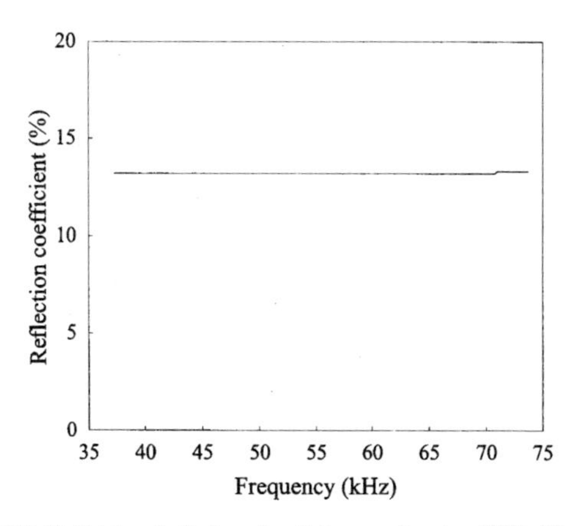

FIG. 22. Variation of reflection ratio with frequency for a step of 20% of the thickness of the pipe. Results are for $T(0,1)$ incident in 3 inch pipe using signal with center frequency of 55 kHz.

$$R_{B1}=\frac{\alpha-1}{1+\alpha}, \tag{5}$$

where R_{A1} is the reflection coefficient from the start of the notch, T_{A1} is the transmission coefficient past the start of the notch and R_{B1} is the signal reflection coefficient from the end of the notch. The reflection described by (5) is time delayed by L/V compared to that of (3) where L is the axial extent of the notch and V is the velocity of the torsional wave, therefore the total phase shift between the two reflections is $\pi+2fL\pi/V$, where the addition of π takes into account the sign reversal in the numerator between (3) and (5), and f is the frequency of the wave. The transmission coefficient at the end of the notch, T_{B1}, is

$$T_{B1}=\frac{2\alpha}{1+\alpha} \tag{6}$$

also with time delay L/V compared to T_{A1}. Similarly, the second reflection from the start of the notch, R_{A2} is

$$R_{A2}=T_{A1}\cdot R_{B1}\cdot T_{B1}, \tag{7}$$

with time delay $2L/V$;

$$R_{A3}=T_{A1}\cdot(R_{B1})^3\cdot T_{B1}, \tag{8}$$

with time delay $4L/V$;

$$R_{Aj}=T_{A1}\cdot(R_{B1})^{2j-3}\cdot T_{B1}, \tag{9}$$

with time delay $j\cdot 2\cdot L/V$. The total reflection coefficient is given by

$$R_{\mathrm{TOT}}=R_{A1}+R_{A2}+\cdots+R_{Aj}. \tag{10}$$

The total reflection coefficient has a maximum when L is equal to a quarter wavelength ($\lambda/4$) and a minimum when L is equal to $\lambda/2$. The maximum is about twice the value of the reflection from a step down of the same depth of the notch. The FE predictions of Fig. 14 and the experiments of Fig. 18 confirm that the maximum is obtained at about $\lambda/4$ and the minimum at about $\lambda/2$. In practice, the reflection maxima and minima are not at axial extents of exactly 25% and 50% of the wavelength. This is because the phase delay at a transition between two different thicknesses is not zero or π as predicted from equations (3) and (5), respectively. It should be stressed here that such a large increase in reflectivity when the defect length is a quarter wavelength long is unlikely in practice since real defects would not have a sharp, rectangular profile. Nevertheless, maxima and minima can occur in practical inspection and it is wise to test at more than one frequency.

VIII. CONCLUSIONS

A quantitative study of the reflection of the $T(0,1)$ mode from defects in pipes in the frequency range 10–300 kHz has been carried out, finite element predictions being validated by experiments on selected cases. Both cracklike defects with zero axial extent and notches with varying axial extents have been considered. The predictions have largely been done on part-thickness, axisymmetric defects and on full-wall-thickness, part-circumference defects, both of which can be modelled with two-dimensional analyses. However, it has also been shown that it is reasonable to use a combination of these two approaches to predict the reflection from part-thickness, part-circumference defects.

It has been shown that the reflection coefficient from axisymmetric cracks increases monotonically with depth at all frequencies and increases with frequency at a given depth. In the frequency range of interest, $T(0,1)$ is the only propagating axisymmetric mode so there is no mode conversion at axisymmetric defects. With nonaxisymmetric cracks, the reflection coefficient is a roughly linear function of the circumferential extent of the defect at relatively high frequencies, the reflection coefficient at low circumferential extents falling below the linear prediction at lower frequencies. With nonaxisymmetric defects, mode conversion to the $F(1,2)$ mode occurs at all but the lowest frequencies, the amplitude of the mode converted signal being a maximum when the circumferential extent is 50%; at low circumferential extents, the amplitudes of the mode converted and direct reflections are similar. Some mode conversion to $F(1,3)$ as well as $F(1,2)$ is seen at lower frequencies.

The depth and circumferential extent of the defect are the parameters controlling the reflection from cracks; when notches having finite axial extent, rather than cracks, are considered, interference between the reflections from the start and the end of the notch causes a periodic variation of the reflection coefficient as a function of the axial extent of the notch, maxima occurring when the notch width is 25% and 75% of the wavelength, and minima appearing when the width is zero (the crack case), 50% or 100% of the wavelength. This paper has only considered square sided defects; real defects will not have such regular shapes so the interference effects will be less severe. However, some frequency dependence is likely to be seen.

The results have been explained in terms of the wavenumber-defect size product, ka. Low frequency scattering behavior is seen when $ka<0.1$, high frequency scattering characteristics being seen when $ka>1$.

The results demonstrate that the torsional wave is attractive for the practical testing of pipes. The sensitivity to defects is improved at higher frequencies, though this conflicts with the desire to use lower frequencies to increase the propagation range, and hence the length of pipe that can be inspected in a single test. When testing for defects that may have significant axial extent, it is wise to test at more than one frequency in order to avoid missing defects due to destructive interference of the reflections from the two ends of the defect.

The 3 inch and 24 inch pipes have been considered in this paper. It would be valuable to generalize the results to other pipe sizes and crack dimensions without the need to analyze each specific case. This will be the subject of a future paper.

[1] R. Thompson, G. Alers, and M. Tennison, "Application of direct electromagnetic Lamb wave generation to gas pipeline inspection," Proceedings of the 1971 IEEE Ultrasonic Symposium, (IEEE, New York, 1972), pp. 91–94.

[2] W. Mohr and P. Holler, "On inspection of thin-walled tubes for transverse and longitudinal flaws by guided ultrasonic waves," IEEE Trans. Sonics Ultrason. **SU-23**, 369–378 (1976).

[3] M. Silk and K. Bainton, "The propagation in metal tubing of ultrasonic wave modes equivalent to Lamb waves," Ultrasonics **17**, 11–19 (1979).

[4] M. Brook, T. Ngoc, and J. Eder, "Ultrasonic inspection of steam generator tubing by cylindrical guided waves," in *Review of Progress in Quantitative NDE*, edited by D. Thompson and D. Chimenti (Plenum, New York, 1990), pp. 243–249.

[5] J. Ditri, J. Rose, and A. Pilarski, "Generation of guided waves in hollow cylinders by wedge and comb type transducers," in *Review of Progress in Quantitative NDE*, edited by D. Thompson and D. Chimenti (Plenum, New York, 1993), pp. 211–218.

[6] J. Ditri, "Utilization of guided elastic waves for the characterization of circumferential cracks in hollow cylinders," J. Acoust. Soc. Am. **96**, 3769–3775 (1994).

[7] M. Lowe, D. Alleyne, and P. Cawley, "The mode conversion of a guided wave by a part-circumferential notch in a pipe," J. Appl. Mech. **65**, 649–656 (1998).

[8] D. Alleyne, M. Lowe, and P. Cawley, "The reflection of guided waves from circumferential notches in pipes," J. Appl. Mech. **65**, 635–641 (1998).

[9] M. Lowe, "Characteristics of the reflection of lamb waves from defects in plates and pipes," in *Review of Progress in Quantitative NDE*, edited by D. Thompson and D. Chimenti (Plenum, New York, 1998), p. 113.

[10] D. Alleyne, B. Pavlakovic, M. Lowe, and P. Cawley, "Rapid, long range inspection of chemical plant pipework using guided waves," Insight **43**, 93–96, 101 (2001).

[11] H. Shin and J. Rose, "Guided wave tuning principles for defect detection in tubing," J. Nondestruct. Eval. **17**, 27–36 (1998).

[12] G. Kino, *Acoustic Waves: Devices, Imaging and Analogue Signal Processing* (Prentice-Hall, New Jersey, 1987).

[13] B. Pavlakovic, M. Lowe, D. Alleyne, and P. Cawley, "DISPERSE: A general purpose program for creating dispersion curves," in *Review of Progress in Quantitative NDE*, edited by D. Thompson and D. Chimenti (Plenum, New York, 1997), Vol. 16, pp. 185–192.

[14] D. Alleyne and P. Cawley, "Long range propagation of Lamb waves in chemical plant pipework," Mater. Eval. **55**, 504–508 (1997).

[15] D. Alleyne and P. Cawley, "The interaction of Lamb waves with defects," IEEE Trans. Ultrason. Ferroelectr. Freq. Control **39**, 381–397 (1992).

[16] P. Wilcox, M. Lowe, and P. Cawley, "A signal processing technique to remove the effect of dispersion from guided wave signals," in *Review of Progress in Quantitative NDE*, edited by D. Thompson and D. Chimenti (American Institute of Physics, New York, 2001), Vol. 20, pp. 555–562.

[17] W. Zhuang, A. Shah, and S. Datta, "Axisymmetric guided wave scattering by cracks in welded steel pipes," J. Pressure Vessel Technol. **119**, 401–406 (1997).

[18] C. Valle, M. Niethammer, J. Qu, and L. Jacobs, "Crack characterization using guided circumferential waves," J. Acoust. Soc. Am. **110**, 1282–1290 (2001).

[19] B. Pavlakovic, "Leaky guided ultrasonic waves in NDT," Ph.D. thesis, University of London, 1998.

[20] D. Alleyne and P. Cawley, "The excitation of Lamb waves in pipes using dry-coupled piezoelectric transducers," J. Nondestruct. Eval. **15**, 11–20 (1996).

[21] D. Alleyne, "The nondestructive testing of plates using ultrasonic lamb waves," Ph.D. thesis, University of London, 1991.

[22] M. Koshiba, H. Hasegawa, and M. Suzuki, "Finite-element solution of horizontally polarized shear wave scattering in an elastic plate," IEEE Trans. Ultrason. Ferroelectr. Freq. Control **34**, 461–466 (1987).

[23] F. Moser, L. Jacobs, and J. Qu, "Modelling elastic wave propagation in waveguides with finite element method," NDT & E Int. **33**, 225–234 (1999).

[24] N. Kishore, I. Sridhar, and N. Iyengar, "Finite element modelling of the scattering of ultrasonic waves by isolated flaws," NDT & E Int. **33**, 297–305 (2000).

[25] M. Lowe and O. Diligent, "Low frequency reflection characteristics of the S0 Lamb wave from a rectangular notch in a plate," J. Acoust. Soc. Am. **111**, 64–74 (2002).

[26] E. Kraut, "Review of theories of scattering of elastic waves by cracks," IEEE Trans. Sonics Ultrason. **SU-23**, 162 (1972).

[27] S. Datta, "Scattering of elastic waves," Mechanics Today **4**, 149–208 (1978).

[28] J. Achenbach, A. Gautesen, and H. McMaken, *Ray Methods for Waves in Elastic Solids* (Pitman Books Limited, New York, 1982).
[29] D. Alleyne and P. Cawley, "The interaction of Lamb waves with defects," IEEE Trans. Ultrason. Ferroelectr. Freq. Control **39**, 381–397 (1992).
[30] Y. Wang and J. Shen, "Scattering of elastic waves by a crack in a isotropic plate," Ultrasonics **35**, 451–457 (1997).
[31] J. Wang, K. Lam, and G. Liu, "Wave scattering of interior vertical crack in plates and detection of the crack," Eng. Fract. Mech. **59**, 1–6 (1998).
[32] H. Bai, A. Shah, N. Popplewell, and S. Datta, "Scattering of guided waves by circumferential cracks in steel pipes," J. Appl. Mech. **68**, 619–631 (2001).
[33] B. Auld, *Acoustic Fields and Waves in Solids* (Krieger, Malabar, FL, 1990), Vol. 2.

Damage detection with auxiliary subsystems

Fabrizio Vestroni* and Stefano Vidoli*
* Dipartimento di Ingegneria Strutturale e Geotecnica, Università di Roma La Sapienza, Rome, Italy

Abstract The small sensitivity to local variations of mechanical characteristics turns out to be the major limit of indirect identification techniques based on frequency response measurements. To overcome this limit, the use of sensitivity enhancement techniques has recently proposed: the monitored structure is coupled to an auxiliary system, the constitutive parameters of which are then suitably tuned to enhance the sensitivities relevant to the identification process. The damage identification problem in these "augmented" structures is here introduced, and its main advantages and drawbacks are discussed. A beam-like structure coupled to a network of piezoelectric patches supplies an enlightening application of the proposed technique.

1 Limits of frequency-response based identification techniques

The measurements of natural frequencies, displacement, strain or stress values in selected sites of a structure belong to the simplest solutions worth of interest for health monitoring purposes. These methods, in spite of significant advantages, are associated with important difficulties. As a matter of fact, whilst a punctual damage detection is necessary to prevent abnormal functioning or structural failures, damage often results in local changes of the structural characteristics. On the other hand, global structural characteristics, as the natural vibration frequencies, manifest a small sensitivity to lumped variations of the mass or stiffness properties (Bendat and Piersol (1980) or Ljung (1986)). A well-know argument is presented here to convice the reader of the above mentioned difficulties, Morassi (1993). To this aim the Rayleigh quotient:

$$R(u) := \frac{\int_\Omega K[D(u)] \cdot D(u)\, d\Omega}{\int_\Omega M(u) \cdot u\, d\Omega}, \tag{1.1}$$

is used to estimate the natural structural pulsation. In equation (1.1) u is an approximation to the structural modal shape, $D(u)$ is the associated strain field and K and M represent the stiffness and mass operators respectively. A well known theorem guarantees that the Rayleigh quotient is stationary when the displacement function u equals the k-th modal shape, say ϕ_k; moreover in this case its value corresponds to the square of the k-th natural pulsation, say ω_k:

$$R(\phi_k + \epsilon\, \eta) = R(\phi_k) + o(\epsilon^2), \qquad \epsilon > 0,\ \ \|\eta\| = 1, \tag{1.2}$$

$$\omega_k^2 = R(\phi_k) = \frac{\int_\Omega K[D(\phi_k)] \cdot D(\phi_k)\, d\Omega}{\int_\Omega M(\phi_k) \cdot \phi_k\, d\Omega} \tag{1.3}$$

This theorem substantially means that the Rayleigh quotient gives good frequency estimates even when the error on the tentative modal shape is "not so small". Now suppose that moderate variations of the stiffness (or mass) properties occurs in a small subset, say $\tilde{\Omega}$, of the structural domain Ω. Let ϵ be the distance between the *damaged* modal shape, due to the assumed parameter variation, and the *undamaged* modal shape ϕ_k; up to an error of the order $o(\epsilon^2)$ the *damaged* structural pulsation $\tilde{\omega}_k$ can be estimated using the *undamaged* modal shape as follows:

$$\tilde{\omega}_k^2 \overset{\circ}{=} \frac{\int_\Omega \tilde{K}[D(\phi_k)] \cdot D(\phi_k)\, d\Omega}{\int_\Omega M(\phi_k) \cdot \phi_k\, d\Omega} \overset{\circ}{=} \omega_k^2 + \frac{\int_{\tilde{\Omega}} (\tilde{K} - K)[D(\phi_k)] \cdot D(\phi_k)\, d\Omega}{\int_\Omega M(\phi_k) \cdot \phi_k\, d\Omega}, \tag{1.4}$$

where $\tilde{K}$ represent the *damaged* stiffness operator which is equal to K outside the region $\tilde{\Omega}$. Equation (1.4) implies that the difference of the eigenfrequencies is bounded in norm by the percentage ratio between the volume of the damaged region with respect to the total volume:

$$\|\tilde{\omega}_k^2 - \omega_k^2\| \leq c_k \frac{vol(\tilde{\Omega})}{vol(\Omega)}. \tag{1.5}$$

Hence it can be concluded that frequency variations are poorly affected by moderate lumped ($vol(\tilde{\Omega}) \ll vol(\Omega)$) damages.

A small sensitivity with respect to the observed quantities translates into a relevant uncertainty on the identified parameters, since it implies a small curvature of any reasonable identification functional build on these quantities. Moreover, a similar reasoning can be repeated when the observed quantities are displacements or strains in selected points of the structure; again the sensitivity to local variations of the constitutive parameters reveals often to be too small for an effective identification. Of course a detailed knowledge of the displacement (or strain) fields carries information valuable in damage identification process and can notably improve the situation, but its actual achievement can be sometimes too expensive. To avoid the discussed difficulties, the basic problem, to be solved, is how to increase the sensitivity of selected global measurements up to a level enabling an effective damage detection and identification; this problem is addressed with different perspectives in Lew and Juang (2002), Ray and Tian (1999), Hongbing et al. (2001), Mróz and Lekszycki (1998). In particular the ideas proposed in Hongbing et al. (2001) follow the observation that, by a proper modification of energy distribution in a structure, one can significantly magnify the effects of damage.

Consider, for instance, the five different testing configurations, each one characterized by a different location of the support π_2, shown in Figure 1; which one is optimal in order to detect a possible damage? Of course the correct answer depends on the position π_1^* where the damage is located; since the distance between the maximum of the elastic energy distribution (shown on the right column of Figure 1) and the damage location should be minimized to enhance the damage detection sensitivity, the fourth configuration

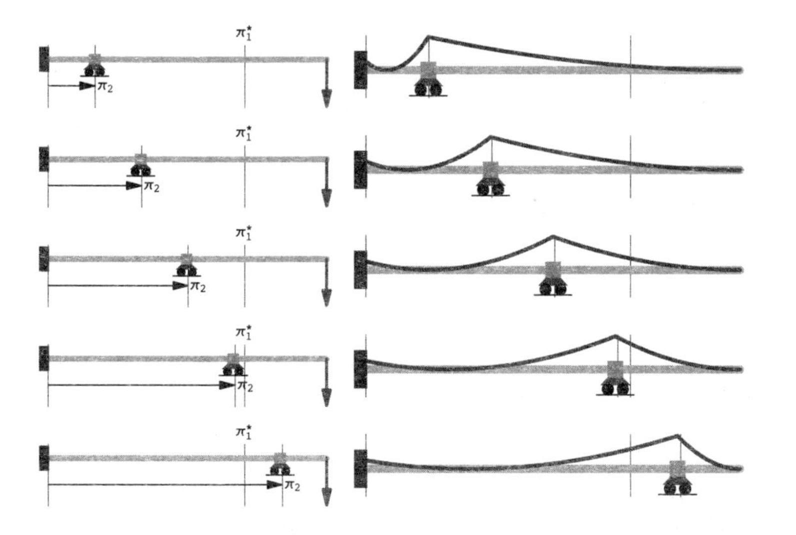

Figure 1. Five testing configurations: on the right the elastic energy per unit length.

is the optimal one. The effect of local structural stiffness variations can be enlarged even by a factor of 10.

Clearly to add moving supports or other constraints to an existing structure is often unpractical, but this is not the only mean to change the elastic energy distribution. More practical realizations of the proposed concept can be achieved by coupling the structure to auxiliary systems whose parameters are tuned according to suitable optimal conditions. To this end concentrated masses, vibration absorbers with variable position, piezoelectric patches connected to circuits with variable impedances have been used and tested in case of simple structural elements as beams and plates.

2 Standard test configuration: general setting

The general identification problems under consideration can be settled in the following mathematical framework. Let $x_1 \in X_1 \subset I\!R^P$ be a P-dimensional vector describing the system constitutive parameters, and let the function $z : X_1 \rightarrow I\!R^O$ be a model of the system mapping each admissible set of constitutive parameters into an O-dimensional vector of observed quantities. Moreover let z^* be an O-dimensional vector representing the actual observed quantities. The subscript '1' is used to refer at the standard testing configuration (Figure 2); in this case the relationship between applied inputs and measured outputs depend only on the constitutive parameters of the system under consideration.

The actual value x_1^* of the system parameters is found solving the minimization problem:

$$\min_{x_1 \in X_1} l(x_1), \quad l(x_1) := \|z(x_1) - z^*\|^2 \tag{2.1}$$

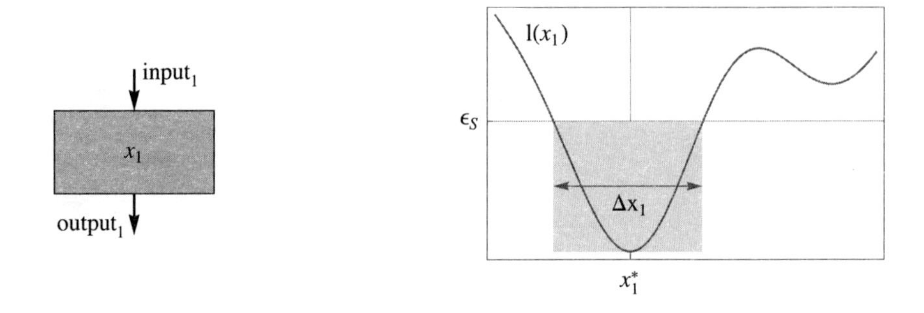

Figure 2. Standard testing configuration: the input-output ratio depends only on the constitutive parameters x_1. The parameter ϵ_S means the experimental sensitivity.

where the function l is the *identification functional.*

When dealing with damage identification, both $z(x_1)$ and z^* physically represent differences of observed quantities between different time instants; usually what must be identified are the actual variations $\Delta x_1^* := x_1^* - x_1^0$ of the actual (*damaged*) constitutive parameters x_1^* from a reference initial (*undamaged*) value x_1^0. In damage detection problems the set of undamaged parameters is usually assumed to be known.

In general the identification functional $l(x_1)$ is not convex; hence global minimization techniques must be used to identify its global minimum. However, in order to have a good starting point for the minimization procedure and to evaluate the sensitivities of the identification functional, it is a standard practice to compute a linear approximation of the model $z(x_1)$ near the *undamaged* state $x_1 = x_1^0$. This is particularly useful in damage identification, where the norm of $\|x_1^* - x_1^0\|$ is usually small and such an approximation is faithfull.

Under this assumption, since a linear model leads to a quadratic approximation $\tilde{l}(x_1)$ of the identification functional $l(x_1)$, the first tentative point for the global minimization procedure is found solving a linear system of equations. In particular one obtains:

$$\tilde{l}(x_1) = \|S_0\, x_1 - c_0^*\|^2, \quad S_0 := \left[\frac{\partial z(x)}{\partial x}\right]_{x=x_1^0}, \quad c_0^* := z(x_1^0) - S_0\, x_1^0 - z^*, \tag{2.2}$$

where S_0 is a $O \times P$ matrix representing the sensitivities of the observed quantities with respect to the parameters, while c_0^* is an O-dimensional constant vector depending not only on the measurements but also on the estimate for the undamaged parameters.

The approximated functional $\tilde{l}$ is minimized when:

$$\frac{\partial \tilde{l}(x_1)}{\partial x_1} = 2\,(S_0^\top\, S_0\, x_1 - S_0^\top\, c_0^*) = 0; \tag{2.3}$$

hence the identification problem, when using the linearized model of the observed quantities, is reduced to the solution of the linear system (2.3). From a geometrical point of view, the $P \times P$ matrix $S_0^\top\, S_0$ measures the curvature in $I\!R^P$ of the functional $\tilde{l}$ near x_1^0.

The relative error when solving Eq. (2.3) is given by:

$$\frac{\Delta x_1^*}{x_1^*} = \kappa \frac{\Delta c_0^*}{c_0^*}, \qquad \kappa = \frac{\lambda_{\max}}{\lambda_{\min}} \tag{2.4}$$

where Δc_0^* is a measure of the variance to which the data c_0^* can be obtained, κ is the condition number of the matrix $S_0^\top S_0$, and $\lambda_{\max}$ ($\lambda_{\min}$) are the maximum (minimum) eigenvalue of the curvature matrix $S_0^\top S_0$. Note that the eigenvalues λ_i of $S_0^\top S_0$, in other words the principal curvatures, are always non-negative[1]; their square roots $\mu_i = \sqrt{\lambda_i}$ are the singular values of the sensitivity matrix S_0. Equation (2.4) is of fundamental importance in damage detection problems since reveals the magnification factor between the uncertainty on the measured data and the uncertainty on the identified parameters. Once the maximum allowed variance $(\Delta x_1)/(x_1)$ on the identified parameters has been declared and the minimal variance on the data $(\Delta c)/(c)_{\min}$ has been estimated, equation (2.4) gives the maximal condition number for a well-defined identification problem.

In order to obtain efficient damage detection procedures, at each step in the iterative process to minimize l, we are led to increase the smallest sensitivities of the problem under consideration; this is equivalent to increase the curvature of the identification functional l measured by the smallest singular values in the list $\{\mu_i\}_{i=1,2,\ldots}$ evaluated at the current value of the parameters x_1.

3 Test configuration with auxiliary subsystems

When the structure to be monitored (also referred as main system) is coupled to an auxiliary dynamical system, the relations between inputs and outputs depend on both the structural and auxiliary constitutive parameters. Besides, the introduction of an auxiliary subsystem allows to monitoring the structure without actually applying inputs and measuring outputs on the structure. Clearly the observability of the structural characteristics depends essentially on the nature of the coupling between the two systems. In Figure 3 a scheme of this testing configuration is presented; the subscript '1' is used to refer to structural quantities while the subscript '2' will refer to the auxiliary subsystem.

Now the general identification problems is settled as follows: let $x_1 \in X_1$ be a P-dimensional vector describing the structural constitutive parameters, $x_2 \in X_2$ a E-dimensional vector describing the auxiliary constitutive parameters, and let the function $z : X_1 \times X_2 \rightarrow I\!R^O$ be a model of the whole system mapping each admissible set of parameters into an O-dimensional vector of observed quantities. Moreover let $z^*(\overline{x}_2)$ be an O-dimensional vector representing the actual measure of the response observed quantities; its dependence on the actual value $\overline{x}_2$ of the auxiliary parameters is explicitly indicated.

For each choice of $\overline{x}_2$, the actual value x_1^* of the system parameters is found solving the following minimization problem:

$$\min_{x_1 \in X_1} l(x_1, \overline{x}_2), \quad l(x_1, \overline{x}_2) := \|z(x_1, \overline{x}_2) - z^*(\overline{x}_2)\|^2 \tag{3.1}$$

[1]Indeed $S_0^\top S_0\, v = \lambda\, v$ implies $\lambda\, v \cdot v = S_0^\top S_0\, v \cdot v = S_0\, v \cdot S_0\, v \geq 0$, i.e. $\lambda \geq 0$

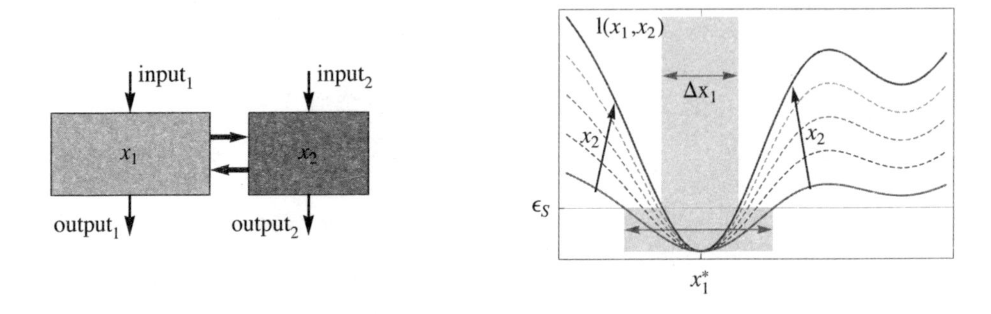

Figure 3. Testing configuration with an auxiliary system: due to the coupling the input-output ratios depend on both the principal (x_1) and the auxiliary (x_2) parameters.

The dependence of *identification functional* $l(x_1, \overline{x}_2)$ on the actual value $\overline{x}_2$ of the auxiliary parameters can now be exploited to minimize the variance of the identified data, as shown in Figure 3, increasing the functional curvature with respect to variations of the structural parameters. To this end, recalling the results of the previous section, one is actually led to maximize the minimum (in module) eigenvalue of the Hessian matrix $H := \partial^2 l/(\partial x_1)^2$: since the maximum eigenvalues is usually bounded, this is equivalent to minimizing the condition number defined in Eq. (2.4). More precisely, given an estimate $\overline{x}_1$ for the actual structural parameters, the optimal value x_2^* for the auxiliary parameters is found solving the following maximization problem:

$$\max_{x_2 \in X_2} \lambda_{\min}(x_2, \overline{x}_1), \quad \lambda_{\min}(x_2, \overline{x}_1) := \min_j |\lambda_j \left(\frac{\partial^2 l(x_1, x_2)}{\partial x_1^2} \right)_{x_1 = \overline{x}_1} |, \tag{3.2}$$

where $\lambda_j(H)$ means the j-th eigenvalue of the matrix H.

As in the case of the beam with moving support in Figure 1, the optimal auxiliary parameters do depend on the estimate $\overline{x}_1$ of the main system parameters; moreover the gained sensitivity enhancement leads to a better estimate for x_1. Hence an iterative min – max procedure is adopted to effectively solve the problem:

$$\begin{cases} \overline{x}_{1,k+1} \leftarrow \min\limits_{x_1 \in X_1} l(x_1, \overline{x}_{2,k}) \\ \\ \overline{x}_{2,k+1} \leftarrow \max\limits_{x_2 \in X_2} \lambda_{\min}(x_2, \overline{x}_1) \end{cases} \tag{3.3}$$

where the minimization steps update the x_1-estimates at the given values of the auxiliary parameters, the maximization steps update the x_2-values to improve the conditioning of the minimization problems.

Figure 4 depicts a good scenario for the identification functional $l(x_1, x_2)$: the functional is convex with respect to x_1 parameters whilst its curvature is concave with respect to x_2 parameters. The scheme (3.3) will converge to the correct (and unique!) solution from every point in the box $X_1 \times X_2$. However, both the convexity of l with respect to

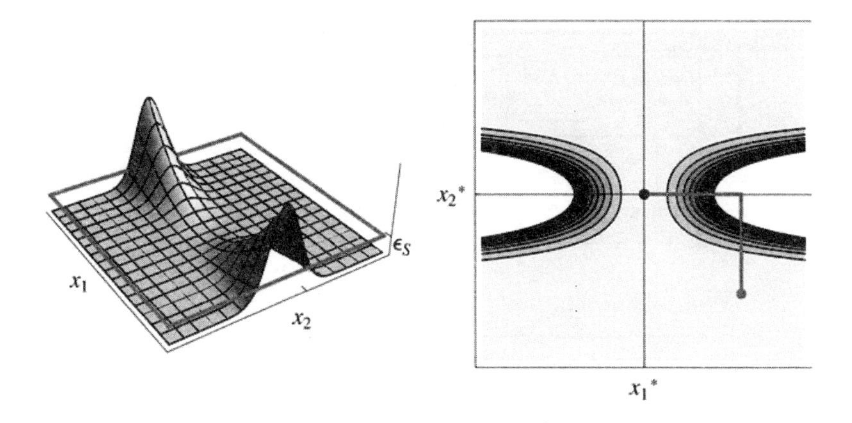

Figure 4. Identification functional as function of mechanical (x_1) and auxiliary (x_2) parameters. The gray region on the right corresponds to all the points where the functional is lower than the threshold ϵ_S.

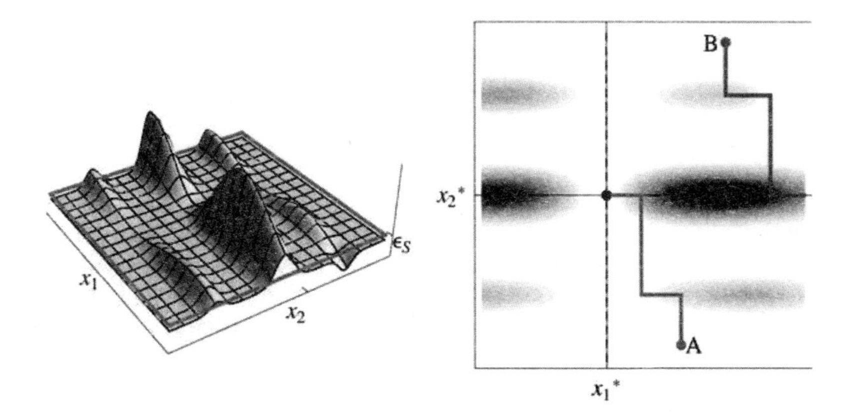

Figure 5. Identification functional as function of mechanical (x_1) and auxiliary (x_2) parameters.

x_1 and the concavity of $\lambda_{\min}$ with respect to x_2 are not guaranteed *a-priori*. Figure 5 shows a case in which the scheme (3.3) can locally converges to the correct solution (path starting from point A) but fails to converge globally (path from point B).

Critical remarks to the described procedure are as follows:

- the procedure is experimentally more expensive, since involves the acquisition of the system response for several values of the auxiliary parameters;
- the set X_2, *i.e.* the values of auxiliary parameters at which the undamaged response is measured, must be discrete in most practical situations; seeking the best value

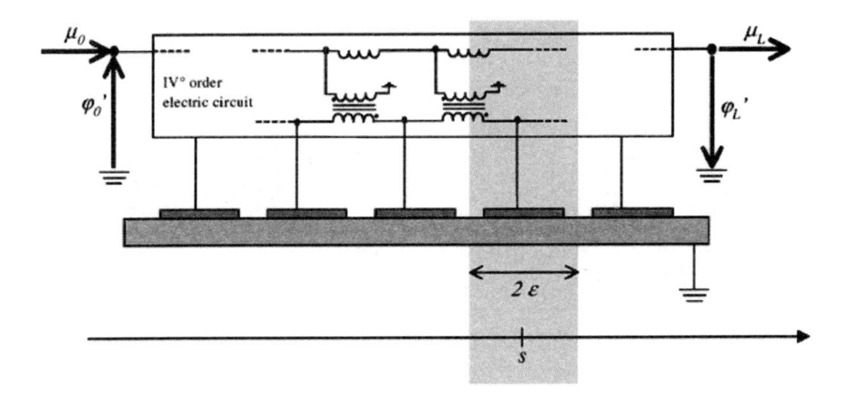

Figure 6. An electric circuit as auxiliary subsystem

of auxiliary parameters is then reduced to a discrete maximization problem;

- the procedure is computationally more expensive, since a bigger amount of data must be stored and processed in the min-max procedure.

4 A numerical example: sensitivity enhancement by auxiliary electric circuits

A practical realization of the testing configuration with an auxiliary subsystem (Figure 3) can be obtained using piezoelectric actuators and a standard electric circuitry. A set of piezoelectric patches distributed along beam-like or plate-like structural members can provide the necessary coupling between the structure and an auxiliary dynamical system constituted by an electric circuit interconnecting them. In particular the electromechanical systems sketched in Figure 6, characterized by the presence of several distributed and interconnected actuators, have been recently introduced and used to control structural vibrations (see *e.g.* Vidoli and Dell'Isola (2000), Dell'Isola et al. (2003)). In particular the circuit shown, composed by a two-port inductance and a 4-ports transformer in each module, is analog to a flexible Euler-Bernouilli beam. In these integrated systems, based on the concept of electric analog, the circuit is designed to have the same modal behavior of the structure to be controlled. Hence an optimal tuning of the circuital impedances allows to efficiently transduce the mechanical energy into electric energy through internal resonance phenomena: an electric "tuned mass damper" is then realized.

As observed quantities we choose the frequency response function of the circuital nodes, *i.e.* the matrix collecting the ratios between the input current at node i and the resulting electric potential drop at node j. This matrix is the electric analog of the frequency response function for a mechanical system with finite degrees of freedom and is called the impedance matrix: for a set of N_P piezoelectric patches and a non-dissipative circuit it is given by $N_P \times N_P$ symmetric frequency-dependent matrix. Due to the presence of the piezoelectric patches, the circuit is intrinsically coupled to the

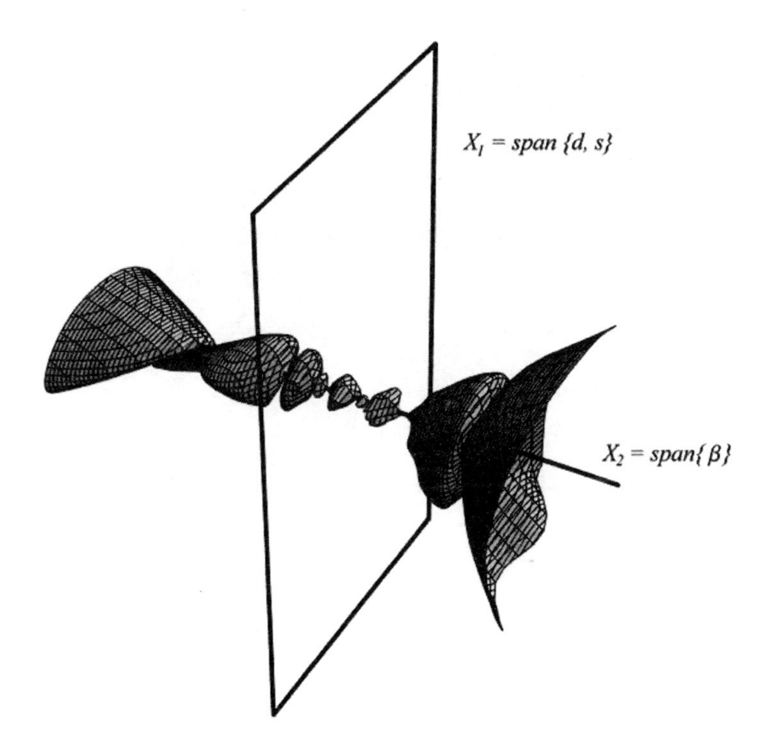

Figure 7. Level plot of $l\,(\{d,s\},\,\{\beta\}) = \epsilon_S$ in the space $X_1 \times X_2$.

structure and the impedance matrix does depend on the structural parameters.

Hence measuring only electric input and outputs (auxiliary systems), variations of the mechanical constitutive parameters should be sensed; moreover by changing the values of the circuital elements the sensitivity to this variations could be increased. These expectations are successfully confirmed by numerical simulations in Dell'Isola et al. (2005): the case of a lumped damage is considered for a simply supported Euler beam endowed with piezoelectric patches and an auxiliary circuit. The damage is defined by two scalar parameters, namely its location s and its intensity d, meaning the percent drop in the beam bending stiffness for a given extension 2ε; the damaged area is denoted by a gray shadow in Figure 6. Concerning the circuit, four modules connect $N_P = 5$ piezoelectric patches uniformly distributed along the beam length; for seek of simplicity we limit our attention to the optimal tuning of the line inductance value supposed constant in all the circuital modules. Hence, within the framework delineated in the previous section, monitored and auxiliary parameters are chosen to be respectively:

$x_1 = \{s,\, d\}$	damage position and damage intensity
$x_2 = \{\beta\}$	electric inductance

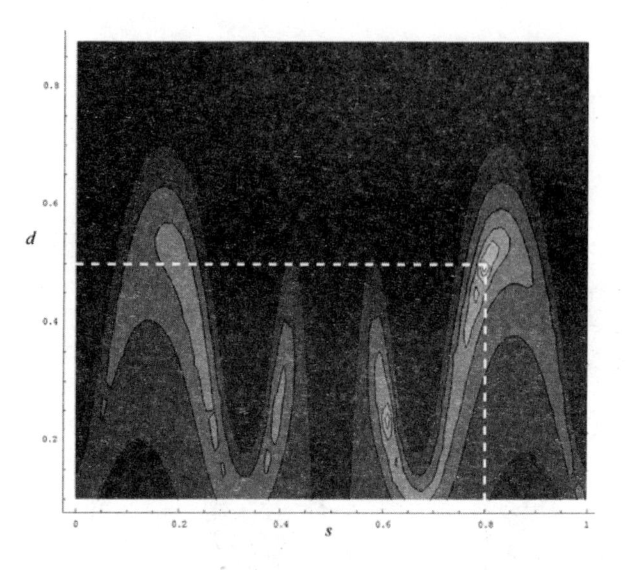

Figure 8. Contour plot $l(\{d, s\}, \{\beta^*\})$ as function of s and d.

The observed quantities $z(x_1, x_2)$ are the differences between the undamaged and damaged values of the $N_P(N_P+1)/2 = 15$ independent entries of the impedance matrix sampled at nine frequency values around the first three structural eigenfrequencies. The identification functional l is then defined as in Eq. (3.1).

Figure 7 shows the level plot of the function $l(\{d, s\}, \{\beta\}) = \epsilon_S$ or in other words all the points in the augmented parameters space $X_1 \times X_2$ for which the identification functional value equals the experimental sensitivity ϵ_S. The points inside this "foil-wrapped candy" shape are admissible estimates for the constitutive parameters s and d; clearly there are values of the auxiliary parameter β where the shape shrinks and the uncertainty on the constitutive parameters is sensibly reduced. As evident from Figure 7 there are several local optima for the auxiliary parameter β and this means that the functional λ_{min} is not concave with respect to it; methods for global optimizations are then required to select the best value for the electric impedance. The plane and the line drawn respectively correspond to the global optimal solutions for both the auxiliary parameter and the monitored parameters *i.e.* $\beta = \beta^*$ and $s = s^*, d = d^*$ respectively; in this case the uncertainty is sensibly reduced with respect to the initial value of the line inductance which was set the optimal value for vibration damping.

As final step of the min-max procedure, the identification functional $l(\{d, s\}, \{\beta = \beta^*\})$, evaluated at the optimal value of line inductance, is minimized with respect to s and d; its contour plot (Figure 8) shown several local minima; the Nelder-Mead simplex algorithm has been successfully used from a set of random points in X_1 to find the correct global minimum.

5 Conclusions

The use of auxiliary systems to enhance the sensitivity of standard damage identification techniques has been discussed. In particular, the case of a Euler beam coupled to a set of piezoelectric patches distributed along the beam length and interconnected by lumped electric impedances has been examined; the impedance matrix between the nodes of the electric circuit, which also depends on the underlying constitutive parameters of the beam, is measured and monitored to sense the occurence of mechanical damages. In this respect the optimal tuning of the impedances has numerically proven to affect the curvature of a suitable damage detection functional and to increase the sensitivity of the impedance matrix to local variations of mechanical characteristics. Hence the mechanical characteristics are identified through an iterative procedure which at each step minimize the objective function and maximize a measure of its curvature near the current minimum. Whilst a relevant magnification of the sensitivity has been obtained, the main drawbacks of the proposed technique are the increased computational cost and the need of a relevant set of *undamaged* data to compare with.

Bibliography

J. S. Bendat and A. G. Piersol. *Engineering applications of correlation and spectral analysis.* New York, Wiley-Interscience, 1980.

F. Dell'Isola, Vestroni F., and S. Vidoli. Structural damage detection by distributed piezoelectric transducers and tuned electric circuits. *Research in Nondestructive Evaluation*, 16(3):101–118, 2005.

F. Dell'Isola, M. Porfiri, and S. Vidoli. Piezo-electromechanical (pem) structures: Passive vibration control using distributed piezoelectric transducers. *Comptes Rendus - Mecanique*, 331(1):69–76, 2003.

Z. Hongbing, Y. Gu, and T. Lekszycki. Identification of structural damage based on energy and optimization methods. In *WCSMO4 Proceedings*, 2001.

J.S. Lew and J.N. Juang. Structural damage detection using virtual passive controllers. *Journal of Guidance, Control, and Dynamics*, 25(3):419–424, 2002.

L. Ljung. *System identification: theory for the user.* Prentice-Hall, Inc., Upper Saddle River, NJ, USA, 1986. ISBN 0-138-81640-9.

A. Morassi. Crack-induced changes in eigenparameters of beam structures. *Journal of Engineering Mechanics*, 119(9):1798–1803, 1993.

Z. Mróz and T. Lekszycki. Identification of damage in structures using parameter dependent modal response. In *ISMA23 Proceedings*, 1998.

L.R. Ray and L. Tian. Damage detection in smart structures through sensitivity enhancing feedback control. *Journal of Sound and Vibration*, 227(5):987–1002, 1999.

S. Vidoli and F. Dell'Isola. Modal coupling in one-dimensional electromechanical structured continua. *Acta Mechanica*, 141(1):37–50, 2000.

HENRY
AND THE
YETI
A RUSSELL AYTO PRODUCTION

BLOOMSBURY
NEW YORK LONDON OXFORD NEW DELHI SYDNEY

For my mother and father

First published in Great Britain in February 2017 by Bloomsbury Publishing Plc
Published in the United States of America in August 2018
by Bloomsbury Children's Books
www.bloomsbury.com

Bloomsbury is a registered trademark of Bloomsbury Publishing Plc

For information about permission to reproduce selections from this book, write to
Permissions, Bloomsbury Children's Books, 1385 Broadway, New York, New York 10018
Bloomsbury books may be purchased for business or promotional use. For information on bulk purchases please contact Macmillan Corporate and Premium Sales Department at specialmarkets@macmillan.com

Library of Congress Cataloging-in-Publication Data
available upon request
ISBN 978-1-68119-683-1 (hardcover) • ISBN 978-1-68119-869-9 (e-book) • ISBN 978-1-68119-868-2 (e-PDF)

Typeset in Amasis MT Std
Hand lettering by Russell Ayto
Book design by Goldy Broad
Printed in China by Leo Paper Products, Heshan, Guangdong
2 4 6 8 10 9 7 5 3 1

All papers used by Bloomsbury Publishing, Inc., are natural, recyclable products made from wood grown in well-managed forests. The manufacturing processes conform to the environmental regulations of the country of origin.

Henry loves yetis.

Yes, *yetis*.

Yet nobody knows if
yetis actually exist.

"Yetis?" says
Henry's father. "Hmm,
nobody actually knows."

But Henry is sure yetis do exist . . .

so he will go on an expedition to find one.

Henry asks his principal if he
can miss school to go on the expedition.

"Yetis?" says
the principal.
"They don't exist."

"This is a school announcement.
Henry is going on an
expedition to find a yeti!"

Everybody laughs.

"And if you do happen to see one," says the principal,
"don't forget to bring back some evidence."

Henry packs all the equipment
he needs for the expedition.

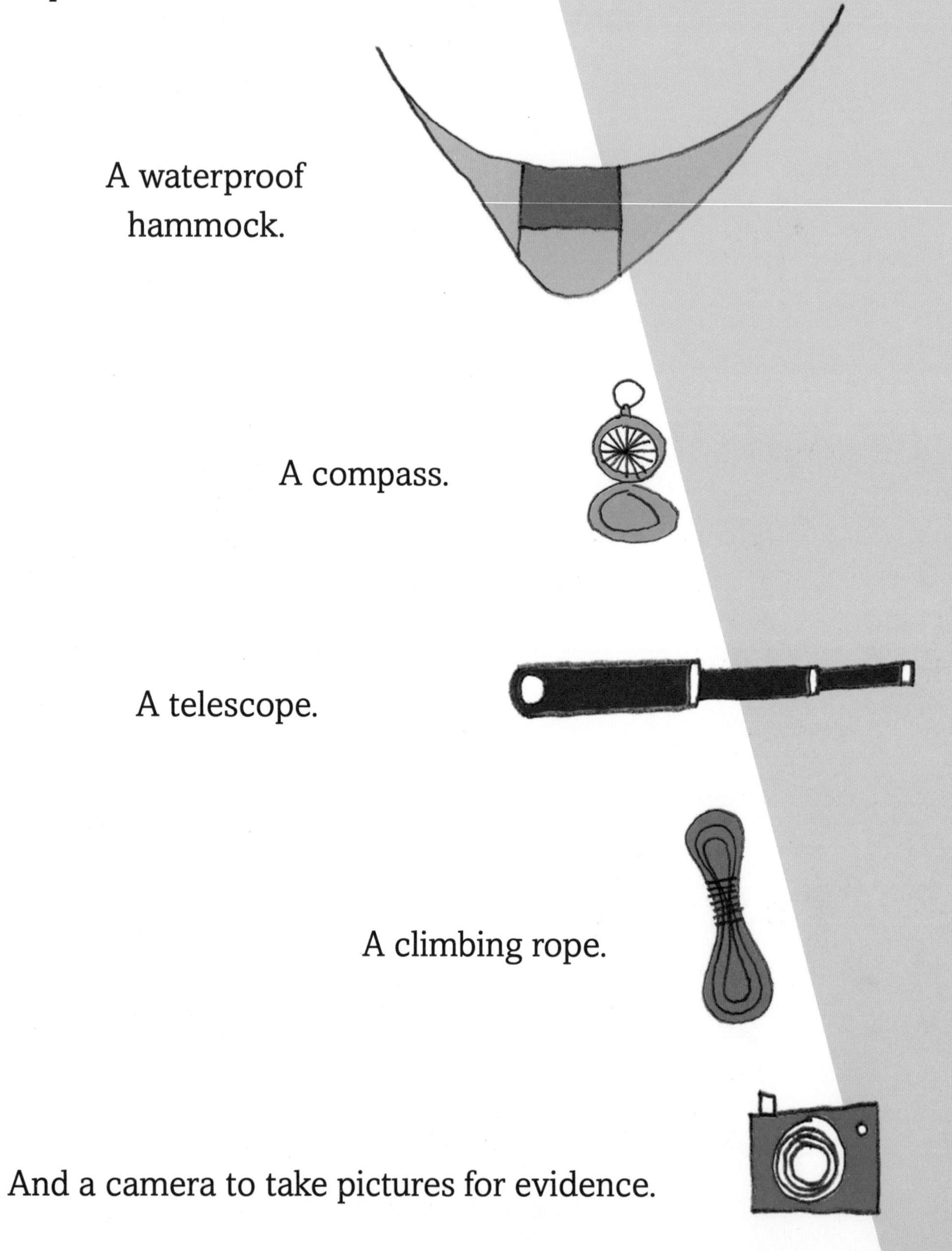

A waterproof
hammock.

A compass.

A telescope.

A climbing rope.

And a camera to take pictures for evidence.

Now Henry is ready.

“Remember, no staying up late,” says Henry’s father.

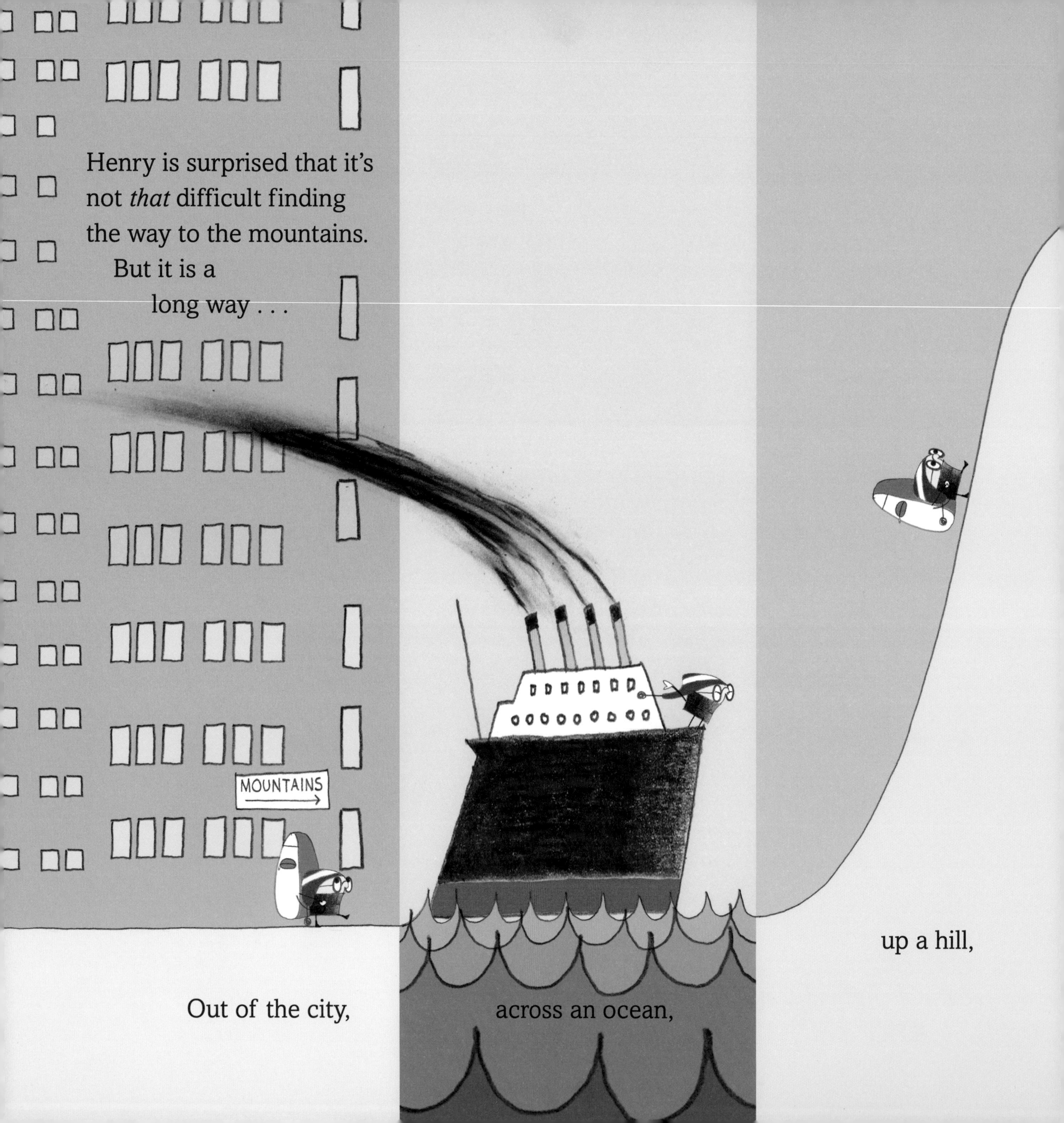

Henry is surprised that it's not *that* difficult finding the way to the mountains. But it is a long way . . .

Out of the city,

across an ocean,

up a hill,

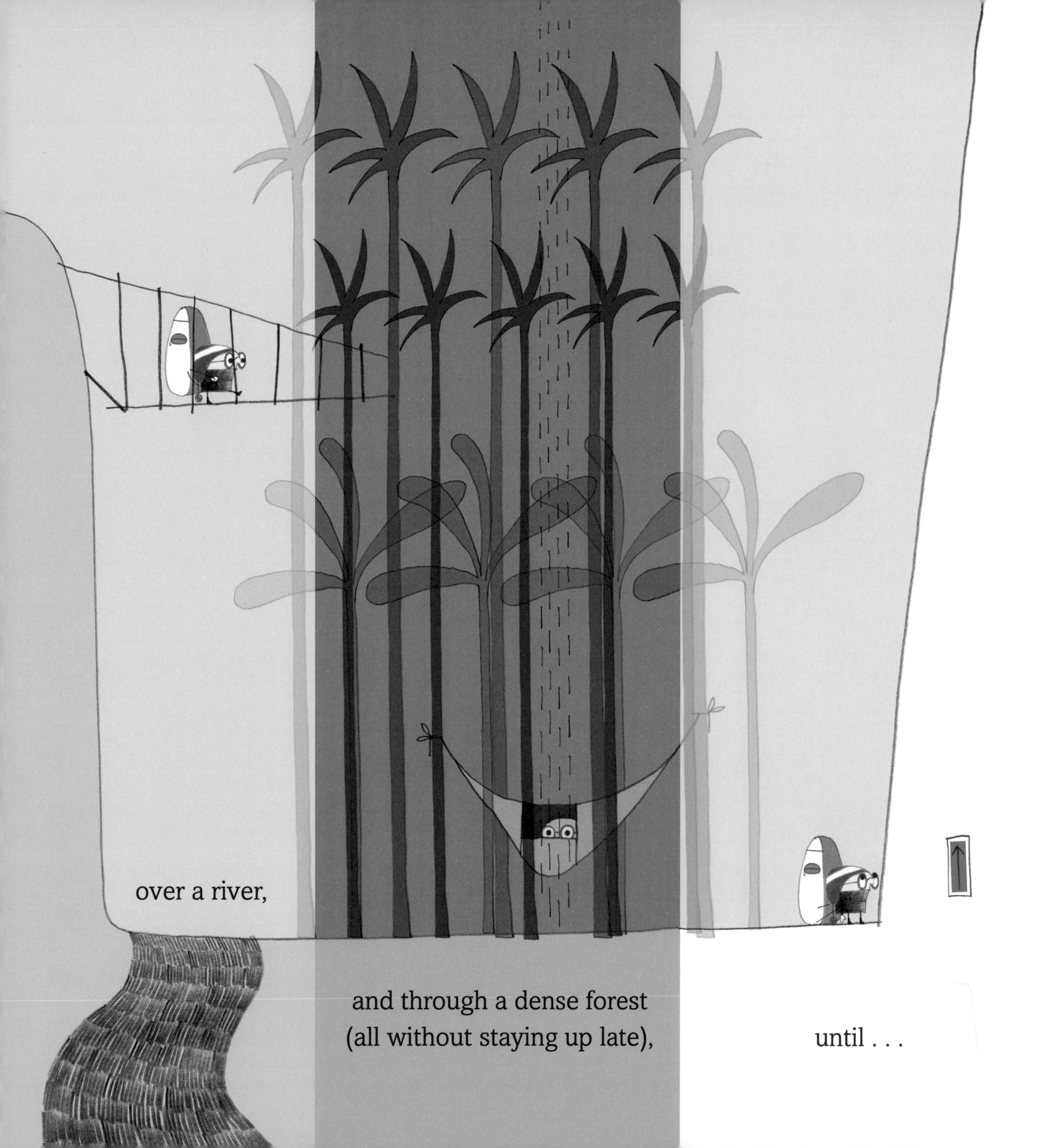

over a river,

and through a dense forest
(all without staying up late),

until . . .

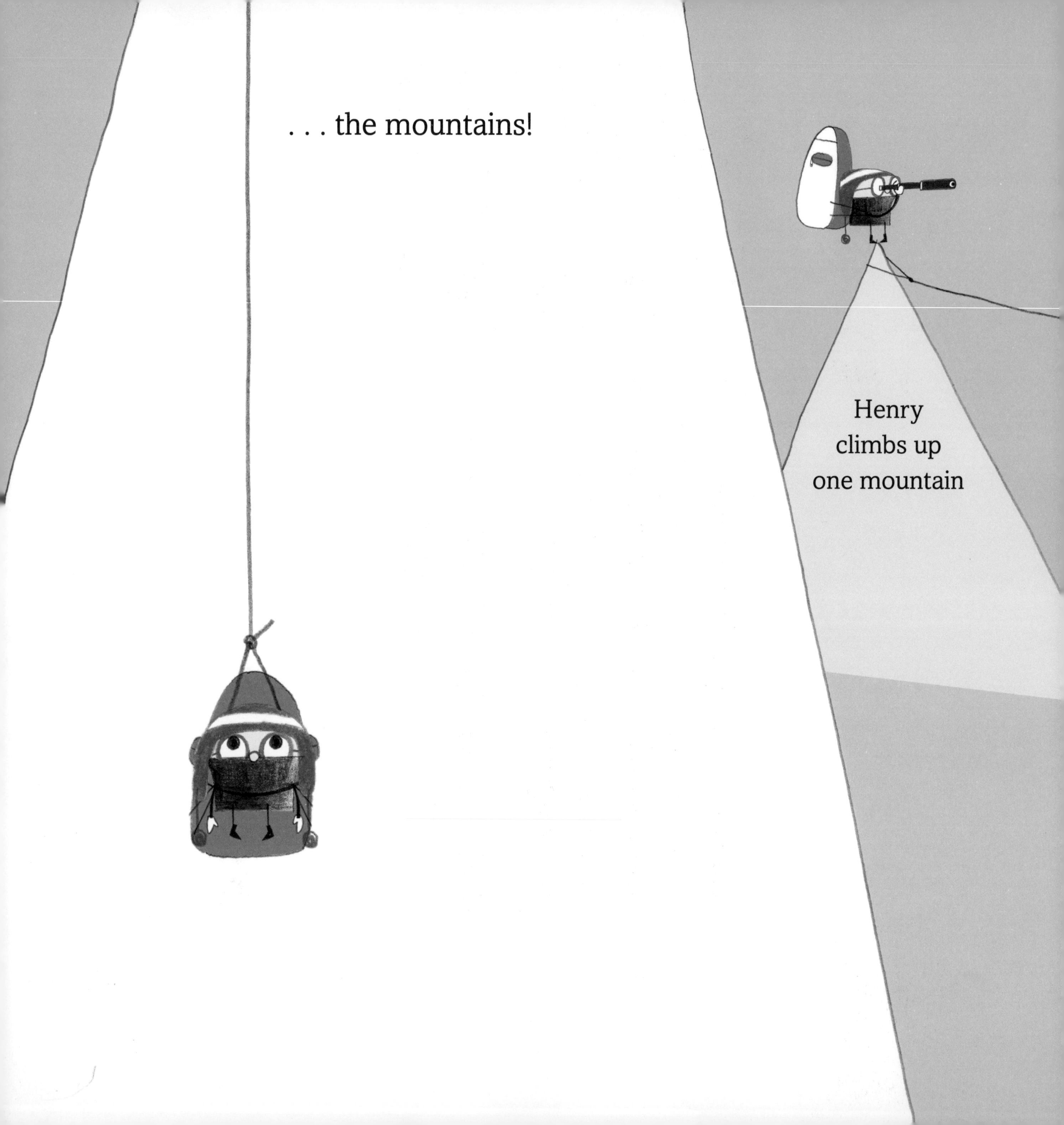
. . . the mountains!
Henry
climbs up
one mountain

after another,

searching **everywhere** for a yeti.

But Henry finds nothing.
There is no sign of a yeti anywhere.
Not even a suspicious-looking footprint.

Henry was **sure** yetis do exist,
but now he isn't so sure.

Maybe he should just turn around and go right back . . .

Oh!

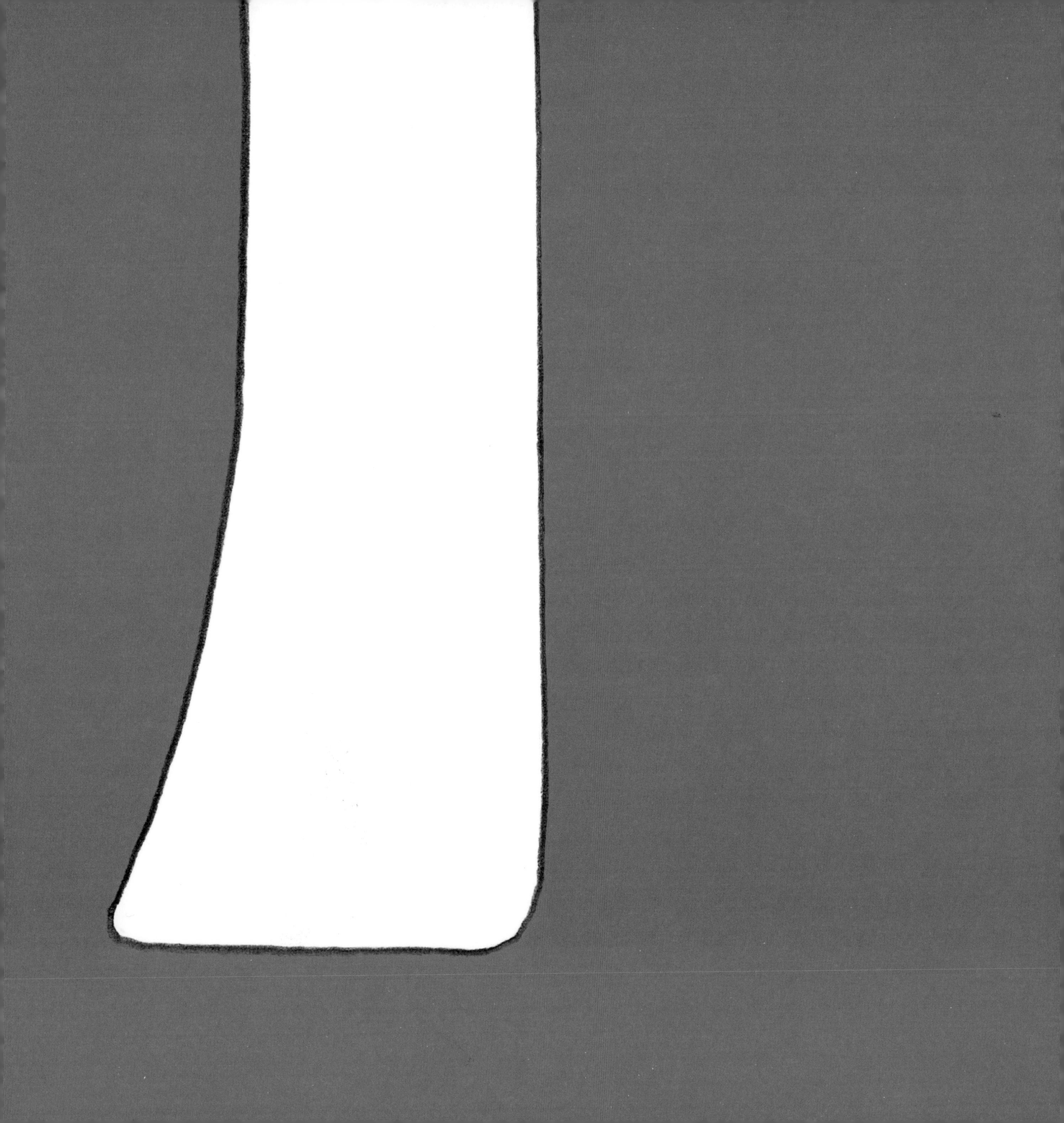

Henry sees a yeti.
The yeti sees Henry.

The yeti is slightly **bigger** than Henry expects.
And friendlier.

Henry takes pictures of
the yeti for evidence.

Then they
play hide and seek.
5
4
3
2
1

Now it is time for Henry to go home.

Henry is surprised that it's more
difficult knowing the way back . . .

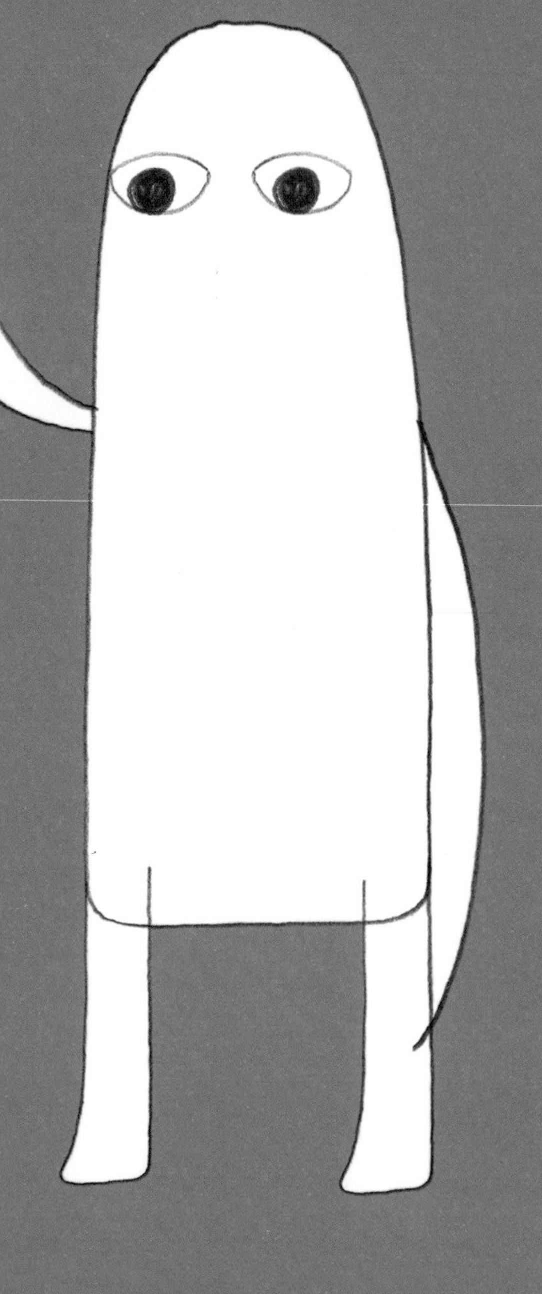

MOUNTAINS

so he uses his compass to find the right way home.

"Well?" says Henry's father.

"I didn't stay up late once," says Henry.

"No," says Henry's father. "Did you see a yeti?"

"Oh yes!" says Henry. "Yetis do exist! And I've brought back the evidence."

Henry unpacks all the equipment.

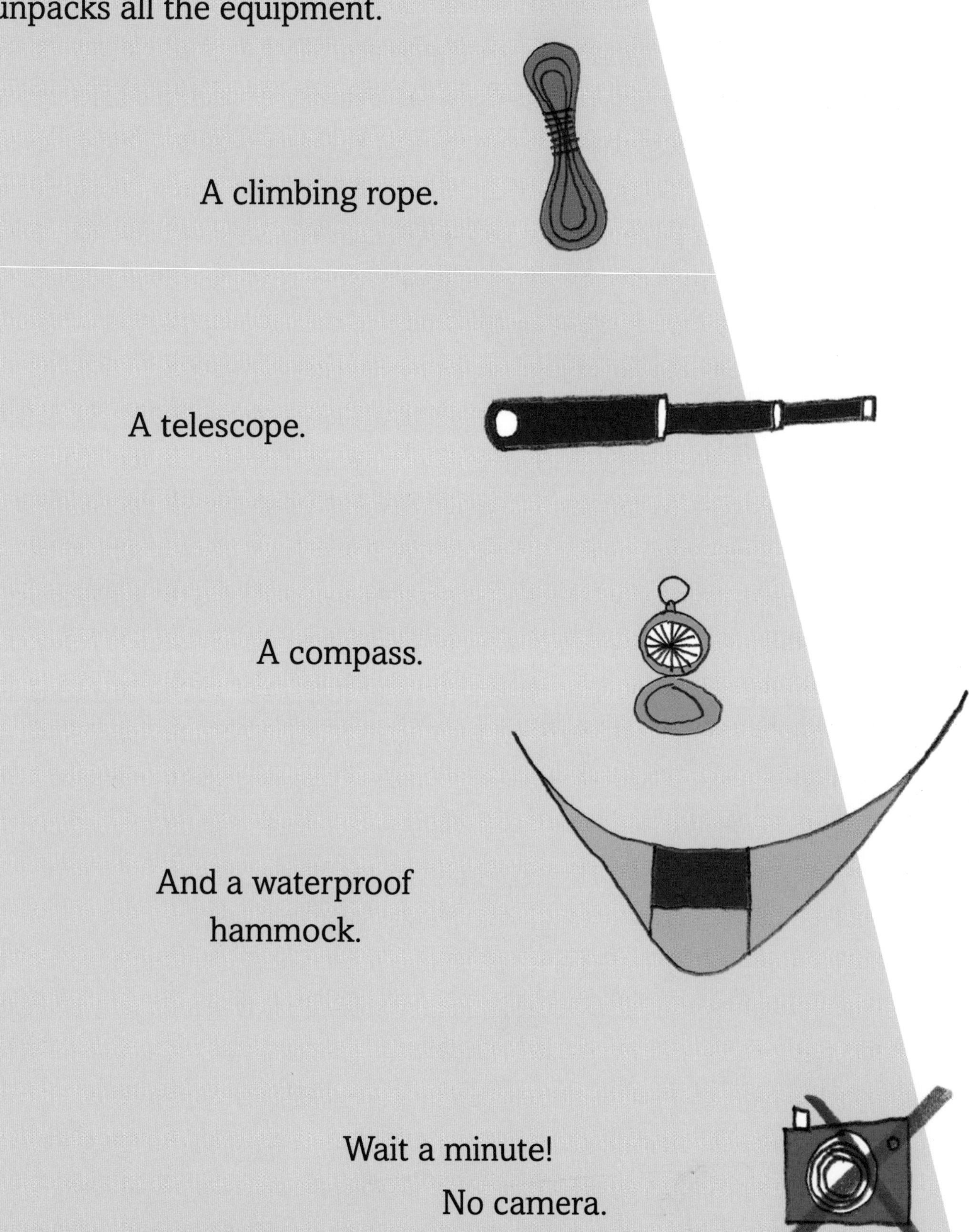

A climbing rope.

A telescope.

A compass.

And a waterproof
hammock.

Wait a minute!

No camera.

“No camera!”
says Henry.

“No camera, no evidence,” says Henry’s father.

"No evidence," says Henry.

"No evidence!" says the principal.

"Write me ten million lines
for making things up.
Yetis indeed!"
Everybody laughs.
Ha ha ha!
Ha ha ha!

What can Henry do now?
He is **not** making things up. He **did** see a yeti. Yetis **do** exist.

But nobody, except his own father, believes him.

Oh!
Henry sees the yeti again.
The yeti sees Henry.
The principal sees the yeti.
And everybody stops laughing.

Now the principal is
lying down.

The yeti gives Henry back his camera.

Henry is thinking he will probably not have to write ten million lines after all.

Henry loves yetis.

Yes, *yetis*.